Ordered Groups and Infinite Permutation Groups

Mathematics and Its Applications

Volume 354

Ordered Groups and Infinite Permutation Groups

edited by

W. Charles Holland
Bowling Green State University

KLUWER ACADEMIC PUBLISHERS
DORDRECHT / BOSTON / LONDON

A C.I.P. Catalogue record for this book is available from the Library of Congress.

ISBN-13: 978-1-4613-3445-3 e-ISBN-13: 978-1-4613-3443-9
DOI: 10.1007/978-1-4613-3443-9

Published by Kluwer Academic Publishers,
P.O. Box 17, 3300 AA Dordrecht, The Netherlands.

Kluwer Academic Publishers incorporates
the publishing programmes of
D. Reidel, Martinus Nijhoff, Dr W. Junk and MTP Press.

Sold and distributed in the U.S.A. and Canada
by Kluwer Academic Publishers,
101 Philip Drive, Norwell, MA 02061, U.S.A.

In all other countries, sold and distributed
by Kluwer Academic Publishers Group,
P.O. Box 322, 3300 AH Dordrecht, The Netherlands.

Printed on acid-free paper

Contents

Preface

The subjects of ordered groups and of infinite permutation groups have long enjoyed a symbiotic relationship. Although the two subjects come from very different sources, they have in certain ways come together, and each has derived considerable benefit from the other.

My own personal contact with this interaction began in 1961. I had done Ph. D. work on sequence convergence in totally ordered groups under the direction of Paul Conrad. In the process, I had encountered "pseudo-convergent" sequences in an ordered group G, which are like Cauchy sequences, except that the differences between terms of large index approach not 0 but a convex subgroup C of G. If C is normal, then such sequences are conveniently described as Cauchy sequences in the quotient ordered group G/C. If C is not normal, of course G/C has no group structure, though it is still a totally ordered set. The best that can be said is that the elements of G permute G/C in an order-preserving fashion. In independent investigations around that time, both P. Conrad and P. Cohn had showed that a group admits a total right ordering if and only if the group is a group of automorphisms of a totally ordered set. (In a right ordered group, the order is required to be preserved by all right translations, unlike a (two-sided) ordered group, where *both* right and left translations must preserve the order.)

It was at about that time that I had the good fortune to be able to attend the lectures of Helmut Wielandt and Olaf Tamaschke in Tübingen. Wielandt's lecture notes on infinite permutation groups had just become available, and I became aware of the powerful methods of that theory. In particular, the most natural context for my problem on Cauchy sequences is that of a group acting on a totally ordered set. Such a group is not itself necessarily totally ordered (two-sided), but is (two-sided) *lattice* ordered by the naturally inherited pointwise order. Indeed, G. Birkhoff had observed, in his book on lattice theory, that the group of automorphisms of a totally ordered set is a natural example of a (usually, non-Abelian) lattice-ordered group, and F. Šik had investigated the structure of such automorphism groups. Using ideas based on my Cauchy sequence problem, I was able to show that *every* lattice-ordered group is embeddable in the lattice-ordered group of automorphisms of a totally ordered set. I then found the obvious analogues in the context of lattice-ordered groups for many of the basic results about general permutation groups.

Since then, these methods have become crucial in the study of non-Abelian lattice-ordered groups. From the other direction, it was already pointed out in Wielandt's lecture notes that certain permutation groups (the "primitive but not strongly primitive" ones) can be viewed as automorphism groups of a certain partial order on the permuted set. The pursuit of similar ideas has led a portion of the theory of permutation groups in the direction of viewing them as automorphism groups of certain structures on the permuted set.

In the summer of 1993, a conference at Luminy in France, organized by M. Giraudet, F. Lucas, and D. Gluschankof, provided the opportunity for many experts in ordered groups and in infinite permutation groups to explore the areas of overlap between the two subjects. The present volume is not, in the usual sense, a proceedings of that conference; not all of the authors represented here were participants at the conference, and not all participants are represented. Rather, I have tried to choose a sample of articles by various authorities to represent a spectrum of ideas ranging from almost purely ordered groups at one end to almost purely permutation groups at the other. I hope that in this way the reader can get a good feeling for the interaction of the two subjects.

The chapter in the present volume by V. M. Kopytov and N. Ya. Medvedev on quasivarieties of lattice-ordered groups, while mainly of a purely algebraic nature, still contains many results whose proofs rely on the representation as ordered permutation groups.

The structure theory of lattice-ordered permutation groups, much of it due to S. H. McCleary, is surveyed in McCleary's chapter in the present volume.

In the chapter in this volume on Jordan groups by S. A. Adeleke, the permutation groups become automorphism groups of various order-like structures.

From another standpoint, if G is the group of all automorphisms of some structure, then certain properties of the structure are reflected in G itself. In the best cases, one can "reconstruct" the structure from G. This is the subject of the chapters in this volume by J. K. Truss (the structure is a totally ordered set and the group of interest is a certain quotient of the full automorphism group), M. Droste and R. Göbel (the structure is a McLain group), and especially the very general approach in the chapter of M. Rubin; to some extent, the same ideas are present also in the chapter by J. L. Alperin, J. Covington, and D. Macpherson (the structure is a quotient of an infinite symmetric group).

The chapters of C. Praeger and of D. Macpherson are essentially of a permutation-group-theoretic nature, but still use methods that are strongly reminiscent of those used in the more order-theoretic parts of the subject.

To all of these authors, to the three organizers of the Luminy conference, and to several anonymous referees, I express my sincerest gratitude.

W. Charles Holland
In the Black Swamp
August, 1995

Quasivarieties and Varieties of Lattice-Ordered Groups *

V. M. Kopytov

Mathematical Institute
Siberian Academy of Science
Novosibirsk 630090
Russia

N. Ya. Medvedev

Altai State University
Barnaul 656099
Russia

1 Introduction

Many properties and statements of the theory of lattice-ordered groups (ℓ-*groups*) can be formulated and proved in terms of first order logic. Special mention should be made of properties expressed by universal sentences such as identities and implications, which can be referred to as the theory of varieties and quasivarieties, respectively, of ℓ-groups. The theory of varieties of ℓ-groups has been developed for more than two decades and contributions to it have been included in books and survey articles (Anderson and Feil [1], Kopytov and Medvedev [44, 45], Reily [63]). It was not until the mid-80s that the systematic investigation of the theory of quasivarieties of ℓ-groups began and the results obtained in this area are not available for wide audience yet. It is clear that the theory of quasivarieties is more general than that of varieties. Nevertheless, there have been obtained a number of results on quasivarieties of ℓ-groups asserting that helpful and non-trivial properties of ℓ-groups can be defined by means of implications, and the theory of quasivarieties of ℓ-groups itself is exciting and profound.

This paper consists of two parts which are different in style and goal. The first part (sections 2 through 6) contains an introduction to the theory of quasivarieties of ℓ-groups and presents the statements and proofs of those results which are, in the authors' opinion, the most interesting. The second part (sections 7 through 11) is a survey of results in the theory of ℓ-varieties obtained after the publication of N. Reilly's paper [63](1989). The account in the present paper is "semiclosed" since we assume the standard definitions and results on ℓ-groups from the well-known

*This work was done with financial support of the Russian Fund of Fundamental Research (project code 93-011-1524)

W. C. Holland (ed.), Ordered Groups and Infinite Permutation Groups, 1–28.

books and survey articles, while the notions not previously defined are explained in the article. We employ standard notation and terminology of group theory (Hall [28], Kargapolov and Merzljakov [38]) and ℓ-group theory (Reily [63]).

PART 1: ℓ-Quasivarieties

2 Generalities

Let us recall that a formula φ of the first-order language and of the signature σ is an *implication* (or *quasi-identity*) if φ is of the form

$$(\forall x_1, \ldots, x_n)((A_1 = B_1 \& A_2 = B_2 \& \ldots \& A_k = B_k) \Rightarrow (A = B))$$

where $A_1, \ldots, A_k$, $B_1, \ldots, B_k$, A, B are terms of signature σ in free variables $x_1, \ldots, x_n$. The class $\mathcal{K}$ of algebraic systems of signature σ is a *quasivariety* if there is a set Φ of implications such that an algebraic system G belongs to $\mathcal{K}$ iff all implications of Φ are valid in G.

An implication of signature $\ell = \{\cdot\ , \ ^{-1},\ e\ , \vee, \wedge\}$ is a formula φ of the predicate calculus of the form

$$(\forall x_1) \ldots (\forall x_n)((w_1(x_1, \ldots, x_n) = e \& \ldots \& w_k(x_1, \ldots, x_n) = e) \Rightarrow$$
$$\Rightarrow w_{k+1}(x_1, \ldots, x_n) = e),$$

where $w_1(x_1, \ldots, x_n), \ldots, w_{k+1}(x_1, \ldots, x_n)$ are ℓ-group words. An ℓ-group G satisfies the implication φ if whenever

$$w_1(g_1, \ldots, g_n) = e, \ldots, w_k(g_1, \ldots, g_n) = e,$$

then $w_{k+1}(g_1, \ldots, g_n) = e$.

A *quasivariety* of ℓ-groups (or, simply, ℓ-quasivariety) is the class $\mathcal{X}$ of all ℓ-groups which satisfy a given set Φ of implications of the signature $\ell = \{\cdot, ^{-1}, e, \vee, \wedge\}$. The set Φ is called a *basis* of the implications of the ℓ-quasivariety $\mathcal{X}$. From standard results in universal algebra (see Burris and Sankappanavar [7], Theorem 2.25.) we have the following characterization of ℓ-quasivarieties:

Theorem 2.1 *Let $\mathcal{X}$ be a class of ℓ-groups. Then the following statements are equivalent:*

(1) *$\mathcal{X}$ is a quasivariety of ℓ-groups,*

(2) *$\mathcal{X}$ is closed under ℓ-isomorphisms, ℓ-subgroups and reduced products, and contains a trivial ℓ-group,*

(3) *$\mathcal{X}$ is closed under ℓ-isomorphisms, ℓ-subgroups, cartesian products and ultraproducts, and contains a trivial ℓ-group.*

It is clear that the intersection of ℓ-quasivarieties is an ℓ-quasivariety. Hence, given any class $\mathcal{K}$ of ℓ-groups, there is a unique minimal ℓ-quasivariety that contains $\mathcal{K}$. This ℓ-quasivariety is called the ℓ-quasivariety generated by $\mathcal{K}$ and is denoted by $\mathrm{q}_\ell(\mathcal{K})$.

For any class $\mathcal{K}$ of ℓ-groups let $P(\mathcal{K})$, $S(\mathcal{K})$ and $P_U(\mathcal{K})$ denote the classes of ℓ-groups which are cartesian products, ℓ-subgroups and ultraproducts, respectively, of elements of $\mathcal{K}$.

From Theorem 2.1 the following useful observation follows.

Theorem 2.2 *If $\mathcal{K}$ is any class of ℓ-groups, then the ℓ-quasivariety $\mathrm{q}_\ell(\mathcal{K})$ generated by $\mathcal{K}$ is $SPP_U(\mathcal{K})$.*

The following result can be proved by direct verification.

Proposition 2.3 *The conjunction of a finite number of implications in the signature $\ell = \{\cdot\,,\ ^{-1}\,,\ e\ ,\vee,\wedge\}$ is equivalent to a single implication in the class of ℓ-groups $\mathcal{L}$; namely, if we denote $|g| = g \vee g^{-1}$ and $\mathbf{x} = (x_1, \ldots, x_n)$, then*

$$\forall \mathbf{x}(\&_{k=1}^{m} w_k(\mathbf{x}) = e \rightarrow \mathbf{x} = e)$$

is equivalent to

$$\forall \mathbf{x}((\bigvee_{k=1}^{m} |w_k(\mathbf{x})| = e) \rightarrow w(\mathbf{x}) = e).$$

Corollary 2.4 *If an ℓ-quasivariety $\mathcal{X}$ has a finite basis of implications, then $\mathcal{X}$ can be defined by one implication.*

It is easy to see that many properties of ℓ-groups can be formulated as implications in the signature $\ell = \{\cdot\,,\ ^{-1}\,,\ e\ ,\vee,\wedge\}$. Each ℓ-variety is an ℓ-quasivariety, but there are ℓ-quasivarieties which are not ℓ-varieties. One such ℓ-quasivariety is the class of all orderable ℓ-groups. That this is an ℓ-quasivariety follows from Theorem 2.1. To see that it is not an ℓ-variety, we observe the following. Let D_ℓ^0 be the ℓ-group of all bounded piecewise linear order-preserving permutations of the totally ordered set R of real numbers.

Then D_l^0 is o-2-transitive on R and orderable. Thus, D_l^0 generates the ℓ-variety $\mathcal{L}$ of all ℓ-groups (Holland [32](1985)). If the orderable ℓ-groups formed an ℓ-variety, this would imply that all ℓ-groups are orderable. But this is not true since, for example, the ℓ-group $A(R)$ of all order permutations of the real numbers is not orderable, because there exist non-trivial $f, g \in A(R)$ such that $f^{-1}gf = g^{-1}$. So the set Λ of all ℓ-quasivarieties properly contains the set $\mathbf{L}$ of all ℓ-varieties. The set Λ of all ℓ-quasivarieties is a lattice where, for any ℓ-quasivarieties $\mathcal{X}, \mathcal{Y}$, the greatest lower bound, or meet, of $\mathcal{X}$ and $\mathcal{Y}$ is their set-theoretical intersection:

$$\mathcal{X} \bigwedge^{\Lambda} \mathcal{Y} = \mathcal{X} \cap \mathcal{Y}$$

and their least upper bound, or join, is the intersection of all ℓ-quasivarieties containing both $\mathcal{X}$ and $\mathcal{Y}$:

$$\mathcal{X}\overset{\Lambda}{\bigvee}\mathcal{Y}=\bigcap\{\mathcal{W}:\mathcal{X}\cup\mathcal{Y}\subseteq\mathcal{W}\}.$$

Let $\mathcal{X}$ and $\mathcal{Y}$ be varieties of ℓ-groups and $\mathcal{X}\overset{\mathbf{L}}{\bigvee}\mathcal{Y}$ be the join of $\mathcal{X}$ and $\mathcal{Y}$ in the lattice $\mathbf{L}$ of ℓ-varieties. Then by (Martinez [48](1974)),

$$G\in\mathcal{X}\overset{\mathbf{L}}{\bigvee}\mathcal{Y}$$

if and only if there exist ℓ-ideals M and N in G such that $M\cap N=E$ and $G/M\in\mathcal{X}, G/N\in\mathcal{Y}$. In this case, therefore, G is isomorphic to some ℓ-subgroup of the direct product $G/M\times G/N$ and so $G\in\mathcal{X}\overset{\Lambda}{\bigvee}\mathcal{Y}$. Hence, the lattice $\mathbf{L}$ of all ℓ-varieties is a sublattice of the lattice Λ of all ℓ-quasivarieties.

The lattice Λ of ℓ-quasivarieties is more diverse and complicated than the lattice $\mathbf{L}$ of ℓ-varieties.

For a discussion of the relationship between free ℓ-groups and and ℓ-quasivarieties, see V. M. Kopytov [42](1979).

3 The lattice of ℓ-quasivarieties

The following result establishes a difference between the lattice $\mathbf{L}$ of ℓ-varieties and the lattice Λ of ℓ-quasivarieties.

Theorem 3.1 (S. A. Gurchenkov [private communication]) *The lattice Λ of ℓ-quasivarieties is not modular and therefore is not distributive.*

Proof. Consider the group

$$G=\mathrm{gp}\langle a,b\mid[a^{b^j},\ a^{b^i}]=[a,b,b,b]=e,\ (i,j\in\mathbf{Z})\rangle.$$

It is clear that G is metabelian and torsion-free nilpotent of class ≤ 3. Now we define on G two total orders. The total order P_1 is uniquely defined by the inequalities $b\gg a\gg[a,b]\gg[a,b,b]>e$. Let G_1 be the group G totally ordered by P_1. The total order P_2 is uniquely defined by the inequalities: $b\gg a\gg[a,b]\gg[a,b,b]^{-1}>e$. The group G totally ordered by P_2 is denoted by G_2. Let G_3 be the ℓ-subgroup of the direct product $G_1\times G_2$ generated by the elements $\bar{b}=(b,b)$, $\bar{a}=(a,a)$. Now consider the following implications (we omit quantifiers):

t_1 : $$[x,y]\geq e\Rightarrow[x,y,y]\geq e;$$

t_2 : $$[x,y] \geq e \Rightarrow [x,y,y] \leq e;$$

t_4 : $$[x,y] \geq e \Rightarrow [x,y,y] = e;$$

$t_{3,1}$: $$[x,y,y] \geq e \Rightarrow [x,y,y] = e;$$

$t_{3,2}$: $$[x,y,y] \leq e \Rightarrow [x,y,y] = e,$$

where $x = |z| \wedge |w|$, $y = |z| \vee |w|$.

Direct verification shows that G_1 satisfies t_1 but not t_2, while G_2 satisfies t_2 but not t_1. Moreover, G_3 satisfies both $t_{3,1}$ and $t_{3,2}$ but neither t_1 nor t_2. Let $\mathcal{M}$ be the ℓ-quasivariety of nilpotent ℓ-groups of class ≤ 3 defined by the implication t_4, and

$$\mathcal{M}_1 = \mathrm{q}_\ell(G_1) \overset{\Lambda}{\bigvee} \mathcal{M}, \; \mathcal{M}_2 = \mathrm{q}_\ell(G_2) \overset{\Lambda}{\bigvee} \mathcal{M}, \; \mathcal{M}_3 = \mathrm{q}_\ell(G_1) \overset{\Lambda}{\bigvee} \mathcal{M} \overset{\Lambda}{\bigvee} \mathrm{q}_\ell(G_3).$$

Then $\mathcal{M}_1, \mathcal{M}_2, \mathcal{M}_3$ are different. Direct verification shows that $\mathcal{M}_2 \neq \mathcal{M}_1 \vee \mathcal{M}_2 = \mathcal{M}_2 \vee \mathcal{M}_3 \neq \mathcal{M}_3$ and $\mathcal{M}_1 \vee \mathcal{M}_2 = \mathcal{M}$. We claim that $\mathcal{M}_3 \wedge \mathcal{M}_2 = \mathcal{M}$. Let $H \in \mathcal{M}_3 \wedge \mathcal{M}_2$ and $z, w \in H$ such that $[x,y] \geq e$. As $H \in \mathcal{M}_2$, then $[x,y,y] \leq e$. Since $H \in \mathcal{M}_3$, then H is a subdirect product of ℓ-groups $H_1 \in M$, $H_2 \in \mathrm{q}_\ell(G_1)$, and $H_3 \in \mathrm{q}_l(G_3)$. Then

$$[x,y,y] = [x_1,y_1,y_1] \cdot [x_2,y_2,y_2] \cdot [x_3,y_3,y_3],$$

where $[x_i,y_i,y_i] \in H_i$ $(i = 1,2,3)$. Since $[x,y] \geq e$ and $[x,y,y] \leq e$ in H, then $[x_i,y_i] \geq e, [x_i,y_i,y_i] \leq e$ in H_i. Since the implications $t_{3,1}, t_{3,2}$ are valid in H_3, the inequalities are valid in H_3, and the inequalities $[x_3,y_3] \geq e$, and $[x_3,y_3,y_3] \leq e$ imply $[x_3,y_3,y_3] = e$. Since the implication t_1 holds in H_2, then the inequality $[x_2,y_2] \geq e$ implies $[x_2,y_2,y_2] = e$. Finally, the validity of the implication t_4 and the inequality $[x_1,y_1] \geq e$ implies $[x_1,y_1,y_1] = e$. Hence the sublattice of the lattice Λ of ℓ-quasivarieties generated by the ℓ-quasivarieties $\mathcal{M}_1$, $\mathcal{M}_2$, and $\mathcal{M}_3$ is a 5-element non-modular lattice.

It is also known that the sublattice of all those ℓ-quasivarieties which can be defined by implications of the signature of the group theory $\{\ \cdot\ ,\ ^{-1}\ ,\ e\ \}$ is not modular (N. Ya. Medvedev [51](1984)).

It is well known that the ℓ-variety $\mathcal{N}$ of normal-valued ℓ-groups is the greatest element in the lattice $\mathbf{L}$ of ℓ-varieties (Holland [31](1976)). In the lattice Λ of ℓ-quasivarieties this is not true, as we see in the following.

For each positive integer n, let $\mathcal{R}_n$ be the ℓ-quasivariety defined by the implication $(x^n = y^n) \Rightarrow (x = y)$. Let $\mathcal{R}_*$ denote the ℓ-quasivariety of all ℓ-groups which admit unique extraction of all roots, i.e.,

$$\mathcal{R}_* = \bigcap_{n=2}^{\infty} \mathcal{R}_n.$$

It is clear that the ℓ-quasivariety $\mathcal{O}$ of all orderable ℓ-groups is contained in $\mathcal{R}_*$. As usual, we denote the Scrimger ℓ-groups as G_n (see, for example, Reilly [63]).

Lemma 3.2 (Arora [3](1985)) *$G_n \in \mathcal{R}_m$ if and only if $(m,n) = 1$ where (m,n) denotes the greatest common divisor of m and n.*

Corollary 3.3 *If p and q are distinct prime numbers, then $G_p \in \mathcal{R}_q \setminus \mathcal{R}_p$ and $G_q \in \mathcal{R}_p \setminus \mathcal{R}_q$. Hence, $\mathcal{R}_p$ and $\mathcal{R}_q$ are incomparable ℓ-quasivarieties if p and q are distinct prime numbers.*

Let n be a positive integer ≥ 2 and $n = p_1^{n_1} \cdots p_k^{n_k}$ where $p_1, \cdots p_k$ are distinct prime divisors of n and $n_1, \cdots, n_k > 0$. Then an easy induction on k shows that $\mathcal{R}_n = \mathcal{R}_{p_1} \cap \cdots \cap \mathcal{R}_{p_k}$. By the above arguments, we conclude

Corollary 3.4 *Let m and n be positive integers. Then $\mathcal{R}_n \supseteq \mathcal{R}_m$ if and only if every prime divisor of m is a prime divisor of n.*

Let P be the set of positive prime integers and let $P' \subseteq P$ be any non-empty subset of P. We define a quasivariety (for each P') as follows:

$$R_{P'} = \bigcap_{p \in P'} \mathcal{R}_p.$$

Let P_1 and P_2 be two incomparable subsets of P; that is, $P_1 \not\supseteq P_2$ and $P_2 \not\supseteq P_1$. Then, $\mathcal{R}_{P_1}$ and $\mathcal{R}_{P_2}$ are two incomparable ℓ-quasivarieties. In fact, for $p \in P_1 \setminus P_2$ and $q \in P_2 \setminus P_1$ we have $G_p \in \mathcal{R}_{P_2} \setminus \mathcal{R}_{P_1}$, and $G_q \in \mathcal{R}_{P_1} \setminus \mathcal{R}_{P_2}$.

Since there is a continuum of pairwise incomparable subsets of P, we have exhibited a collection of continuum many pairwise incomparable ℓ-quasivarieties.

As before, let D_l^0 be the ℓ-group of all bounded piecewise linear order permutations of the real numbers R. Let $H = D_l^0 \mathbf{Wr}(Z,Z)$. For each $p \in P$, define H_p as follows:

$$H_p = \mathrm{gp}\langle(\{g_n\}, m) \in H \mid g_i = g_j \text{ if } i \equiv j \pmod{p}\rangle.$$

Since D_l^0 is o-2-transitive, it follows that D_l^0 is not normal-valued. Thus, for all $p \in P$, H_p is not normal-valued either.

Lemma 3.5 (Arora [3](1985)) *$H_p \in \mathcal{R}_q \setminus \mathcal{R}_p$ if p and q are different prime numbers.*

Lemma 3.6 *Let $\{H_{ij} \mid i \in I,\ 1 \leq j \leq n < \infty\}$ be a collection of ℓ-groups. Let $\mathcal{D}$ be an ultrafilter over I. Then*

$$\prod_{i\in I}(H_{i1} \times H_{i2} \times \ldots \times H_{in})/\mathcal{D} \cong \prod_{i\in I} H_{i1}/\mathcal{D} \times \prod_{i\in I} H_{i2}/\mathcal{D} \times \ldots \times \prod_{i\in I} H_{in}/\mathcal{D}.$$

Lemma 3.7 *Let $\mathcal{X}_j (j \in J)$ be a finite collection of ℓ-quasivarieties and let G be an ℓ-group. Then $G \in \bigvee_{j\in J} \mathcal{X}_j$ if and only if there exist $M_j \in \mathcal{X}_j$ $(j \in J)$ such that G can be ℓ-embedded in the ℓ-group $\prod_{j\in J} M_j$.*

Lemma 3.8 $H_p \notin \mathcal{N} \overset{\wedge}{\bigvee} \mathcal{R}_p$.

Theorem 3.9 (Arora [3](1985)) *There exists a continuum of incomparable ℓ-quasivarieties, each properly containing the ℓ-variety $\mathcal{N}$ of normal-valued ℓ-groups.*

Lemma 3.10 (Arora [3](1985)) *Let G be a finitely subdirectly irreducible ℓ-group. Let $\{\mathcal{X}_j \mid j \in J\}$ be a finite collection of ℓ-quasivarieties. Then $G \in \overset{\wedge}{\bigvee_{j \in J}} \mathcal{X}_j$ if and only if $G \in \mathcal{X}_{j_0}$ for some $j_0 \in J$.*

Lemma 3.11 (Arora [3](1985)) $q_\ell(A(R)) = \mathcal{L}$.

Theorem 3.12 (Arora [3](1985)) *The ℓ-variety $\mathcal{L}$ of all ℓ-groups is finitely join-irreducible in the lattice Λ of ℓ-quasivarieties.*

4 The universal theory of abelian ℓ-groups

Let us recall that a first-order formula Φ is a *universal formula* if it is in prenex form and all the quantifiers are universal. An ℓ-group G_1 is *universally equivalent* to an ℓ-group G_2 iff the sets $\mathrm{Th}_\forall G_1$ and $\mathrm{Th}_\forall G_2$ of universal sentences of the signature $l = \{\cdot\, , \,^{-1}\, , \, e\, , \vee, \wedge\}$ which are satisfied by G_1 and G_2, respectively, are the same (Burris and Sankappanavar [7], p.215). For abelian groups we use the additive notation.

Theorem 4.1 (Gurevich and Kokorin [27](1963)) *All nontrivial totally ordered abelian groups are universally equivalent.*

Theorem 4.2 (Khisamiev and Kokorin [40](1966), Khisamiev [39](1966))
(1) Let G and G' be abelian ℓ-groups of finite orthogonal ranks r and r', respectively. Then G is universally equivalent to G' if and only if $r = r'$.
(2) Let G and G' be abelian ℓ-groups of infinite orthogonal rank. Then G and G' are universally equivalent.

It is clear that an ℓ-group of finite orthogonal rank is not universally equivalent to any ℓ-group of infinite orthogonal rank.

Corollary 4.3 *The ℓ-variety $\mathcal{A}$ of abelian ℓ-groups is the smallest non-trivial ℓ-quasivariety in the lattice of ℓ-quasivarieties.*

5 Covers of $\mathcal{A}$ in the lattice of ℓ-quasivarieties

The structure of the lattice Λ of ℓ-quasivarieties is much more complicated than that of the lattice $\mathbf{L}$ of ℓ-varieties. For example, the list of covers of $\mathcal{A}$ in Λ is more diverse than the same in $\mathbf{L}$. As usual we denote by G_p the Scrimger ℓ-group for any prime number p, and by W^+, W^-, N_0 the Medvedev ℓ-groups (see Reilly [63]).

Proposition 5.1 *The ℓ-variety* $\mathrm{var}_\ell(G_p) = \mathcal{S}_p = \mathrm{q}_\ell(G_p)$ *covers* $\mathcal{A}$ *in the lattice of ℓ-quasivarieties.*

The proof of Proposition 5.1 follows from the work of Gurchenkov [16](1984), and Holland, Mekler and Reilly [34](1986).

Proposition 5.2 *The ℓ-quasivarieties* $\mathrm{q}_\ell(W^+)$ *and* $\mathrm{q}_\ell(W^-)$ *cover* $\mathcal{A}$ *in the lattice of ℓ-quasivarieties.*

The proof of Proposition 5.2 follows immediately from Medvedev [49](1977).

By Proposition 5.1 the ℓ-variety $\mathcal{S}_p = \mathrm{var}_\ell(G_p)$ generated by the Scrimger ℓ-group G_p coincides with the ℓ-quasivariety $\mathrm{q}_\ell(G_p)$ generated by G_p. Therefore, we have the following classification of the covers of $\mathcal{A}$ in the lattice of ℓ-quasivarieties.

Theorem 5.3 (Gurchenkov and Kopytov [26](1987), Van Rie [64](1991)) *If $\mathcal{X}$ is a cover of $\mathcal{A}$ in the lattice of ℓ-quasivarieties, then one of the following is true:*
(1) $\mathcal{X} = \mathcal{S}_p = \mathrm{q}_l(G_p)$ *for some prime p where G_p is a Scrimger l-group ;*
(2) $\mathcal{X} \subseteq \mathrm{q}_l(\mathbf{Zwr}(\mathbf{Z},\mathbf{Z}))$;
(3) $\mathcal{X} \subseteq \mathcal{R}$.

We now consider nilpotent ℓ-quasivarieties. Let K_0 be the group having the presentation

$$\begin{aligned} K_0 &= \mathrm{gp}\langle a, b, c, c^+, c^- \mid c = [a,b] = c^+c^-,\ e = [c^+,\ c^-] = \\ &= [c^+, a] = [c^-, a] = [c^-, b] = [c^+, b]\rangle \end{aligned}$$

and lattice-ordered by the rule: $a^k b^l (c^+)^m (c^-)^n \geq e$ iff $k > 0$ or $k = 0$ and $l > 0$ or $k = l = 0$ and $m \geq 0,\ n \leq 0$.

Theorem 5.4 (Medvedev [52](1985)) *Each non-abelian nilpotent ℓ-quasivariety contains K_0. Hence, the ℓ-quasivariety $\mathrm{q}_\ell(K_0)$ is the unique cover of $\mathcal{A}$ in the lattice of nilpotent ℓ-quasivarieties.*

Recalling that N_0 is the nilpotent Medvedev o-group which generates the unique nilpotent ℓ-variety covering $\mathcal{A}$ in the lattice of ℓ-varieties, it is natural to ask whether $q_\ell(N_0)$ and $\mathrm{q}_l(K_0)$ are equal. The answer is negative since the implication

$$([x,y] \vee e = [x,y]) \Rightarrow ([x,y] = e)$$

holds in K_0 and violated in the o-group N_0. Moreover, the representable ℓ-quasivariety $\mathrm{q}_l(K_0)$ does not contain any non-abelian o-group.

Since the ℓ-variety $\mathcal{A}$ is defined by a single identity, by the compactness theorem (see Burris and Sankappanavar [7]), it follows that each member of $\mathcal{A}$ in the lattice $\mathbf{L}$ of varieties of ℓ-groups which properly contains $\mathcal{A}$ contains a cover of $\mathcal{A}$ in the lattice Λ (see, for example, Reilly [63]). All these covers of $\mathcal{A}$ are very close to those of $\mathcal{A}$ in Λ . Now, however, we will construct an infinite set of covers of $\mathcal{A}$ in the lattice Λ which are very far from the above covers.

Let $G = \mathbf{Zwr}(\mathbf{Z},\mathbf{Z})$ and $a = (\{a_t\},0)$, $b = (\{o_t\},1)$ where $o_t = 0 \in \mathbf{Z}$ for all $t \in \mathbf{Z}$ and $a_t = 0$ for $t \neq 0$, $a_0 = 1$. Then $\mathrm{gp}\langle a,b\rangle = G$ and the lower central series

$$G = \gamma_1(G) \geq \gamma_2(G) \geq \dots \gamma_n(G) \geq \dots$$

has the following properties (Kargapolov and Merzljakov [38]):

(4) $\bigcap_{n\in\mathbf{N}} \gamma_n(G) = E$;

(5) the quotient group $\gamma_{i+1}(G)/\gamma_{i+2}(G) = ([a,\underbrace{b,\dots,b}_{i}]\gamma_{i+2}(G))$ is an infinite cyclic group generated by the element $[a,\underbrace{b,\dots,b}_{i}]\gamma_{i+2}(G)$.

Now for each integer $n \geq 2$ and each finite set $\{\varepsilon_0,\varepsilon_1,\dots,\varepsilon_{n-1}\}$, where $\varepsilon_0 = +1, \varepsilon_l = \pm 1 (l = 1,2,\dots,n-1)$, let $Q_n(\varepsilon_0,\varepsilon_1,\dots,\varepsilon_{n-1})$ be the total order of $G = \mathbf{Zwr}(\mathbf{Z},\mathbf{Z})$ uniquely defined by the following inequalities:

(6) $b \gg a \gg |[a,b]| \gg |[a,b,b]| \gg \dots |[a,\underbrace{b,\dots,b}_{k}]| \gg \dots > e$;

(7) $$[a,\underbrace{b,\dots,b}_{k}]^{\varepsilon_{\bar{k}}} > e;$$

where $k \equiv \bar{k} \pmod{n}$ and $\bar{k} \in \{0,1,\dots,n-1\}$.

Direct verification shows that if $[x,y] \neq e$ in the o-group

$$(G, Q(\varepsilon_0,\varepsilon_1,\dots,\varepsilon_{n-1}))$$

for some $x,y \in G$, then $|x| \vee |y| = b^s\varphi$, where $s > 0$ and φ belongs to the abelian radical $\mathcal{A}(G)$ (the largest abelian convex ℓ-subgroup). Let

$$\begin{aligned}
v^{(i)}(x,y) &= v^{(i)}(x,y,\varepsilon_0,\varepsilon_1,\dots,\varepsilon_{n-1}) \\
&= (u_0^{(i)}(x,y)^{\varepsilon_0} \vee e) \wedge (u_1^{(i)}(x,y)^{\varepsilon_1}) \wedge \dots \wedge (u_{n-1}^{(i)}(x,y)^{\varepsilon}_{n-1} \vee e), \\
w(x,y) &= w(x,y,\varepsilon_0,\dots,\varepsilon_{n-1}) \\
&= \bigvee_{i=0}^{n-1} v^i(x,y,\varepsilon_0,\dots,\varepsilon_{n-1}), \\
\tilde{w}(x,y) &= |[w(x,y), |x| \vee |y|]|,
\end{aligned}$$

where

$$\begin{aligned}
u_0^{(0)}(x,y) &= |[x,y]|, \\
u_0^{(i)}(x,y) &= |[u_0^{(i-1)}(x,y), |x| \vee |y|]|, \quad i = 1,2,\dots,n-1; \\
u_j^{(i)}(x,y) &= [u_{j-1}^{(i)}(x,y), |x| \vee |y|], \quad j = 1,2,\dots,n-1.
\end{aligned}$$

Let $\varepsilon_0 = +1, \varepsilon_i = -1,\ i \in \{1, 2, \ldots, n-1\}$ and let (G, Q_n) be the group $G = \mathbf{Zwr}(\mathbf{Z}, \mathbf{Z})$ totally ordered by the order

$$Q_n = Q_n(1, -1, -1, \ldots, -1).$$

Lemma 5.5 *Let $x, y \in G$ and $[x, y] \neq e$. Then*

$$w(x, y) = [a, \underbrace{b, \ldots, b}_{nt-1}]^r \varphi' > e \quad \text{in } (G, Q_n),$$

where $\varphi' \in \gamma_{nt+1}(G)$, and r is a negative integer.

Lemma 5.6 *Let $x, y \in G$ and $[x, y] \neq e$. Then the totally ordered subgroup $H = \mathrm{gp}\langle \tilde{w}(x, y), |x| \vee |y| \rangle$ of (G, Q_n) generated by the elements $\tilde{w}(x, y)$ and $|x| \vee |y|$ is order isomorphic to the o-group (G, Q_n).*

Lemma 5.7 *Let $V = (G, Q_n)^I / \mathcal{F}$ be an ultrapower and $x, y \in V$ such that $[x, y] \neq e$. Then the subgroup $H = \mathrm{gp}(\tilde{w}(x, y), |x| \vee |y|)$ of V is order isomorphic to the o-group (G, Q_n).*

Lemma 5.8 *Let $x, y \in \prod_{j \in J} V_j$, where $V_j = (G, Q_n)^{I_j} / \mathcal{F}$ is an ultrapower of the o-group (G, Q_n) for each $j \in J$ and $[x, y] \neq e$. Then the subgroup $H = \mathrm{gp}\langle \tilde{w}(x, y), |x| \vee |y| \rangle$ of $\prod_{j \in J} V_j$ is totally ordered and is order isomorphic to (G, Q_n).*

The proofs of Lemmas 2, 3 and 4 are all similar.

By Theorem 2.2, $\mathrm{q}_\ell(G, Q_n) = SPP_U(G, Q_n)$, where S, P, P_U are the operations of taking ℓ-subgroups, cartesian products and ultraproducts, respectively. Let H be any non-abelian ℓ-group and $H \in \mathrm{q}_\ell(G, Q_n)$. Then by Lemma 5.8, H contains an o-subgroup which is order isomorphic to (G, Q_n). This yields the following result.

Theorem 5.9 (Isaeva and Medvedev [36](1992)) *Each ℓ-quasivariety*

$$\mathrm{q}_\ell(G, Q_n), \quad n \in \mathbf{Z}, n \geq 2$$

has the following properties:

(8) *$\mathrm{q}_\ell(G, Q_n)$ covers $\mathcal{A}$ in the lattice Λ of ℓ-quasivarieties,*

(9) *$\mathrm{q}_\ell(G, Q_n)$ is contained in the ℓ-variety $\mathcal{W}a$ of weakly abelian l-groups,*

(10) *each nilpotent ℓ-group of $\mathrm{q}_\ell(G, Q_n)$ is abelian.*

It is evident that $\mathrm{q}_\ell(G, Q_m) \neq \mathrm{q}_\ell(G, Q_n)$ if $n \neq m$. In fact, let us assume, on the contrary, that $\mathrm{q}_\ell(G, Q_m) = \mathrm{q}_\ell(G, Q_n)$ for $m < n$. From the definition of the total orders Q_n and Q_m and Lemma 5.5 it follows that the implication $((w(x, y) = e) \Rightarrow ([x, y] = e))$ holds in (G, Q_n) and is violated in (G, Q_m) if $x = a$ and $y = b$.

Here we construct a new cover of $\mathcal{A}$ in the lattice Λ with the following properties:

(11) it does not contain non-abelian o-groups or nilpotent ℓ-groups;
(12) it is contained in the ℓ-variety $\mathcal{W}a$ of weakly-abelian ℓ-groups;
(13) it is metabelian.

Recall that the group $G = \mathbf{ZwrZ}$ described earlier has the properties (2) and (3). Now let I be the set of all infinite sequences $\alpha = (\alpha_0, \alpha_1, \ldots, \alpha_n, \ldots)$, where $\alpha_n \in \{+1, -1\}$. For every sequence $\alpha \in I$, let us consider the linear order P_α on group G, defined uniquely by the following relations:

$$b \gg a^{\alpha_0} \gg [a,b]^{\alpha_1} \gg \ldots \gg [a, \underbrace{b, \ldots, b}_{n}]^{\alpha_n} \gg \ldots > e.$$

For each n, let $(\bar{G}_n, \bar{P}_{\alpha(n)})$ be the quotient group $G/\gamma_n G$ with the naturally induced order. It is not difficult to verify the following properties of (G, P_α):

(14) for sequences $\alpha = (\alpha_0, \alpha_1, \ldots,)$ and $-\alpha = (-\alpha_0, -\alpha_1, \ldots,)$ the o-groups (G, P_α) and $(G, P_{-\alpha})$ are order isomorphic under an isomorphism φ such that $\varphi(b) = b$ and $\varphi(a^{\alpha_0}) = a^{-\alpha_0}$;
(15) the o-groups $(\bar{G}_n, \bar{P}_{\alpha(n)})$ and $(\bar{G}_n, \bar{P}_{-\alpha(n)})$ are also order isomorphic;
(16) the o-group (G, P_α) is a subdirect product of the o-groups $(\bar{G}_n, \bar{P}_{\alpha(n)})$ because $\bigcap_{n=1}^{\infty} \gamma_n G = e$;
(17) if for some elements $x, y \in (G, P_\alpha)$, $|[x,y]| > e$, then $|x| \vee |y| = b^s\varphi$ for some integer $s \in \mathbf{N}$ and $\varphi \in \oplus \sum_{k\in\mathbf{Z}} (a_{b^k}) = \mathrm{fun}((b),(a))$, the base subgroup of G.

We denote by $\overline{\alpha(n)}$ the final segment $(\alpha_n, \ldots, \alpha_{n+l}, \ldots)$ of the sequence $\alpha = (\alpha_0, \alpha_1, \ldots, \alpha_{n+l}, \ldots)$.

Lemma 5.10 *Let $\alpha = (\alpha_0, \alpha_1, \ldots, \alpha_{n-1}, \ldots, \alpha_n, \ldots)$ be a sequence from I; let $s, t \in \mathbf{Z}$ be such that $s > 0$ and $t\alpha_n > 0$; let $\varphi, \psi \in \mathrm{fun}((b),(a))$ with $\psi \in \gamma_{n+2}G$; and let*

$$x = b^s\varphi, \qquad y = [a, \underbrace{b, \ldots, b}_{n}]^t\psi.$$

Then the subgroup $\mathrm{gp}\langle x, y\rangle$ of the o-group (G, P_α) is order isomorphic to the o-group $(G, P_{\overline{\alpha(n)}})$ or $(G, P_{-\overline{\alpha(n)}})$.

Let us denote by $\hat{I} \subseteq I$ the set of all those sequences α such that $\alpha_0 = +1$. Now consider the group $H = \prod_{\alpha\in\hat{I}}(G, P_\alpha)$, and let $\hat{G}$ be the ℓ-subgroup of H generated by $\hat{a}, \hat{b}$, where $\hat{a}(\alpha) = a$ and $\hat{b}(\alpha) = b$ for all α.

Lemma 5.11 *If for some elements $x, y \in \hat{G}$ $|[x,y]| \neq e$, then ℓ-$gp\langle |[x,y]|, |x| \vee |y|\rangle$ (the ℓ-subgroup of $\hat{G}$ generated by the elements $|[x,y]|$ and $|x| \vee |y|$) is ℓ-isomorphic to $\hat{G}$ under an ℓ-isomorphism which maps $\hat{a} \mapsto |[x,y]|$ and $\hat{b} \mapsto |x| \vee |y|$.*

Let $H_1 = \hat{G}^J/\mathcal{F}$ be an ultrapower of the ℓ-group $\hat{G}$. The following statements follow immediately from the definitions and Lemma 5.11.

Lemma 5.12 *Let u, v be elements of $H_1 = \hat{G}^J/\mathcal{F}$ and $|[u,v]| \neq e$. Then the ℓ-subgroup ℓ-$gp\langle|[u,v]|, |u| \vee |v|\rangle$ is ℓ-isomorphic to $\hat{G}$.*

Lemma 5.13 *Let X be a non-abelian ℓ-subgroup of the cartesian product $\prod_{\gamma\in\Gamma} V_\gamma$ where each V_γ is an ultrapower of the ℓ-group $\hat{G}$ .Then for every pair of elements $u, v \in X$ such that $[u,v] \neq e$, ℓ-$gp\langle|u| \vee |v|, |[u,v]|\rangle$ is ℓ-isomorphic to $\hat{G}$.*

By Theorem 2.2, the ℓ-quasivariety $\mathrm{q}_\ell(\hat{G})$ generated by the ℓ-group $\hat{G}$ coincides with the class of ℓ-groups $SPP_U(\hat{G})$, where S, P, and P_U are the closure operations of forming ℓ-subgroups, cartesian products and ultraproducts. Therefore, if X is a non-abelian ℓ-group from $\mathrm{q}_\ell(\hat{G})$, then by Lemma 5.13, X contains an ℓ-subgroup which is ℓ-isomorphic to the ℓ-group $\hat{G}$. It follows that every non-abelian ℓ-quasivariety $\mathcal{Q}_\ell \subseteq \mathrm{q}_\ell(\hat{G})$ contains the ℓ-group $\hat{G}$ and so $\mathcal{Q}_\ell = \mathrm{q}_\ell(\hat{G})$.

Theorem 5.14 (Medvedev [56](1992)) *The ℓ-quasivariety $\mathrm{q}_\ell(\hat{G})$ has the following properties:*

(18) *$\mathrm{q}_\ell(\hat{G})$ covers the ℓ-variety $\mathcal{A}$ of abelian ℓ-groups in the lattice Λ of ℓ-quasivarieties;*

(19) *every o-group in $\mathrm{q}_\ell(\hat{G})$ is abelian;*

(20) *every nilpotent ℓ-group in $\mathrm{q}_\ell(\hat{G})$ is abelian;*

(21) *$\mathrm{q}_\ell(\hat{G})$ is contained in the ℓ-variety $\mathcal{W}a$ of weakly abelian ℓ-groups.*

The following series of covers of $\mathcal{A}$ was discovered by S. V. Varaksin [67](1993). Let $F = F(x,y)$ be the free group on x and y, and

$$F = \gamma_1(F) \geq \gamma_2(F) \geq \ldots \geq \gamma_n(F) \geq \gamma_{n+1}(F) \geq \ldots$$

be the lower central series of F. It is well known (see, e.g., Hall [28]) that each element $\overline{a}$ of the factor-group $\gamma_n(F)/\gamma_{n+1}(F)$ can be uniquely represented in the form $\overline{a} = \alpha_1^{k_1} \cdots \alpha_l^{k_l} \gamma_{n+1}(F)$, where $\alpha_1, \ldots, \alpha_l$ are basic commutators of weight n and $\alpha_1 \prec \alpha_2 \prec \ldots \prec \alpha_l$ in the standard order on the set of basic commutators of the weight n.

Now we fix some infinite sequence $\varphi = \varepsilon_1, , \ldots, \varepsilon_l, \ldots$ with each $\varepsilon = \pm 1$, and define a total lexicographic order P_n on the factor-group $\gamma_n(F)/\gamma_{n+1}(F)$ by setting:

$$\alpha_1^{\varepsilon_1}\gamma_{n+1}(F) > \gamma_{n+1}(F), \ldots, \alpha_l^{\varepsilon_l}\gamma_{n+1}(F) > \gamma_{n+1}(F),$$

and $\alpha_1^{\varepsilon_1}\gamma_{n+1}(F) \ll \ldots \ll \alpha_l^{\varepsilon_l}\gamma_{n+1}(F)$. Since $\bigcap_{n=1}^{\infty} \gamma_n(F) = E$, we can define a total lexicographic order Q_φ on$F$ by setting $a > e, a \in F$ if and only if $a\gamma_{n_0+1}(F) > \gamma_{n_0+1}(F)$ in the total order P_{n_0}, where n_0 is the smallest integer n such that $a\gamma_{n+1}(F) \neq \gamma_{n+1}(F)$. It is clear that there exist $2^{\aleph_0}$ different lexicographic total orders Q_φ. Let I be the set of all sequences of ± 1.

Theorem 5.15 (Varaksin [67](1993)) *There exists a subset $I_l \subseteq I$ of cardinality $2^{\aleph_0}$ and lexicograpically totally ordered free groups $(F, Q_\varphi), (\varphi \in I_1$ such that:*

1) $q_\ell(F, Q_\varphi)$ covers $\mathcal{A}$ in the lattice Λ of ℓ-quasivarieties;

2) all these covers $q_\ell(F, Q_\varphi)$ are different and are contained in the ℓ-variety $\mathcal{W}a$ of weakly-abelian ℓ-groups

6 Covers in the lattice of ℓ-quasivarieties

Here we will exhibit some ℓ-quasivarieties which have no covers in the lattice Λ of ℓ-quasivarieties.

Let P be the set of all prime numbers and $M \subseteq P$. The ℓ-quasivariety $\mathcal{Q}_M$ is defined by the following system of implications Σ_M:

$$((|x| \vee |y|)^{-1}|[x,y]|(|x| \vee |y|) = |[x,y]|^{p^k}) \to ([x,y] = e),$$

where $k \in N, k \geq 2, p \in M \subseteq P$. Let $G_{p^k} = \text{gp}\langle c, d \mid c^{-1}dc = d^{p^k}\rangle$. Then the normal closure $D = \text{gp}\langle d^g \mid g \in G_{p^k}\rangle$ of the element d is isomorphic to the subgroup of the rational numbers $\mathbf{Q}$ which consists of the elements $\frac{a}{b}$ with $a \in \mathbf{Z}$ and $b = p^{kn}$, $n \in Z$. Thus, G_{p^k} is the semidirect product of D and the infinite cyclic group $\langle c\rangle$. Since D is isomorphic to a subgroup of the rational numbers Q, then D has only two different total orders. Let P be a total order on D such that $d > e$. Now we define a total order on G_{p^k} by the rule: $c^i d \geq e$ iff $i > 0$ or $i = 0$ and $d \in D$, $d \geq e$ in o-group (D, P).

Theorem 6.1 (Medvedev [52](1985)) *For each non-empty set M of prime numbers there is no ℓ-quasivariety covering the ℓ-quasivariety $\mathcal{Q}_M$ in the lattice Λ of ℓ-quasivarieties.*

Corollary 6.2 *There are $2^{\aleph_0}$ different ℓ-quasivarieties each of which has no covers in the lattice Λ.*

In the theory of quasivarieties (or varieties) of algebraic systems there is a well-known connection between independent bases of implications (identities) of a quasivariety (variety) and covers in the lattice of quasivarieties (varieties) of this quasivariety (variety). We formulate this connection for ℓ-quasivarieties and ℓ-varieties.

Proposition 6.3 (Gorbunov [14](1977)) *Let $\mathcal{V} \subseteq \mathcal{W} \subseteq \mathcal{L}$ be ℓ-varieties (ℓ-quasivarieties) and let $\mathcal{W}$ be finitely based. Let $\mathcal{V}$ have an infinite independent basis of identities (implications). Then there is an infinite set of ℓ-varieties (ℓ-quasivarieties) $\mathcal{U}_i$, $i = 1, 2, \ldots, n, \ldots$ such that each $\mathcal{U}_i$ covers $\mathcal{V}$ in the lattice of ℓ-varieties (ℓ-quasivarieties) and $\mathcal{U}_i \subseteq \mathcal{W}$, $i = 1, 2, \ldots, n, \ldots$.*

Proof. Let $\Phi = \{\varphi_j \mid j \in \mathbf{N}\}$ be an infinite independent basis of identities (implications) of $\mathcal{V}$. Let

$$\Phi_k = \Phi \setminus \{\varphi_k\}, \ k \in \mathbf{N}$$

and let $\mathcal{V}_k$ be the ℓ-variety (ℓ-quasivariety) defined by the identities Φ_k, $k \in \mathbf{N}$. It is clear that $\mathcal{V}_{k_1} \cap \mathcal{V}_{k_2} = \mathcal{V}$ for all $k_1, k_2 \in \mathbf{N}$, $k_1 \neq k_2$. From the compactness theorem (see Burris and Sankappanavar [7] , page 212), it follows that there exists a finite subset $K \subseteq \mathbf{N}$ such that $\mathcal{V}_k \subseteq \mathcal{W}$ for all $k \in \mathbf{N} \setminus K$. We claim that each $\mathcal{V}_k$ ($k \in \mathbf{N} \setminus K$) contains a cover of $\mathcal{V}$. Assume, on the contrary, that $\mathcal{V}_k$ has no cover of $\mathcal{V}$. Then there exists a descending chain of ℓ-varieties (ℓ-quasivarieties) $\mathcal{V} \subsetneq \ldots \subsetneq \mathcal{X}_\alpha \subsetneq \ldots \subsetneq \mathcal{V}_k \subseteq \mathcal{W}$ for which $\mathcal{W} = \bigcap_{\alpha \in I} \mathcal{X}_\alpha$. By our assumption, $\mathcal{V}$ is defined in the ℓ-variety (ℓ-quasivariety) $\mathcal{V}_k$ by one identity (implication) φ_k. By the compactness theorem, the identity φ_k follows from the identities defining the ℓ-variety (ℓ-quasivariety) $\mathcal{X}_{\alpha_0}$. Thus, $\mathcal{V} = \mathcal{X}_{\alpha_0}$, in contradiction to our assumption.

Corollary 6.4 *Each ℓ-quasivariety $\mathcal{Q}_M$ has no independent basis of implications.*

The proof follows immediately from Proposition 6.3.

7 ℓ-Varieties

We begin by recalling a fundamental theorem due to G. Birkhoff (see Burris and Sankappanavar [7]).

Theorem 7.1 (Birkhoff [5](1935)) *For a non-empty class $\mathcal{K}$ of ℓ-groups the following statements are equivalent:*

(a) *$\mathcal{K}$ is closed under the formation of Cartesian products, homomorphic images and ℓ-subgroups;*

(b) *$\mathcal{K}$ is equationally defined (that is, for some family of identities $\Sigma, \mathcal{K}$ is the class of all ℓ-groups which satisfy all identities of Σ).*

For any class $\mathcal{K}$ of ℓ-groups let $H(\mathcal{K})$ denote the class of ℓ-groups which are homomorphic images of elements of $\mathcal{K}$. Let $\mathrm{var}_\ell \mathcal{K}$ be the ℓ-variety generated by $\mathcal{K}$ (that is, the smallest ℓ-variety containing $\mathcal{K}$). From Theorem 7.1 it follows that $\mathrm{var}_\ell \mathcal{K} = HSP(\mathcal{K})$.

By a result of G. Birkhoff's [6](1942), the lattice of ideals of any ℓ-group is Brouwerian and so distributive. Therefore there is another description due to B. Jónsson, using ultraproducts, of the ℓ-variety generated by a class $\mathcal{K}$.

Theorem 7.2 (Jónsson [37](1967)) *Let $\mathcal{V} = \mathrm{var}_\ell(\mathcal{K})$ be the ℓ-variety generated by some class $\mathcal{K}$ of ℓ-groups. If A is a finitely subdirectly irreducible ℓ-group in $\mathcal{V}$, then $A \in HSP_U(\mathcal{K})$ where H, S, and P_U are the operations of ℓ-homomorphic images, ℓ-subgroups and ultraproducts, respectively. Hence,*

$$\mathcal{V} = \mathrm{var}_\ell(\mathcal{K}) = SPHSP_U(\mathcal{K}).$$

We will denote the ℓ-variety of all ℓ-groups by $\mathcal{L}$ and the ℓ-variety of trivial ℓ-groups by $\mathcal{E}$. Some ℓ-varieties are defined by identities in the signature of group theory alone, $\{\ \cdot\ ,\ ^{-1}\ ,\ e\ \}$. We call them the *group ℓ-varieties.* For group ℓ-varieties we use the following notation:

(1) $\mathcal{A}$ is the ℓ-variety of all abelian ℓ-groups;
(2) $\mathcal{N}_k$ is the ℓ-variety of all nilpotent ℓ-groups of class $\leq k$;
(3) $\mathcal{A}^k$ is the ℓ-variety of all ℓ-soluble ℓ-groups of ℓ-soluble length $\leq k$;

As an example of an ℓ-variety which is not a group ℓ-variety, let $\mathcal{X}$ be the ℓ-variety defined by the identities $[x, y, z] = e$ and $[|[x, y]| \wedge |z|, t] = e$. It is easy to see that $G \in \mathcal{X}$ iff the ℓ-ideal of G generated by the derived subgroup G' of G is central. It is not difficult to construct an example of a totally ordered nilpotent group G of class 2, such that the convex subgroup generated by G' is not central. Therefore, $\mathcal{X} \neq \mathcal{N}_2$. On the other hand, it is well known that every torsion-free nilpotent group G has a total order such that the centre Z of G is a convex subgroup of G. Applying this observation to a free nilpotent group of class 2, it is evident that every identity of signature $\{\ \cdot\ ,\ ^{-1}\ ,\ e\ \}$ which is valid in all ℓ-groups from $\mathcal{X}$ is valid in $\mathcal{N}_2$, too. Hence, the ℓ-variety $\mathcal{X}$ cannot be defined by identities of signature $\{\ \cdot\ ,\ ^{-1}\ ,\ e\ \}$.

We next recall some very important ℓ-varieties:

(4) $\mathcal{N}$ is the ℓ-variety defined by the identity $|x| \cdot |y| \wedge |y^2| \cdot |x^2| = |x| \cdot |y|$;
(5) $\mathcal{R}$ is the ℓ-variety defined by the identity $(x \wedge y^{-1}x^{-1}y) \vee e = e$;
(6) $\mathcal{W}a$ is the ℓ-variety defined by the identity $x^{-1}|y|x|y|^{-2} \vee e = e$.

It is clear that the set $\mathbf{L}$ of all ℓ-varieties is a complete lattice where for any ℓ-varieties $\mathcal{X}, \mathcal{Y}$ the greatest lower bound, or meet, of $\mathcal{X}$ and $\mathcal{Y}$ is their set-theoretic intersection and the least upper bound, or join, is the intersection of all ℓ-varieties containing both $\mathcal{X}$ and $\mathcal{Y}$.

8 The lattice of nilpotent ℓ-varieties

For a long time there was the conjecture that the lattice of ℓ-varieties of nilpotent ℓ-groups might be in some sense simpler than the lattice $\mathbf{L}$ of all ℓ-varieties. These hopes were dashed by the results of S. A. Gurchenkov [17, 19] and N. Ya. Medvedev [54].

Let us consider the following $2n$ ($n \in \mathbf{N}$) terms of the signature ℓ;

$$\begin{aligned} u_1 &= (|[x, y]| \vee t_1) \wedge |[x, y]|^2, & w_1 &= [|x| \vee |y|, u_1], \\ u_2 &= (|w_1| \vee t_2) \wedge |w_1|^2, & w_2 &= [|x| \vee |y|, u_2], \\ &\cdots & &\cdots \\ u_n &= (|w_{n-1}| \vee t_n) \wedge |w_{n-1}|^2, & w_n &= [|x| \vee |y|, u_n]. \end{aligned}$$

Now for every positive integer $k \geq 3$ we define the following identity

$$f_k = w_1^+ \wedge \cdots \wedge w_k^+ \wedge [u_1, u_2]^+ \wedge [u_2, u_3]^+ \wedge \cdots \wedge [u_{k-1}, u_k]^+ \wedge [u_1, u_k]^+ = e.$$

Let $\mathcal{U}_n$ be the ℓ-variety defined by the identities $f_k = e$, for $k = 3, 4, \ldots, n$ and $[x, y, z] = e$. It is evident that

$$\mathcal{U}_3 \supseteq \mathcal{U}_4 \supseteq \ldots \supseteq \mathcal{U}_n \supseteq \mathcal{U}_{n+1} \supseteq \ldots .$$

N. Ya. Medvedev proved that each containment is proper.
Therefore, we have the following result.

Theorem 8.1 (Medvedev [54](1989))

a) *The ℓ-variety $\mathcal{U} = \bigcap\limits_{n \in \mathbf{N}} \mathcal{U}_n$ of nilpotent ℓ-groups of class ≤ 2 is not finitely based.*

b) *The set of identities $\{f_n = e \mid n \in \mathbf{N}\}$ is independent.*

c) *The lattice of nilpotent ℓ-varieties of class ≤ 2 has uncountable cardinality.*

Corollary 8.2 *The ℓ-variety $\mathcal{U} = \bigcap_{n \in \mathbf{N}} \mathcal{U}_n$ has infinite axiomatic rank.*

Now let us consider the following terms of the signature ℓ:

$$\begin{aligned}
u_1 &= |[x, y]| \wedge |z_1|, \\
u_2 &= (|[x, y_1]| \vee |[y, u_1]|) \wedge |z_2|, \\
u_3 &= (|[x, u_2]| \vee |[y, u_2]|) \wedge |z_3|, \\
u_4 &= |[x, u_3]| \vee |[y, u_3]|, \\
v &= |[u_1, x]|^{-1} u_3 \wedge |[u_2, y]|^{-1} u_4 \wedge x \wedge y, \\
u &= v \wedge x^2 y^{-1} \wedge y^2 x^{-1}.
\end{aligned}$$

For each pair of positive integers m, n let

$$\begin{aligned}
w_{m,n} &= (v \wedge x^2 y^{-1} \wedge |[x, u_3]|^{-2n} |[y, u_3]|^{m}) \vee e, \\
u_{m,n} &= (v \wedge x^2 y^{-1} \wedge |[x, u_3]|^{n} |[y, u_3]|^{-2m}) \vee e.
\end{aligned}$$

Now for each positive real number $\alpha > 0$ let

$$\begin{aligned}
\mathcal{U}_\alpha &= \mathcal{A}^2 \overset{\mathbf{L}}{\bigwedge} \mathcal{N} \overset{\mathbf{L}}{\bigwedge} \mathrm{var}_l\{u_{n,m} = e \mid \ n, m \in \mathbf{N}, \ \ n/m \leq \alpha\}, \\
\mathcal{U}^\alpha &= \mathcal{A}^2 \overset{\mathbf{L}}{\bigwedge} \mathcal{N} \overset{\mathbf{L}}{\bigwedge} \mathrm{var}_l\{w_{n,m} = e \mid \ n, m \in \mathbf{N}, \ \ n/m \leq \alpha\}, \\
\mathcal{B} &= \mathcal{A}^2 \overset{\mathbf{L}}{\bigwedge} \mathcal{N} \overset{\mathbf{L}}{\bigwedge} \mathrm{var}_l\{u \vee e = e\}.
\end{aligned}$$

Theorem 8.3 (Gurchenkov [20](1990))

a) *The ℓ-variety $\mathcal{B}$ has $2^{\aleph_0}$ different covers in the lattice of nilpotent ℓ-varieties.*

b) *For each positive irrational real number α, neither of the ℓ-varieties $\mathcal{U}_\alpha$ and $\mathcal{U}^\alpha$ has a finite basis of identities.*

c) *The lattice of nilpotent ℓ-varieties is not Brouwerian.*

It is known (see [Gurchenkov [19]) that the lattice $\mathbf{L}$ of all ℓ-varieties has the covering condition (i.e., each non-trivial ℓ-variety has a cover in the lattice $\mathbf{L}$) and (see Medvedev [50]) the lattice $\mathbf{L}_{\mathcal{R}}$ of all representable ℓ-varieties does not have the covering condition.

The following results were proved by S. A. Gurchenkov.

Theorem 8.4 (Gurchenkov [19](1990))

a) *The lattice of all weakly abelian ℓ-varieties does not have the covering condition.*

b) *The lattice of all nilpotent ℓ-varieties does not have the covering condition.*

c) *For each $k \geq 5$, the lattice $\mathbf{L}_{n_k}$ of nilpotent ℓ-varieties of class $\leq k$ does not have the covering condition.*

By Proposition 6.3 each non-trivial ℓ-variety without covers in the lattice of the ℓ-varieties contained in some finitely based ℓ-variety has no independent basis of identities. Therefore the following result is connected with the covers in the lattice $\mathbf{L}_{n_k}$.

Theorem 8.5 (Gurchenkov [18](1988),[19](1990)) *There exists an ℓ-variety $\mathcal{X}$ of nilpotent ℓ-groups of class ≤ 3 without an independent basis of identities.*

Let us recall that an ℓ-variety $\mathcal{X}$ is *divisible* if each ℓ-group $G \in \mathcal{X}$ can be embedded in a divisible ℓ-group $G^* \in \mathcal{X}$. It is known that the ℓ-variety $\mathcal{L}$ of all ℓ-groups is divisible (Holland [30](1963)).

Theorem 8.6 (Gurchenkov [21](1991))

a) *There are uncountably many nilpotent class 2 divisible ℓ-varieties with finite axiomatic rank.*

b) *There are uncountably many nilpotent class ≤ 3 ℓ-varieties each of which is not divisible.*

In [23](1994), S. A. Gurchenkov constructed an infinite family of weakly abelian ℓ-varieties $\mathcal{M}_p$, (p is a prime number) which are not generated by their nilpotent ℓ-groups.

Let us recall that a group G is called an *Engel group* if for each pair of elements $x, y \in G$ there is an integer $t = t(x, y) > 0$ such that

$$[x, \underbrace{y, \ldots, y}_{t}] = e.$$

If the identity holds in a group G for some fixed positive integer t independent of x and y, then G is called a t-Engel group.

For any pair $x, y \in G$ with $x > e$, let

$$M(x, y) = \{x \wedge y^{-i_1} x y^{i_1} \wedge \cdots \wedge y^{-i_k} x y^{i_k} \mid i_1, \ldots, i_k \in \mathbf{Z},\ k \geq 1\}.$$

The following results give necessary and sufficient conditions for representability of Engel ℓ-groups.

Theorem 8.7 (Medvedev [53](1988)) *Let G be an Engel ℓ-group. Then G is representable if and only if for every $x, y \in G$, $x > e$, there exists a positive integer $t = t(x, y)$ such that the equality $[u, \underbrace{y, \ldots, y}_{t}] = e$ holds in G for all $u \in M(x, y)$.*

Theorem 8.8 (Gurchenkov [25](1994)) *Let G be an Engel ℓ-group. Then* $\mathrm{var}_\ell(G) \subset \mathcal{N}$ *if and only if* $\mathrm{var}_\ell(G) \subset \mathcal{R}$.

Corollary 8.9 *Let G be a bounded t-Engel ℓ-group. Then G is representable.*

The following surprising result was obtained by Y. K. Kim and A. H. Rhemtulla with the use of this Corollary.

Theorem 8.10 (Kim and Rhemtulla [41](1992)) *A bounded t-Engel lattice-ordered group is nilpotent.*

9 Covers in the lattice of ℓ-varieties

The smallest proper ℓ-variety in the lattice $\mathbf{L}$ is the abelian ℓ-variety $\mathcal{A}$. Because $\mathcal{A}$ is defined by a single equation, it follows from general principles that every ℓ-variety which properly contains $\mathcal{A}$ contains a cover of $\mathcal{A}$ in the lattice $\mathbf{L}$ of ℓ-varieties. All the covers of $\mathcal{A}$ containing non-representable ℓ-groups are known; they are the prime Scrimger varieties (see Reilly [63]). The covers of $\mathcal{A}$ consisting of only representable ℓ-groups may not be completely known. Medvedev [49](1977) found three solvable covers of $\mathcal{A}$ which are representable (generated by o-groups), and proved that they are the only solvable representable covers. Kopytov [43](1985) and Bergman [4](1984), independently, discovered two more representable covers.

Then Holland and Medvedev [33](1994) produced an uncountable collection of representable covers of $\mathcal{A}$. Each of these ℓ-varieties is generated by a free group, totally ordered so that conjugation by certain elements produces large shifts of the archimedean classes, either up or down, according to the pattern of any prescribed infinite sequence of ± 1's. Specifically, if a and b are free generators and $e \ll a \ll b$, then there is generated a sequence of ± 1's (corresponding to "up" and "down", respectively) by considering the effect of conjugation first of b on a, then by the larger of a and a^b on the smaller, etc. The Kopytov-Bergman groups correspond to the constant sequences of all $+1$'s and all -1's, respectively. Holland and Medvedev were able to order the free group so that *any* given sequence of ± 1's can be produced. Furthermore, the action of conjugating up or down can be captured by equations. From this, the following theorem was proved.

Theorem 9.1 (W. C. Holland and N. Ya. Medvedev [33](1994)) *There are an uncountable number of representable covers of the abelian ℓ-variety $\mathcal{A}$ in the lattice of ℓ-varieties.*

Indeed, using a combinatorial argument of G. Bergman (personal communication), the cardinality in Theorem 9.1 can be shown to be $2^{\aleph_0}$.

Medvedev [50](1982) defined a variety containing all representable covers of $\mathcal{A}$. Sometime later, M. Anderson, M. Darnel and T. Feil [2] also found a variety containing all representable covers of $\mathcal{A}$.

Let $\mathcal{C}$ be the ℓ-variety defined by the infinite system of identities

$$(([|x| \wedge |y|, |x|] \vee e) \wedge (|x| \wedge |y|))^n \wedge ((|x| \wedge |y|) \vee |x|^{-1}(|x| \wedge |y|)|x|) =$$
$$= (([|x| \wedge |y|, |x|] \vee e) \wedge (|x| \wedge |y|))^n \quad (n \in \mathbf{N}).$$

Theorem 9.2 (Anderson, Darnel, and Feil [2](1991)) *The ℓ-variety $\mathcal{C}$ contains all covers of $\mathcal{A}$ contained in $\mathcal{R}$.*

We recall that a formula φ is *universal* if it is of the form $\forall x_1, \ldots, \forall x_n\ \psi$, where ψ is quantifer-free. A class $\mathcal{K}$ is *universal class* if for some family of universal formulas Σ, $\mathcal{K}$ is the class of all ℓ-groups which satisfy all formulas of Σ. The question about covers of ℓ-varieties in the lattice $\mathbf{T}$ of universal classes of ℓ-groups was considered by Medvedev [57].

Let $\mathcal{U}$ be the ℓ-variety defined by the following system of identities Σ_1 :

a) $(|x| \vee |y|)^{-1}(|[x,y]| \wedge |t|)(|x| \vee |y|) \wedge (|[x,y]| \wedge |t|)^n =$
$= (|[x,y]| \wedge |t|)^n, \quad (n \in \mathbf{N}, \quad n \geq 3),$

b) $(x \wedge y^{-1}x^{-1}y) \vee e = e,$

c) $[|[x,y]| \wedge |t|, |\,[x_1,y_1]| \wedge |t_1\,|\,|] = e.$

By $\hat{\mathcal{U}}$ we denote the ℓ-variety defined by the system of identities $\hat{\Sigma}$:

a) $(|x| \vee |y|)^{-1}(|[x,y]| \wedge |t|)(|x| \vee |y|) \wedge (|[x,y]| \wedge |t|)^3 = (|[x,y]| \wedge |t|)^3,$

b) $(x \wedge y^{-1}xy) \vee e = e,$

c) $[|[x,y]| \wedge |t|, |[x_1,y_1]| \wedge |t|_1] = e.$

It is clear that $\mathcal{U} \subseteq \hat{\mathcal{U}}$ and $\mathcal{U} \neq \hat{\mathcal{U}}$.

Theorem 9.3 (Medvedev [57](1992)) *In the lattice* $\mathbf{T}$ *of universal classes of ℓ-groups there are no universal classes* $\mathcal{W}$ *such that*

a) $\mathcal{W}$ *covers the ℓ-variety* $\mathcal{U}$ *in the lattice* $\mathbf{T}$;

b) $\mathcal{W} \subseteq \hat{\mathcal{U}}$.

Corollary 9.4 *The ℓ-variety* $\mathcal{U}$ *has no independent basis of universal formulas.*

M. R. Darnel [11](1994) studied non-representable covers of the ℓ- variety $\mathcal{R}$ of all representable ℓ-groups. In that paper, a variation of the method used to construct the Scrimger ℓ-varieties is developed that is shown to produce every non-representable cover of any representable ℓ-variety. All non-representable covers of any weakly abelian ℓ-variety and any ℓ-metabelian representable ℓ-variety are described.

It is shown that any non-representable ℓ-variety is a subvariety of a certain quasi-representable ℓ-variety defined by N. Reilly [62](1981).

10 The semigroup of ℓ-varieties

It is known (Reilly [63](1989)) that the lattice-ordered semigroup $\mathbf{L}$ of of ℓ-varieties satisfies the identity

$$\mathcal{U}(\mathcal{V}_1 \bigwedge \mathcal{V}_2) = \mathcal{U}\mathcal{V}_1 \bigwedge \mathcal{U}\mathcal{V}_2.$$

N. Ya. Medvedev [55](1989) proved that the identity

$$\mathcal{U}(\bigwedge_{i \in I} \mathcal{V}_i) = \bigwedge_{i \in I} \mathcal{U}\mathcal{V}_i$$

is not valid in the semigroup $\mathbf{L}$ for an infinite index set I.

Let us recall that for the ℓ-variety $\mathcal{R}$ of representable ℓ- groups and for representable ℓ-varieties $\mathcal{U}, \mathcal{V}$ we set $\mathcal{U} * \mathcal{V} = \mathcal{U} \cdot \mathcal{V} \cap \mathcal{R}$.

N. Ya. Medvedev [54](1989) also proved that the identity

$$\mathcal{U} * (\bigwedge_{i \in I} \mathcal{V}_i) = \bigwedge_{i \in I} \mathcal{U} * \mathcal{V}_i$$

is not valid in the semigroup $\mathbf{L}_{\mathcal{R}}$ of representable ℓ- varieties for an infinite index set I.

As usual, let

$$\mathcal{A}^1 = \mathcal{A}, \quad \mathcal{A}^{n+1} = \mathcal{A}^n\mathcal{A} \quad \text{for } n \in \mathbf{N},$$

$$\mathcal{A}^\omega = \bigvee_{n\in\mathbf{N}} \mathcal{A}^n, \mathcal{A}^{*\,1} = \mathcal{A}, \quad \mathcal{A}^{*\,n+1} = \mathcal{A}^{*\,n} * \mathcal{A} \text{ for } n \in \mathbf{N}$$

and

$$\mathcal{A}^{*\,\omega} = \bigvee_{n\in\mathbf{N}} \mathcal{A}^{*\,n}.$$

Li Si Ze [47](1990) proved the following statements;

$$\mathcal{A}^{*\,\omega} * \mathcal{A}^{*\,n} \subset \mathcal{A}^{*\,\omega} * \mathcal{A}^{*\,n+1} (n = 1, 2, \ldots) \quad (1)$$

$$\bigvee_{n=1}^{\infty} (\mathcal{A}^{*\,\omega} * \mathcal{A}^{*\,n}) = \mathcal{R}. \quad (2)$$

The powers of the ℓ-variety $\mathcal{R}$ of all representable ℓ-groups were considered by M. Darnel [11](1994). He proved the following statements:

a) Let $\mathcal{V}$ be any ℓ-variety such that $\mathcal{V} \subset \mathcal{N}$ and $\mathcal{V} \neq \mathcal{N}$. Then $\mathcal{V} \subseteq \mathcal{R}^n$ for some positive integer n.

b) An ℓ-variety $\mathcal{V}$ is solvable if and only if $\mathcal{V} \cap \mathcal{R}$ is solvable; and $\mathcal{V}$ is ℓ-solvable if and only if $\mathcal{V} \cap \mathcal{R}$ is ℓ-solvable.

For some finitely based varieties of ℓ-groups $\mathcal{U}$ and $\mathcal{V}$ it is possible to prove the existence of a finite basis of identities of the product $\mathcal{U}\mathcal{V}$. For example, we have the following.

Theorem 10.1 (Litvinova [46](1994))*Let $\mathcal{N}_k$ be the ℓ-variety of all nilpotent ℓ-groups of class $\leq k$ and $\mathcal{L}_n$ be the ℓ-variety defined by the identity*

$$[x, y] = e.$$

Then the ℓ-varieties $\mathcal{N}_k \cdot \mathcal{V}$ and $(\mathcal{L}_n \cap \mathcal{A}^2) \cdot \mathcal{V}$ are finitely based for each finitely based ℓ-variety $\mathcal{V}$.

11 Ordered permutation groups and ℓ-varieties

Let Ω be a totally ordered set and $A(\Omega)$ be the ℓ-group of all order preserving permutations of Ω. One can ask which ℓ- varieties can be generated by the ℓ-groups $A(\Omega)$. Such ℓ-varieties are now completely known. This remarkable result is due to W. C. Holland.

Theorem 11.1 (Holland [32](1985)) *Let $A(\Omega)$ be the ℓ-group of all order preserving permutations of the totally ordered set Ω. Then* $\mathrm{var}_\ell(A(\Omega))$ *is either the ℓ-variety $\mathcal{L}$ of all ℓ- groups, the ℓ-variety $\mathcal{N}$ of all normal-valued ℓ-groups, the ℓ-variety $\mathcal{E}$ of trivial ℓ-groups, or $\mathcal{A}^n$ for some positive integer n.*

Let G be a right-ordered group and $R : G \to A(G, \leq)$ be the right regular representation of the right-ordered group $(G, \leq)$. Then $R(G)$ is a partially ordered subgroup of the ℓ-group $A(G, \leq)$ of all order preserving permutations of the totally ordered set $(G, \leq)$. Let G^* be the ℓ-subgroup of $A(G, \leq)$ generated by the subgroup $R(G)$. Some properties of G^* are inherited from G. For example, if a right-ordered group G is Conradian, then G^* is a normal-valued ℓ-group (Kopytov [42](1979)).

S. V. Varaksin [65], M. Darnel and A. M. W. Glass [13], and M. Darnel [8] investigated the relation between the group identities of G and the ℓ-group identities of G^*.

Theorem 11.2 (Varaksin [65](1989)) *Let $(G, \leq)$ be a right-ordered group and suppose there is a finite subnormal series*

$$G = G_n \geq \ldots \geq G_1 \geq G_0 = e$$

of convex subgroups, all factors G_k/G_{k-1} being o-groups $(k = 1, \ldots, n-1)$. *If* $\mathrm{var}_\ell(G_k/G_{k-1}) = \mathcal{X}_k$, *then* $\mathrm{var}_l G^* \subseteq \mathcal{X}_1 \cdot \mathcal{X}_2 \cdots \mathcal{X}_n$.

Corollary 11.3 (Darnel [8](1991)) *Let G be a finitely generated torsion-free nilpotent group of Hirsch length n. Then the free lattice-ordered group $F(G)$ over G is ℓ-solvable of rank at most n.*

Corollary 11.4 (Darnel and Glass [13](1989)) *The free lattice-ordered group over any m-generator torsion-free nilpotent group of class 2 is ℓ-solvable of length at most* $\binom{m}{2} + 1$.

The varieties of lattice-ordered groups generated by simple ℓ-groups were considered by S. V. Varaksin.

Theorem 11.5 (Varaksin [66](1990)) *Let G be an ℓ-simple non-representable ℓ-group. Then the ℓ-variety generated by G is $\mathcal{N}$ or $\mathcal{L}$.*

Corollary 11.6 *Every (group-)solvable ℓ-simple ℓ-group is abelian.*

Let us recall that the ℓ-variety $\mathcal{X}$ has the *amalgamation property* if for all ℓ-groups$A, B, C \in \mathcal{X}$ and embeddings $\sigma : A \to B$, $\mu : A \to C$, there exists an ℓ-group D in $\mathcal{X}$ and embeddings $\varphi : B \to D$, $\psi : C \to D$ such that $\varphi\sigma = \psi\mu$. The quintuple (A, B, C, σ, μ) is called a *V-formation* in $\mathcal{X}$ and the triple (σ, μ, α) is an *amalgamation* in $\mathcal{X}$ of this formation. An algebra $A \in \mathcal{X}$ is said to be an *amalgamation base* for $\mathcal{X}$ if every V-formation (A, B, C, σ, μ) in $\mathcal{X}$ has an amalgamation in $\mathcal{X}$. The *amalgamation class* of $\mathcal{X}$, $Amal\mathcal{X}$, is the class consisting of all amalgamation bases for $\mathcal{X}$.

S. A. Gurchenkov investigated ℓ-varieties with the amalgamation property. He proved the following remarkable result.

Theorem 11.7 (Gurchenkov [24](1994))

a) *Each non-representable ℓ-variety fails the amalgamation property.*

b) *Each ℓ-variety $\mathcal{X}$ such that $\mathcal{X} \supset \mathcal{A}^2 \cap \mathcal{W}a$ fails the amalgamation property.*

c) *If the totally ordered additive group $\mathbf{Z}$ of integers is contained in $Amal\mathcal{X}$ and $\mathcal{X} \cap \mathcal{R} \neq \mathcal{X}$, then $\mathcal{X}$ contains the ℓ-variety $\mathcal{N}$ of all normal-valued ℓ-groups.*

An ℓ-variety $\mathcal{V}$ has the *lex-property* if $\mathcal{V} = \text{var}_\ell(\{G \overleftarrow{\times} \mathbf{Z} | G \in \mathcal{V}\})$. M. E. Hansen in [29](1991) studied ℓ-varieties with the lex-property. She proved that each quasi-representable ℓ-variety defined by N. Reilly in [62](1981) has the lex-property. On the other hand, the Feil ℓ-varieties and the Medvedev ℓ-varieties (for both of these, see Reily [63]) do not have it.

Let A and B be totally ordered groups. The restricted wreath product $A\mathbf{wr}B$ can be totally ordered in two ways. Let $g \in A\mathbf{wr}B$, $g = (\hat{f}, b)$, where $\hat{f} : B \to A$ has finite support. Define $(\hat{f}, b) > (e, e)$ if $b > e$ or both $b = e$ and $\hat{f}(b_0) > e$ where $b_0 = \max(\mathbf{supp}(\hat{f}))$. This gives an o-group denoted by $A \overleftarrow{\mathbf{wr}} B$. The o-group $A \overrightarrow{\mathbf{wr}} B$ is defined analogously, except $(\hat{f}, b) > (e, e)$ if $b > e$ or both $b = e$ and $\hat{f}(b_1) > e$ where $b_1 = \min(\mathbf{supp}\hat{f})$. One solvable Medvedev cover of $\mathcal{A}$, denoted by $\mathcal{M}^+$, is generated by $\mathbf{Z} \overleftarrow{\mathbf{wr}} \mathbf{Z}$, where $\mathbf{Z}$ is the group of integers with the usual order, and the other, denoted by $\mathcal{M}^-$ is generated by $\mathbf{Z} \overrightarrow{\mathbf{wr}} \mathbf{Z}$. Let $\mathcal{V}_t^+ = \mathbf{var}_\ell((\mathbf{Z} \overrightarrow{\mathbf{wr}} \mathbf{Z}) \overleftarrow{\times} \mathbf{Z}$ and $\mathcal{V}_t^- = \mathbf{var}_\ell((\mathbf{Z} \overleftarrow{\mathbf{wr}} \mathbf{Z}) \times \mathbf{Z}$. The following interesting result was obtained by M. Huss and later by Darnel [10](1993).

Theorem 11.8 (Huss [35](1984)) $\mathcal{V}_t^+ = \mathcal{V}_t^-$.

References

[1] M. Anderson and T. Feil, *Lattice-Ordered Groups: An Introduction*, Kluwer Academic Publ., Dordrecht, 1988.

[2] M. Anderson, M. R. Darnel, and T. Feil, *A variety of lattice-ordered groups containing all representable covers of the abelian variety*, Order **7**(1991), 401–405.

[3] A. K. Arora, *Quasi-varieties of lattice-ordered groups*, Algebra Universalis **20**(1985), 34–50.

[4] G. M. Bergman, *Specially ordered groups*, Comm. Algebra **12**(1984), 2315–2333.

[5] G. Birkhoff, *On the structure of abstract algebras*, Proc. Cambridge Phil. Soc. **31**(1935), 433–454.

[6] G. Birkhoff *Lattice-ordered groups*, Ann. Math. **43**(1942), 298–331.

[7] S. Burris and H. P. Sankappanavar, *A Course in Universal Algebra*, Graduate Texts in Mathematics **78**, Springer-Verlag, New York-Heidelberg-Berlin, 1981.

[8] M. R. Darnel, *The free lattice-ordered group over a nilpotent group*, Proc. American Math. Soc. **111** (1991), 301–307.

[9] M. R. Darnel, *Disjoint conjugate chains*, in *Ordered Algebraic Structures (The 1991 Conrad conference)*, ed. J. Martinez and C. Holland, Kluwer Academic Pub., Dordrecht, The Netherlands, (1992), 31–49.

[10] M. R. Darnel, *Varieties minimal over representable varieties of lattice-ordered groups*, Comm. Algebra **21**(1993), 2637–2665.

[11] M. R. Darnel, *Cyclic extension of the Medvedev ordered groups*, Czechoslovak Math. J. **43(118)**(1993), 193–204.

[12] M. R. Darnel, *Powers of the representable variety of lattice-ordered groups*, Algebra Universalis, (to appear).

[13] M. R. Darnel and A. M. W. Glass, *Commutator relations and identities in lattice-ordered groups*, Michigan Math. J. **36**(1989), 203–211.

[14] V. A. Gorbunov, *Covers in the lattice of quasivarieties and independent axiomatizability*, (in Russian), Algebra i Logika, **16**(1977), 507–548.

[15] S. A. Gurchenkov, *Varieties of ℓ-groups with the identity $[x^p, y^p] = e$ have finite basis*, Algebra and Logic **21**(1984), 20–35.

[16] S. A. Gurchenkov, *On covers in the lattice of ℓ-varieties*, (in Russian), Mat. Zametki, **35**(1984), 677–684.

[17] S. A. Gurchenkov, *The lattice of quasivarieties of 2-nilpotent ℓ-groups is not distributive*, (in Russian), Siberian school on varieties of algebraic systems, Barnaul, (1988), 21–24.

[18] S. A. Gurchenkov, *On the theory of varieties of lattice-ordered groups*, (in Russian), Algebra i Logika **27**(1988), 249–273.

[19] S. A. Gurchenkov, *The lattice of varieties of weakly abelian lattice-ordered groups does not have the covering condition*, (in Russian), Mat. Zametki **47**(1990), 35–40.

[20] S. A. Gurchenkov, *The lattice of nilpotent ℓ-varieties is not Browerian and it has the cardinality of the continuum*, (in Russian), Izv. Visch. Uchebn. Zaved. Math. **34**(1990), 17–22.

[21] S. A. Gurchenkov, *On three questions of the theory of ℓ-varieties*, (in Russian), Czechoslovak Math. J. **41**(1991), 405–410.

[22] S. A. Gurchenkov, *On completion of invariant locally nilpotent subgroups of totally ordered groups*, (in Russian), Mat. Zametki, **51**(1992), 35–39.

[23] S. A. Gurchenkov, *About varieties of weakly abelian ℓ-groups*, to appear in Math. Slovaca.

[24] S. A. Gurchenkov, *On amalgamation in ℓ-varieties*, (in Russian), to appear in Algebra i Logika.

[25] S. A. Gurchenkov, *On Engel lattice-ordered groups*, (in Russian), to appear in Algebra i Logika.

[26] S. A. Gurchenkov and V. M. Kopytov, *Description of covers of the variety of Abelian lattice-ordered groups*, Sibirskii Mat. Zh. **28**(1987), 66–69.

[27] Yu. S. Gurevich and A. I. Kokorin, *Universal equivalence of ordered abelian groups*, (in Russian), Algebra i Logika **2**(1963), 37–39.

[28] M. Hall, *The Theory of Groups*, Macmillan, New York, 1959.

[29] M. E. Hansen (M. Huss), *The lex property of varieties of lattice-ordered groups*, Algebra Universalis **28**(1991), 535–548.

[30] W. C. Holland, *The lattice-ordered group of automorphisms of an ordered set*, Michigan Math. J. **10**(1963), 399–408.

[31] W. C. Holland, *The largest proper variety of lattice ordered groups*, Proc. American Math. Soc. **57**(1976), 25–28.

[32] W. C. Holland, *Varieties of automorphism groups of orders*, Trans. American Math. Soc. **288**(1985), 755–763.

[33] W. C. Holland and N. Ya. Medvedev, *A very large class of small varieties of lattice-ordered groups*, Comm. Algebra **22(2)**(1994), 551–578.

[34] W. C. Holland, A. H. Mekler and N. R. Reilly, *Varieties of lattice-ordered groups in which prime powers commute*, Algebra Universalis **23**(1986), 196–214.

[35] M. Huss (M. E. Hansen), *Varieties of Lattice-Ordered Groups*, Ph. D. Dissertation, Simon Fraser University, (1984).

[36] O. V. Isaeva and N. Ya. Medvedev, *Covers in the lattice of quasivarieties of ℓ-groups*, (in Russian), Sibirsk. Mat. Zh. **33**(1992), 102–107.

[37] B. Jònsson, *Algebras whose congruence lattices are distributive*, Math. Scand. **21**(1967), 110–121.

[38] M. I. Kargapolov and Yu. I. Merzljakov, *Fundamentals of the Theory of Groups*, Springer-Verlag, Berlin, 1979.

[39] N. G. Khisamiev, *Universal theory of lattice-ordered abelian groups*, (in Russian), Algebra i Logika **5**(1966), 71–76.

[40] N. G. Khisamiev and A. I. Kokorin, *An elementary classification of lattice-ordered abelian groups with a finite number of fibers*, (in Russian), Algebra i Logika **5**(1966), 41–50.

[41] Y. K. Kim and A. H. Rhemtulla, *Orderable groups satisfying an Engel condition*, in *Ordered Algebraic Structures (The 1991 Conrad conference)*, ed. J. Martinez and C. Holland, Kluwer Academic Pub., Dordrecht, The Netherlands, (1992), 31–49.

[42] V. M. Kopytov, *Free lattice-ordered groups*, Algebra and Logic **18**(1979), 259–270.

[43] V. M. Kopytov, *A non-abelian variety of lattice-ordered groups in which every solvable ℓ-group is abelian*, Mat. Sbornik **126/168**(1985), 247–266.

[44] V. M. Kopytov and N. Ya. Medvedev, *The structure of varieties of lattice-ordered groups*, in *Algebra and Order. Proc. First Int. Symp. Ordered Algebraic Structures, Luminy-Marseilles*, ed. S. Wolfenstein, R&E **14**, Helderman, Berlin, (1984), 35–46.

[45] V. M Kopytov and N. Ya. Medvedev, *Varieties of lattice-ordered groups*, (in Russian), Uspechi Mat. Nauk **40: 6(256)**(1985), 117–128; translated into English in Russian Mathematical Surveys **40: 6(256)**(1985), 97–110.

[46] M. B. Litvinova, *On the product of finitely based varieties of lattice-ordered groups*, (in Russian), to appear in Algebra i Logika.

[47] Li Si Ze, *On the varieties of representable ℓ-groups*, (in Chinese with English summary), J. Math. (Wuhan), **10**(1990), 321–324.

[48] J. Martinez, *Varieties of lattice-ordered groups*, Math. Zeit. **137**(1974), 265–284.

[49] N. Ya. Medvedev, *The lattices of varieties of lattice-ordered groups and Lie algebras*, Algebra and Logic **16**(1977), 27–31.

[50] N. Ya. Medvedev, *On the theory of varieties of lattice ordered groups*, (in Russian), Czechoslovak Math. J. **32**(1982), 364–372.

[51] N. Ya. Medvedev, *Free products of ℓ-groups*, (in Russian), Algebra i Logika **23**(1984), 493–511.

[52] N. Ya. Medvedev, *Quasivarieties of ℓ-groups and groups*, (in Russian), Sibirsk. Mat. Zh. **26**(1985), 111–117.

[53] N. Ya. Medvedev, *On o-approximability of bounded Engel ℓ-groups*, (in Russian), Algebra i Logika **27**(1988), 418–421.

[54] N. Ya. Medvedev, *On nilpotent lattice-ordered groups*, (in Russian), Mat. Zametki **45**(1989), 72–79.

[55] N. Ya. Medvedev, *On infinite distributivity in the lattice of ℓ-varieties*, (in Russian), Sibirsk. Mat. Zh. **30**(1989), 216–220.

[56] N. Ya. Medvedev, *On covers in the lattice of quasivarieties of ℓ-groups*, in *Ordered Algebraic Structures (The 1991 Conrad Conference)*, ed. J. Martinez and C. Holland, Kluwer Academic Pub., Dordrecht, The Netherlands, (1992), 81–98.

[57] N. Ya. Medvedev, *Independent axiomatization of varieties of lattice-ordered groups*, Czechoslovak Math. J. **42**(1992), 53–57.

[58] N. Ya. Medvedev, *HSP≠SHPS for metabelian representable ℓ-groups*, Algebra Universalis **31**(1994), 151–156.

[59] R. T. Botto-Mura and A. H. Rhemtulla, *Notes on Orderable Groups*, University of Alberta, Edmonton, 1975; published as *Orderable Groups*, Lecture Notes in Pure and Applied Math. **27**, Marcel Dekker, New York, 1977.

[60] W. B. Powell and C. Tsinakis, *The failure of the amalgamation property for representable varieties of ℓ-groups*, Proc. Cambridge Phil. Soc. **106**(1989), 439–444.

[61] W. B. Powell and C. Tsinakis, *Covers of the variety of abelian ℓ-groups*, Comm. Algebra **17**(1989), 2461–2468.

[62] N. R. Reilly, *A subsemilattice of the lattice of varieties of lattice-ordered groups*, Canadian J. Math. **33**(1981), 1309–1318.

[63] N. R. Reilly, *Varieties of lattice-ordered groups*, in Lattice-Ordered Groups; Advances and Techniques, ed. A.M.W. Glass and W. C. Holland, Kluwer Academic Publishers, Dordrecht/Boston/London, 1989, 228–277.

[64] D. A. Van Rie, *Quasivarieties of ℓ-metabelian lattice-ordered groups*, Doctoral Thesis, Bowling Green State University, 1991.

[65] S. V. Varaksin, *Lattice-ordered groups constructed from right-ordered groups*, (in Russian), Algebra i Logika **28**(1989), 524–533.

[66] S. V. Varaksin, *Varieties generated by simple ℓ-groups*, (in Russian), Sibirsk. Mat. Zh. **31**(1990),167–180.

[67] S. V. Varaksin, *A continuum of minimal quasivarieties of ℓ-groups*, (in Russian), Sibirsk. Mat. Zh. **34**(1993) N4, 41–49.

Lattice-ordered Permutation Groups: The Structure Theory

Stephen H. McCleary

Bowling Green State University
Bowling Green, Ohio 43403, USA
Email: smcclea@bgnet.bgsu.edu

1 Introduction

This survey of lattice-ordered permutation groups focuses especially on their structure theory and on their relation to unordered infinite permutation groups. However, the survey is essentially self-contained. The reader will need only a little familiarity with unordered permutation groups, and none at all with lattice-ordered groups.

Let Ω be a chain (totally ordered set). The group $A(\Omega)$ of automorphisms (order-preserving permutations) of Ω acquires from Ω the *pointwise order*, i.e., $f \leq g$ iff $\alpha f \leq \alpha g$ for all $\alpha \in \Omega$. It is easily checked that this order makes $A(\Omega)$ a *lattice-ordered group* (*ℓ-group*), i.e., it makes $A(\Omega)$ a lattice and the order is preserved by multiplication on both sides ($f \leq g$ implies $hf \leq hg$ and $fh \leq gh$). The lattice operations are given by $\alpha(f \vee g) = \max\{\alpha f, \alpha g\}$ and dually, and thus are also pointwise; see Figure 1.

A subgroup G of $A(\Omega)$ which is also a sublattice (closed under the lattice operations) is called an *ℓ-subgroup*, and (G, Ω) is called a *lattice-ordered permutation group* (*ℓ-permutation group*). Some parts of the structure theory apply more generally to *ordered permutation groups* (with the order inherited from $A(\Omega)$ but without the sublattice requirement).

For any Ω, the elements of $A(\Omega)$ are homeomorphisms with respect to the order topology on Ω. Especially important is the case in which Ω is the real line $\mathbf{R}$. Here *every* homeomorphism either preserves or reverses order, and of course $A(\mathbf{R})$ consists of those homeomorphisms doing the former. The study of $A(\mathbf{R})$ goes back at least to 1924, when H. Kneser [23] proved that any two positive elements without fixed points must be conjugate. Later, ℓ-permutation groups served as examples of ℓ-groups.

But the main impetus to the study of ℓ-permutation groups has been Charles

W. C. Holland (ed.), Ordered Groups and Infinite Permutation Groups, 29–62.

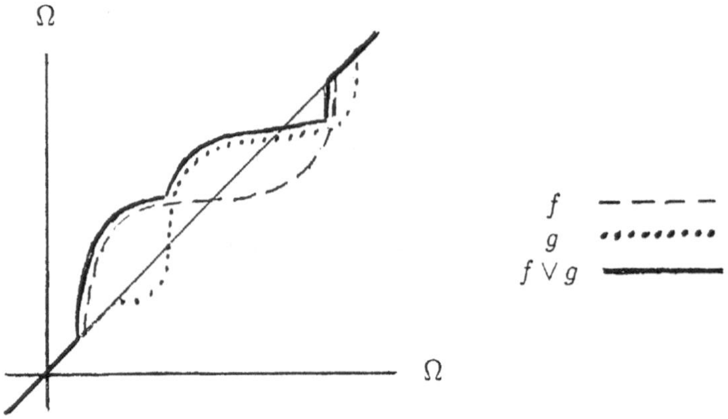

Figure 1: $f \vee g$

Holland's analogue of Cayley's Theorem. The Holland Representation Theorem [14] says that *every* ℓ-group is (isomorphic as an ℓ-group to) an ℓ-subgroup G of some $A(\Omega)$. This made obvious some elementary facts about ℓ-groups which were already known, e.g., that they are torsion free and that the underlying lattice is distributive. The distributivity is acquired from that of the chain Ω since

$$\begin{aligned} \alpha((f \vee g) \wedge h) &= \min\{\max\{\alpha f, \alpha g\}, \alpha h\} \\ &= \max\{\min\{\alpha f, \alpha h\}, \min\{\alpha g, \alpha h\}\} \\ &= \alpha((f \wedge h) \vee (g \wedge h)). \end{aligned}$$

Furthermore, the Representation Theorem led Holland to deeper discoveries, for example that every ℓ-group can be embedded in a *divisible* ℓ-group (extra care being required in the choice of Ω in order to make $A(\Omega)$ divisible).

However, this survey deals directly with ℓ-permutation groups (G, Ω) rather than with abstract ℓ-groups. It includes a detailed description (with proofs) of the structure theory of transitive ℓ-permutation groups, and several important applications. The power of this theory derives from the interplay between its two complementary parts:

(1) The collection of convex congruences (i.e., congruences having convex classes) of a transitive ℓ-permutation group (G, Ω) is totally ordered by refinement. In some arguments this affords a sort of "reduction" to the "o-primitive" case. ("o-primitive" is the ordered analogue of "primitive", and means that there are no proper convex congruences.).
(2) The classification of the o-primitive (G, Ω)'s, which makes the reduction useful.

The survey also offers a brief sketch of the more general intransitive structure theory, with precise statements of just enough of it to enable one to see that many applications of the general theory proceed almost exactly as in the transitive case. This is done in such a way that a reader wishing to consider just the transitive case can easily do so.

Following this is a section indicating which results of the structure theory (e.g., (1)) hold for ordered permutation groups. Even more of the results hold for *coherent* ℓ-permutation groups (G, Ω), meaning that whenever $\alpha < \beta$ and $\alpha g = \beta$ for some $g \in G$ then $\alpha h = \beta$ for some $h \in G^+$. (Here e denotes the identity and $G^+ = \{g \in G \mid e \leq g\}$ is the set of *positive* elements of G.) A word of caution: An ordered permutation group (G, Ω) for which the pointwise order makes G a lattice can fail to be an ℓ-permutation group because the lattice operations can differ from the (pointwise) lattice operations of $A(\Omega)$.

R, **Q**, and **Z** will denote the chains of real numbers, rational numbers, and integers, resp., and **N** the set of natural numbers. The chain Ω will be assumed to contain more than one element.

The standard reference in ℓ-permutation groups is the book [8] by Andrew Glass, and it contains most of the material presented here, along with an extensive annotated bibliography. Also, a shorter survey of the area by Holland appears as a chapter in [9].

2 Examples

Here are some examples of transitive ℓ-permutation groups. Transitivity prevents Ω from having a largest or a smallest point, and in particular forces Ω to be infinite. It also forces Ω to be either dense in itself or discrete. The first three examples will turn out to illustrate the three types of o-primitivity, and the fourth will illustrate a pathological subtype of the first type. The orbits of the *stabilizer subgroup* $G_\alpha = \{g \in G \mid \alpha g = \alpha\}$ of a point $\alpha \in \Omega$ will play an important role, and we precede the examples with a brief discussion of these orbits. A subset Σ of Ω is *convex* if $\sigma_1 \leq \omega \leq \sigma_2$ ($\sigma_i \in \Sigma$, $\omega \in \Omega$) implies $\omega \in \Sigma$.

Proposition 2.1 *Let (G, Ω) be a transitive ℓ-permutation group, and let $\alpha \in \Omega$. The orbits of G_α are convex.*

Proof. Suppose $\sigma_1 \leq \omega \leq \sigma_2$ and that $\sigma_1 g = \sigma_2$ for some $g \in G_\alpha$. By transitivity, $\sigma_1 h = \omega$ for some $h \in G$. Then $g \wedge (h \vee e)$ moves σ_1 to ω and fixes α. See Figure 2.

Because of the previous proposition, the set of orbits of G_α inherits a total order from Ω in the obvious fashion. Any non-singleton orbit of G_α must be infinite, and indeed can contain no largest or smallest point; such orbits are called *long* orbits.

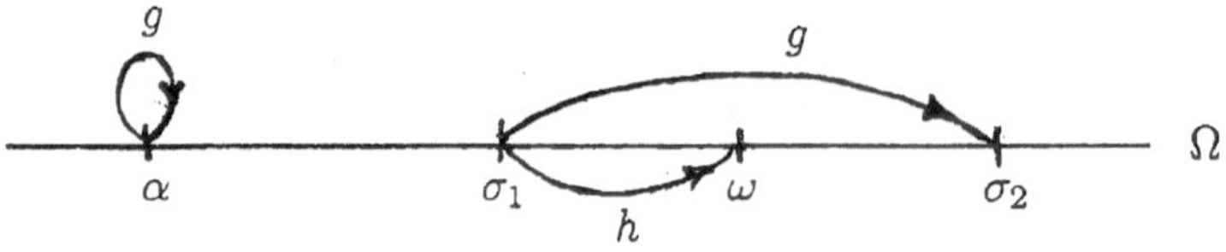

Figure 2: Proof of the convexity of the orbits of G_α

By transitivity this chain, together with the distinction between long orbits and fixed points, is independent (up to isomorphism) of the choice of α. (This provides a chance to emphasize our reliance on the trivial fact that when $g : \sigma_1 k \mapsto \sigma_2 k$ then $k^{-1}gk : \sigma_1 \mapsto \sigma_2$; and of course we have $k^{-1}G_\alpha k = G_{\alpha k}$. We note also that inner (group) automorphisms of ℓ-groups are automatically ℓ-group automorphisms.)

Although we shall not use the information, we mention that the chain of *positive* orbits of G_α (those to the right of $\{\alpha\}$) is anti-order-isomorphic to the chain of *negative* orbits via the map $\Delta \to \Delta'$ from Δ to its paired orbit $\Delta' = \{\alpha g \mid \alpha \in \Delta g\}$. Warning: It is possible for a long orbit of G_α to be paired with a fixed point.

Example 1 $(G, \Omega) = (A(\mathbf{R}), \mathbf{R})$ [o-2-transitive]

This (G, Ω) is *o-2-transitive*, i.e., for all $\alpha_1 < \alpha_2$ and $\beta_1 < \beta_2$ in Ω, there exists $g \in G$ such that $\alpha_1 g = \beta_1$ and $\alpha_2 g = \beta_2$ (and $|\Omega| > 2$). For any o-2-transitive (G, Ω), there are clearly only three orbits of G_α, namely $\{\alpha\}$ itself, a positive long orbit consisting of all points to the right of α, and a similar negative long orbit.

We define o-n-transitivity ($n \in \mathbf{N}$) similarly. In contrast to unordered permutation groups we have

Proposition 2.2 *An o-2-transitive ℓ-permutation group is o-n-transitive for all n.*

Proof. Let (G, Ω) be o-2-transitive. Let $\alpha_1 < \alpha_2 < ... < \alpha_n$ and $\beta_1 < \beta_2 < ... < \beta_n$ in Ω. By induction there exists $g \in G$ such that $\alpha_i g = \beta_i$, $i = 1, ..., n-1$. We suppose $\alpha_n g \geq \beta_n$, the other case being similar. By o-2-transitivity, there exists $h \in G$ such that $\alpha_1 h = \beta_{n-1}$ (forcing $\alpha_i \geq \beta_i$, $i = 1, ..., n-1$) and $\alpha_n h = \beta_n$. Then $\alpha_i(g \wedge h) = \beta_i$, $i = 1, ..., n$.

Example 2 $(G, \Omega) =$ *the right regular representation of a subgroup of the additive reals* $(\mathbf{R}, \leq, +)$ [regular]

Here the orbits of G_a are all singletons. If we take the subgroup to be $\mathbf{Z}$, then each automorphism of the ordered set $\mathbf{Z}$ is simply translation by some integer, so that $G = A(\Omega)$.

In Example 2, we could have been more general, citing the right regular representation of any *totally* ordered group (o-group). However, the example as stated does include all Archimedean o-groups since Hölder's Theorem (1901) tells us that all

Archimedean o-groups are (up to isomorphism) o-subgroups of the additive reals. (G is *Archimedean* if for all $0 < a < b \in G$ there exists $n \in \mathbf{N}$ such that $na \geq b$.)

Example 3 $G = \{g \in A(\mathbf{R}) \mid (\alpha + 1)g = \alpha g + 1\ \forall \alpha \in \mathbf{R}\}$ [periodic]

These are the elements of "period" one (though in a slightly non-standard sense–see Figure 3). Any $g \in G$ maps $[0, 1)$ onto some other interval of length one, and by periodicity that action uniquely determines g throughout $\mathbf{R}$. The fixed points of G_0 are the integers, and the long orbits are the intervals $(n, n+1)$, $n \in \mathbf{Z}$. As we shall prove later, all o-primitive transitive ℓ-permutation groups which are neither o-2-transitive nor regular look very much like this one.

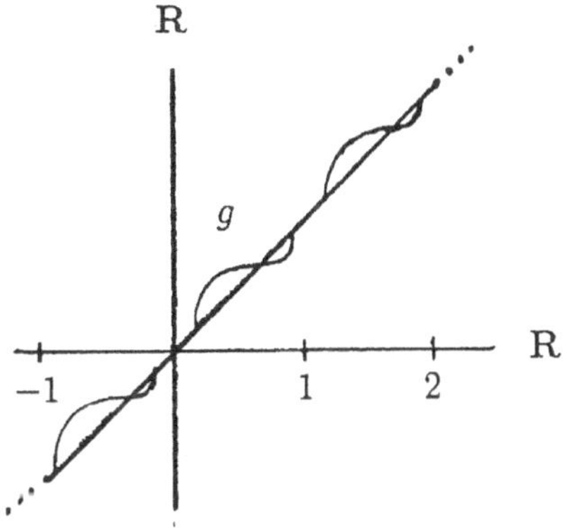

Figure 3: Periodicity

Example 4 $G = \{g \in A(\mathbf{R}) \mid \exists n_g \in \mathbf{N}$ *such that* $(\alpha + n_g)g = \alpha g + n_g\ \forall \alpha \in \mathbf{R}\}$, *with the period* n_g *depending on* g [pathologically o-2-transitive]

Like Example 1, $(G, \mathbf{R})$ is o-2-transitive. However, it is *pathologically o-2-transitive*, meaning that it has no element $\neq e$ of bounded support. (The *support* of g means $\{\alpha \in \mathbf{R} \mid \alpha g \neq \alpha\}$, and is denoted by $\mathrm{supp}(g)$.) Pathologically o-2-transitive (G, Ω)'s cause a considerable amount of trouble.

When (G, Ω) is *non*-pathologically o-2-transitive, *every* non-singleton interval Δ of Ω supports some $e \neq f \in G^+$. For given $e \neq g \in G$, let $h = g \vee g^{-1} \in G^+ \backslash \{e\}$. Picking $\sigma \leq \mathrm{supp}(h) \leq \tau$ and picking $k \in G$ such that $\sigma k, \tau k \in \Delta$, we have $\mathrm{supp}(k^{-1}hk) = (\mathrm{supp}(h))k \subseteq \Delta$.

Example 5 $\Omega = \overleftarrow{\mathbf{Z} \times \mathbf{R}}$, *the lexicographic product ordered from the right;* $G = A(\Omega)$.

The set of local copies of $\mathbf{Z}$ which replace the points of $\mathbf{R}$ in the formation of Ω is "respected" by G. That is, for each $g \in G$ the image Δg of one of these copies is another one of the copies (since two points of Ω lie in the same copy iff there are only finitely many points of Ω between them). Thus each $g \in G$ induces in the obvious fashion an order-preserving permutation $\hat{g}$ of the chain of copies of $\mathbf{Z}$. In effect $\hat{g} \in A(\mathbf{R})$, and of course locally g induces an integer translation t_β for each $\beta \in \mathbf{R}$. For $\alpha = (0,0)$, the positive fixed points of G_α consist of the positive integers in the 0th copy of $\mathbf{Z}$, and the points to the right of the 0th copy form the one other positive orbit.

3 The Holland Representation Theorem

Now we revisit abstract ℓ-groups and the Holland Representation Theorem [14]. A subgroup H of an ℓ-group G is *convex* if $h_1 \leq g \leq h_2$ $(h_i \in H)$ implies $g \in H$; and is *prime* if $g_1 \wedge g_2 \in H$ implies $g_1 \in H$ or $g_2 \in H$.

In an ℓ-permutation group, every G_α is a prime convex ℓ-subgroup of G. Moreover, the orbit αG is in one-to-one correspondence with the set $R(G_\alpha)$ of right cosets $G_\alpha g$ of G_α via $\alpha g \leftrightarrow G_\alpha g$. This becomes an isomorphism of *chains* if we define

$$G_\alpha g_1 \leq G_\alpha g_2 \Leftrightarrow k_1 \leq k_2 \text{ for some } k_i \in G_\alpha g_i.$$

(If $\alpha g_1 \leq \alpha g_2$, take $k_1 = g_1$ and $k_2 = g_1 \vee g_2$.)

An *ℓ-homomorphism* from G_1 into G_2 is a group homomorphism which also preserves the lattice operations The subgroups of G which occur as kernels of ℓ-homomorphisms are precisely the normal convex ℓ-subgroups, known as *ℓ-ideals.*

Theorem 3.1 (Holland Representation Theorem) *Every ℓ-group G is an ℓ-subgroup of $A(\Omega)$ for some chain Ω.*

In outline, Holland's proof goes like this. For each $e \neq g \in G$, pick a convex ℓ-subgroup P_g maximal with respect to not containing g; because of the maximality, P_g is prime. The crux of the argument is that primeness guarantees that setting

$$P_g f_1 \leq P_g f_2 \Leftrightarrow k_1 \leq k_2 \text{ for some } k_i \in P_g f_i$$

gives a total order on $R(P_g)$, just as for G_α above. Let Ω_g be the chain $R(P_g)$, and let G act on Ω_g by right translation. This gives an ℓ-homomorphism $\phi_g : G \to A(\Omega_g)$. The stabilizer subgroup of $P_g \in \Omega_g$ is P_g itself, and $\ker(\phi_g) = \text{core}(P_g)$, the intersection of all conjugates of P_g. In general, no one ϕ_g will be faithful. However, pasting together (disjoint copies of) the various Ω_g's gives a master chain Ω in which the Ω_g's are convex and on which G acts faithfully.

The proof gives some extra information: There is a representation of G which can be obtained by pasting together various transitive actions, i.e., for which the orbits (the Ω_g's above) are convex.

The proof also tells us which subgroups of an ℓ-group G can occur as stabilizer subgroups in representations of G as ℓ-permutation groups, namely the prime convex ℓ-subgroups. (Even if P is not one of the P_g's, $R(P)$ can be adjoined to the end of Ω.) We also discover that the P's which can occur as stabilizer subgroups in representations of G as *transitive* ℓ-permutation groups are those having trivial core (since clearly when (G, Ω) is transitive, $\mathrm{core}(G_\alpha) = \bigcap_{k \in G} G_{\alpha k} = \{e\}$). Such P's are known as *representing subgroups*, and of course the G's having transitive representations are precisely those having representing subgroups.

We mention here an easy identity

$$g = (g \vee e)(g \wedge e) = (g \vee e)(g^{-1} \vee e)^{-1},$$

which is easily verified for ℓ-permutation groups by computing αg, splitting into the cases $\alpha g \geq \alpha$ and $\alpha g < \alpha$. Then by the Holland Representation Theorem this identity holds also for abstract ℓ-groups. In particular, every ℓ-group is generated as a group by G^+.

4 Doubly transitive $A(\Omega)$

Let Ω be an arbitrary chain, let $f \in A(\Omega)$, and let $\delta \in \mathrm{supp}(f)$. The convexification $\mathrm{Conv}\{\delta f^n \mid n \in \mathbf{Z}\}$, i.e.,

$$\{\omega \in \Omega \mid \delta f^{-n} \leq \omega \leq \delta f^n \text{ for some } n \in \mathbf{Z}\},$$

is called a *supporting interval* of f (*positive* if $\delta f > \delta$, *negative* if $\delta f < \delta$). Since $\{\delta f^m \mid m \in \mathbf{Z}\}$ is coterminal in this supporting interval, the interval and its parity are independent of which of its points δ is used to define it. The element f in Figure 1 has one supporting interval of each parity. The set of supporting intervals of any f has a natural total order, namely $\Delta_1 < \Delta_2$ iff $\delta_1 < \delta_2$ for all $\delta_1 \in \Delta_1$, $\delta_2 \in \Delta_2$.

An element b having exactly one supporting interval is called a *bump* or a *convex cycle*. Each $f \in A(\Omega)^+$ is uniquely the supremum of a (possibly infinite) pairwise disjoint set of positive bumps, where $g_1, g_2 \in A(\Omega)^+$ are called *disjoint* if their supports are disjoint, or equivalently, if $g_1 \wedge g_2 = e$.

Any conjugate of a bump is a bump of the same parity. This is a special case of something more general: The *landscape* $\mathcal{L}(f)$ of any $f \in A(\Omega)$ means the chain of supporting intervals and fixed points of f, with positive supporting intervals, negative supporting intervals, and fixed points distinguished; and conjugation by any $h \in A(\Omega)$ induces an isomorphism from $\mathcal{L}(f)$ onto $\mathcal{L}(h^{-1}fh)$. The following conjugacy lemma provides a converse when $A(\Omega)$ is o-2-transitive. The lemma is due to Holland [14], special cases having been proved previously by Kneser [23] and by J. Schreier and S. Ulam [40]. The lemma tells us, for example, that in $A(\mathbf{Q})$ there are exactly nine conjugacy classes of positive bumps (the inf of supp(f) can be rational, irrational, or $-\infty$, and dually).

Lemma 4.1 *Suppose $A(\Omega)$ is o-2-transitive. Then f and g are conjugate in $A(\Omega)$ if and only if $\mathcal{L}(f)$ and $\mathcal{L}(g)$ are isomorphic.*

Proof. Let ψ be a landscape isomorphism from $\mathcal{L}(f)$ onto $\mathcal{L}(g)$. Let Δ_f be any supporting interval of f. Assume Δ_f is positive, the negative case being similar. Let $\Delta_g = \Delta_f\psi$, which is also positive. Pick $\alpha \in \Delta_f$, $\beta \in \Delta_g$. By o-2-transitivity we can pick an order-isomorphism ϕ_0 from the interval $[\alpha, \alpha f)$ onto $[\beta, \beta g)$; see Figure 4.

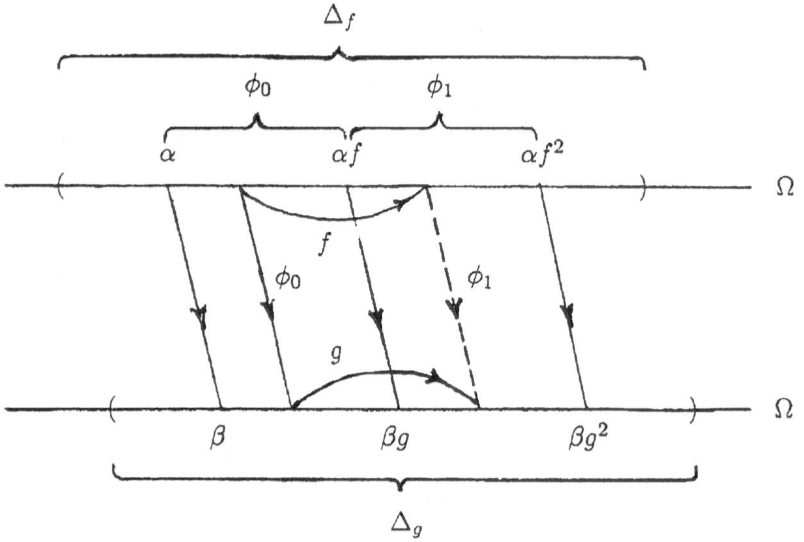

Figure 4: The definition of the conjugator ϕ

We extend ϕ_0 to Δ_f in the only way which can possibly yield the desired conjugator: We define ϕ_{n+1} on $[\alpha f^n, \alpha f^{n+1})$ to be $\phi_{n+1} = f^{-1}\phi_n g$ $(n = 1, 2, \ldots)$, and similarly in the negative direction. Form $\phi_{\Delta_f} : \Delta_f \to \Delta_g$ by patching together the various ϕ_n's. Now form $\phi \in A(\Omega)$ by patching together the ϕ_{Δ_f}'s for the various supporting intervals Δ_f of f and sending each fixed point ω of f to $\omega\psi$. Then $\phi \in A(\Omega)$ and $\phi^{-1}f\phi = g$.

Note that if f and g both have support bounded above (below), the same can be arranged also for the conjugator ϕ.

Let $L(\Omega)$ consist of those elements of $A(\Omega)$ having supports bounded above ("living on the left"). Let $R(\Omega)$ be the dual. Let $B(\Omega) = L(\Omega) \cap R(\Omega)$, the elements of bounded support. Always these are *ℓ-ideals* of $A(\Omega)$.

Various parts of the following theorem were contributed by G. Higman [13], J. T. Lloyd [25], and Holland [14].

Theorem 4.2 *Let $A(\Omega)$ be o-2-transitive. Then every normal subgroup of $A(\Omega)$ is an ℓ-ideal. Moreover, $B(\Omega)$ is a simple group and is the smallest proper normal subgroup of $A(\Omega)$. If Ω has a countable coterminal subset (bounded neither above nor below), then $B(\Omega)$, $L(\Omega)$, and $R(\Omega)$ are the only proper normal subgroups of $A(\Omega)$, and $B(\Omega)$ is the only proper normal subgroup of either $L(\Omega)$ or $R(\Omega)$.*

Proof that $B(\Omega)$ is simple. Let $e \neq b \in B(\Omega)$, and let $g \in B(\Omega)$. Pick a supporting interval Δ of b. Pick $\sigma < \text{supp}(g) < \tau$, and use o-2-transitivity to pick $k \in A(\Omega)$ such that $\sigma k, \tau k \in \Delta$; and arrange in addition that $k \in B(\Omega)$ by using o-4-transitivity (see Proposition 2.2) to make k fix some $\sigma_1 < \sigma$ and some $\tau_1 > \tau$ and then changing k to act like the identity outside $[\sigma_1, \tau_1]$. Then Δ is also a supporting interval of $b(k^{-1}gk)$, and b and $b(k^{-1}gk)$ have the same landscape. By Lemma 4.1 ff., b and $b(k^{-1}gk)$ are conjugate in $B(\Omega)$. It follows that g is the product of a conjugate of b^{-1} by a conjugate of b. Therefore $B(\Omega)$ is simple.

When the countable coterminality hypothesis in Theorem 4.2 is not satisfied, there are necessarily *many* other normal subgroups of $A(\Omega)$—see R. N. Ball [1], Ball and M. Droste [2], and Droste and S. Shelah [6].

Another consequence of Lemma 4.1 is

Corollary 4.3 (Holland [14]) *If $A(\Omega)$ is o-2-transitive, then $A(\Omega)$ is a divisible group.*

Proof. Let $g \in A(\Omega)$ and $n \in \mathbf{N}$. Since g^n and g have the same landscape, $\phi^{-1}g^n\phi = g$ for some $\phi \in A(\Omega)$. Then $(\phi^{-1}g\phi)^n = g$.

Corollary 4.4 (Holland [14]) *Every ℓ-group can be embedded in a divisible ℓ-group.*

Outline of proof. Let G be an ℓ-group. Use the Holland Representation Theorem to embed G in some $A(\Omega_1)$. Then embed Ω_1 in some Ω for which $A(\Omega)$ is o-2-transitive, and do this in a way that embeds $A(\Omega_1)$ in $A(\Omega)$.

5 The transitive structure theory: "Reduction" to the o-primitive case

Throughout this section (G, Ω) will denote a transitive ℓ-permutation group. Without transitivity most of the definitions are valid, but most of the results require some modification (see Section 8).

A *convex congruence* of (G, Ω) is simply a *congruence* $\mathcal{C}$ (an equivalence relation on Ω which is respected by G) for which each congruence class Δ is *convex*. We set $\mathcal{C}_1 \leq \mathcal{C}_2$ iff $\mathcal{C}_1$ refines $\mathcal{C}_2$. The key result of this section is that this partial order on the set of convex congruences is actually a *total* order.

Always there are the two improper convex congruences (the trivial congruence $\mathcal{E}$ whose classes are single points, and the congruence whose only class is Ω). In Example 5 there is also a proper convex congruence; its classes are the local copies of $\mathbf{Z}$. As usual, a *block* of (G, Ω) is a non-empty $\Delta \subseteq \Omega$ such that if $\Delta g \cap \Delta \neq \emptyset$ $(g \in G)$ then $\Delta g = \Delta$. The classes of a convex congruence are convex blocks (*o-blocks*). Conversely, because of transitivity the set $\tilde{\Delta} = \{\Delta g \mid g \in G\}$ of translates of any one o-block Δ forms a convex congruence (we identify congruences with their corresponding partitions of Ω), and each convex congruence is determined by any one of its classes. Thus for any fixed $\alpha \in \Omega$, the set of o-blocks Δ containing α (ordered by inclusion) is in one-to-one order-preserving correspondence with the set of convex congruences of (G, Ω).

Let Δ be an o-block. Then $\tilde{\Delta}$ inherits a natural total order from Ω. The (not necessarily faithful) action of G on $\tilde{\Delta}$ given by $(\Delta g)h \mapsto \Delta(gh)$, $h \in G$, preserves the order on $\tilde{\Delta}$, and the resulting group $\tilde{G}$ of order-preserving permutations of $\tilde{\Delta}$ forms a transitive ℓ-permutation group $(\tilde{G}, \tilde{\Delta})$. $\tilde{G}$ is the quotient of the ℓ-group G by the ℓ-ideal $L(\tilde{\Delta}) = \{g \in G \mid \Sigma g = \Sigma \text{ for all } \Sigma \in \tilde{\Delta}\}$.

Theorem 5.1 (Holland [15]) *Let (G, Ω) be a transitive ℓ-permutation group. Let $\alpha \in \Omega$. Then*

(1) *The collection of o-blocks containing α is totally ordered by inclusion.*

(2) *The collection of convex congruences of (G, Ω) is totally ordered by refinement.*

(3) *An order-isomorphism from the tower of o-blocks containing α onto the tower of convex congruences is given by $\Delta \mapsto \tilde{\Delta}$.*

Proof. We prove (1), which will yield (2) and (3). Suppose that Δ and Σ are o-blocks containing α which are incomparable under inclusion. By transitivity we can pick $g \in G$ such that $\alpha g \in \Sigma \setminus \Delta$. This makes $\Sigma g = \Sigma$. Then Δg meets Δ (see Figure 5), making $\Delta g = \Delta$ and contradicting the fact that $\alpha g \notin \Delta$.

There is another tower of interest here. For $\Delta \subseteq \Omega$, let $G_\Delta = \{g \in G \mid \Delta g = \Delta\}$, the setwise stabilizer of Δ. In the usual one-to-one correspondence $\Delta \mapsto G_\Delta$ between

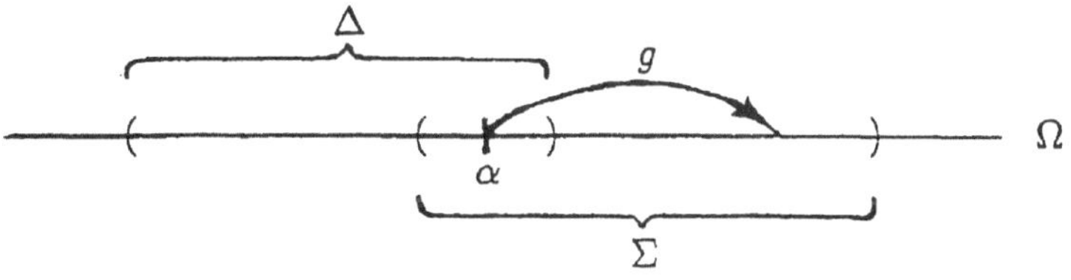

Figure 5: Two o-blocks containing α must be comparable under inclusion.

the set of blocks Δ containing α and the set of subgroups H containing G_α (whose inverse is $H \mapsto \alpha H$), the o-blocks correspond precisely to the convex subgroups. (Let H be a convex subgroup containing G_α, and let $\alpha \leq \beta \leq \alpha h$ for some $h \in H$. Pick $g_1, g_2 \in G^+$ such that $\alpha g_1 = \beta$ and $\beta g_2 = \alpha h$. Then $\alpha g_1 g_2 = \alpha h$, so that $g_1 g_2 h^{-1} \in G_\alpha \subseteq H$, making $g_1 g_2 \in H$. Since $e \leq g_1 \leq g_1 g_2$, $g_1 \in H$, so that $\beta \in \alpha H$.) This gives one version of the next theorem. But clearly these convex G_Δ's are prime ℓ-subgroups of G. Hence all convex subgroups containing G_α are prime ℓ-subgroups as well. We record this in

Theorem 5.2 (Holland [15]) *Let (G, Ω) be a transitive ℓ-permutation group. Let $\alpha \in \Omega$. Then*

(1) *The collection of convex (prime ℓ-) subgroups G which contain G_α is totally ordered under inclusion.*

(2) *An order-isomorphism from the tower of o-blocks containing α onto the tower in (1) is given by $\Delta \mapsto G_\Delta$.*

For any $\beta \neq \alpha \in \Omega$, there exists in the tower of o-blocks Δ containing α a *covering pair* $\Delta_1 \subset \Delta_2$ (i.e., there is no other Δ properly between Δ_1 and Δ_2) such that $\beta \in \Delta_2$ but $\beta \notin \Delta_1$. Namely, Δ_1 is the union of the Δ's omitting β and Δ_2 is the intersection of the Δ's containing β as well as α. Then $C_1 = \tilde{\Delta}_1$ and $C_2 = \tilde{\Delta}_2$ form a covering pair of convex congruences, known as the *value* $\mathrm{Val}(\alpha, \beta)$. The tower of covering pairs of convex congruences of (G, Ω) is denoted by $\Gamma = \Gamma(G, \Omega)$. The covering pair of convex congruences of which a $\boldsymbol{\gamma} \in \Gamma$ consists will be denoted by $C_{\boldsymbol{\gamma}}$ and $C^{\boldsymbol{\gamma}}$. It is easily checked that any convex congruence which is not $C_{\boldsymbol{\gamma}}$ or $C^{\boldsymbol{\gamma}}$ for some $\boldsymbol{\gamma} \in \Gamma$ must be both the union of the $C^{\boldsymbol{\gamma}}$'s below it and the intersection of the $C_{\boldsymbol{\gamma}}$'s above it. To avoid confusion between points of Ω and elements of Γ, *the letter $\boldsymbol{\gamma}$ will be reserved for elements of Γ, i.e., for values, and will be boldfaced.*

Now let $\boldsymbol{\gamma} = (C_{\boldsymbol{\gamma}}, C^{\boldsymbol{\gamma}}) \in \Gamma$, and let $\alpha \in \Omega$. Let $\Omega_{\boldsymbol{\gamma}} = \alpha C^{\boldsymbol{\gamma}} / C_{\boldsymbol{\gamma}}$ (see Figure 6). Let $G_{\boldsymbol{\gamma}}$ consist of the order-preserving permutations of $\Omega_{\boldsymbol{\gamma}}$ induced by the action of $G_{\alpha C^{\boldsymbol{\gamma}}}$ on the chain $\Omega_{\boldsymbol{\gamma}}$. $(G_{\boldsymbol{\gamma}}, \Omega_{\boldsymbol{\gamma}})$ is a transitive ℓ-permutation group, and by the transitivity of (G, Ω), $(G_{\boldsymbol{\gamma}}, \Omega_{\boldsymbol{\gamma}})$ is independent of the choice of α. $G_{\boldsymbol{\gamma}}$ is a quotient of an ℓ-subgroup of G.

Since $C_{\boldsymbol{\gamma}}$ and $C^{\boldsymbol{\gamma}}$ form a covering pair of convex congruences of (G, Ω), it follows

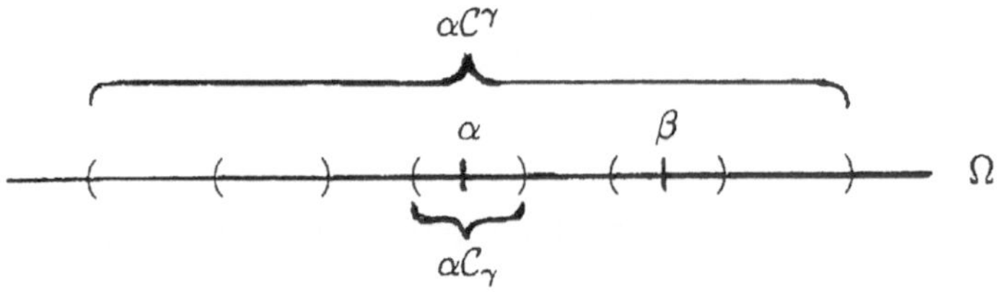

Figure 6: $\Omega_\gamma = \alpha C^\gamma / C_\gamma$

easily that the transitive ℓ-permutation group $(G_\gamma, \Omega_\gamma)$ has no proper convex congruences at all, i.e., that $(G_\gamma, \Omega_\gamma)$ is *o-primitive.* $(G_\gamma, \Omega_\gamma)$ is called an *o-primitive component* of (G, Ω). $\Gamma(G, \Omega)$ is the *spine* of (G, Ω), the $(G_\gamma, \Omega_\gamma)$'s are the *ribs*, and $(\Gamma, \{(G_\gamma, \Omega_\gamma) \mid \gamma \in \Gamma\})$ is the *skeleton.* In Example 5 the spine has two elements. The top rib is $(A(\mathbf{R}), \mathbf{R})$ and the bottom rib is the regular representation $(\mathbf{Z}, \mathbf{Z})$ of the integers.

We recapitulate the preceding discussion, bearing Figure 6 in mind:

Theorem 5.3 *Let (G, Ω) be a transitive ℓ-permutation group.*

(1) *Let $\gamma = (C_\gamma, C^\gamma)$ be a covering pair in the tower of convex congruences of (G, Ω). Let $\alpha \in \Omega$. Then the action of $G_{\alpha C^\gamma}$ on the chain $\Omega_\gamma = \alpha C^\gamma / C_\gamma$ of C_γ-classes contained in αC^γ induces a group G_γ of order-preserving permutations of Ω_γ, and $(G_\gamma, \Omega_\gamma)$ is an o-primitive transitive ℓ-permutation group.*

(2) *Let $\beta \neq \alpha \in \Omega$. Then there exists a unique value $\gamma = (C_\gamma, C^\gamma) = Val(\alpha, \beta) \in \Gamma(G, \Omega)$ such that $\alpha C^\gamma \beta$ but not $\alpha C_\gamma \beta$.*

In many arguments about (G, Ω), this serves to "reduce" the question to the o-primitive case. This puts a premium on understanding o-primitive transitive ℓ-permutation groups. We address this issue in Section 7.

6 Wreath products

Can every conceivable skeleton $(\Gamma, \{(G_\gamma, \Omega_\gamma) \mid \gamma \in \Gamma\})$ actually occur? That is, given a collection $\{(G_\gamma, \Omega_\gamma) \mid \gamma \in \Gamma\}$ of o-primitive transitive ℓ-permutation groups indexed by an arbitrary chain Γ, must there exist a transitive ℓ-permutation group (G, Ω) whose skeleton it is? An affirmative answer is given by constructing the wreath product $(W, \Omega) = \mathrm{Wr}\{(G_\gamma, \Omega_\gamma) \mid \gamma \in \Gamma\}$ [28].

We consider first the wreath produce $(G_{\gamma_0}, \Omega_{\gamma_0})\mathrm{Wr}(G_{\gamma_1}, \Omega_{\gamma_1})$ of just two factors, with $\gamma_0 < \gamma_1$. Thinking of $\Gamma = \{\gamma_0, \gamma_1\}$ as running vertically, take $\Omega = \overleftarrow{\Omega_{\gamma_0} \times \Omega_{\gamma_1}}$, ordered lexicographically from the top (i.e., by greatest difference). Define an equivalence relation $\mathcal{K}$ on Ω by setting

$$(\sigma_{\gamma_0}, \sigma_{\gamma_1})\mathcal{K}(\tau_{\gamma_0}, \tau_{\gamma_1}) \Leftrightarrow \sigma_{\gamma_1} = \tau_{\gamma_1}.$$

Identify each $\mathcal{K}$-class with Ω_{γ_0} in the obvious fashion; and of course the chain of $\mathcal{K}$-classes is a copy of the chain Ω_{γ_1}. Let W consist of those elements w of $A(\Omega)$ such that

(1) w respects $\mathcal{K}$,

(2) The permutation of the chain of $\mathcal{K}$-classes (identified with Ω_1) lies in G_{γ_1}, and

(3) For each $\mathcal{K}$-class Δ, the order-isomorphism induced by w from Δ onto Δw (each identified with Ω_0) lies in G_{γ_0}.

Then (W, Ω) is a transitive ℓ-permutation group whose two ribs are $(G_{\gamma_0}, \Omega_{\gamma_0})$ and $(G_{\gamma_1}, \Omega_{\gamma_1})$. Example 5 is actually the wreath product $(A(\mathbf{Z}), \mathbf{Z})\mathrm{Wr}(A(\mathbf{R}), \mathbf{R})$; in that example it happens that $W = A(\Omega)$.

Now we turn to the general case. In the Cartesian product $\Sigma = \prod_{\gamma \in \Gamma} \Omega_\gamma$, choose any point and label it 0. Let Ω consist of the elements $\sigma \in \Sigma$ having inversely well ordered support, i.e., such that $\{\gamma \in \Gamma \mid \sigma(\gamma) \neq 0(\gamma)\}$ is inversely well ordered. Again order Ω lexicographically from the top.

For each $\gamma \in \Gamma$ define a pair of equivalence relations on the chain Ω by setting

$$\sigma\mathcal{K}^\gamma\tau \quad \Leftrightarrow \quad \sigma(\delta) = \tau(\delta) \text{ for all } \delta > \gamma, \text{ and}$$
$$\sigma\mathcal{K}_\gamma\tau \quad \Leftrightarrow \quad \sigma(\delta) = \tau(\delta) \text{ for all } \delta \geq \gamma.$$

(In the special case $\Gamma = \{\gamma_0, \gamma_1\}$, $\mathcal{K}_{\gamma_1} = \mathcal{K}^{\gamma_0} = \mathcal{K}$, the equivalence relation defined above.) Let W' consist of those elements of $A(\Omega)$ which respect all these equivalence relations. Then W' is a transitive ℓ-permutation group having the $\mathcal{K}_\gamma$'s and $\mathcal{K}^\gamma$'s as convex congruences. For each $w \in W'$, and for each $\alpha \in \Omega$ and $\gamma \in \Gamma$, w induces an order-isomorphism $w_{\alpha\mathcal{K}^\gamma}$ from $\alpha\mathcal{K}^\gamma/\mathcal{K}_\gamma$ onto $(\alpha w)\mathcal{K}^\gamma/\mathcal{K}_\gamma$, which via the obvious identifications with Ω_γ induces in turn an automorphism of Ω_γ. Finally, let W consist of those elements $w \in W'$ for which $w_{\alpha\mathcal{K}^\gamma} \in G_\gamma$ for all $\alpha \in \Omega$, $\gamma \in \Gamma$. The *wreath product* $\mathrm{Wr}\{(G_\gamma, \Omega_\gamma) \mid \gamma \in \Gamma\}$ is the transitive ℓ-permutation group (W, Ω).

The covering pairs (C_γ, C^γ) of convex congruences of (W, Ω) are precisely the pairs $(\mathcal{K}_\gamma, \mathcal{K}^\gamma)$ because of the o-primitivity of the given $(G_\gamma, \Omega_\gamma)$'s (which matters only for this feature of (W, Ω), not for the construction itself). Moreover the o-primitive component $(W_\gamma, \Omega_\gamma)$ is the given $(G_\gamma, \Omega_\gamma)$ when we take the *component* chain Ω_γ to be $0C^\gamma/C_\gamma$ and make the obvious identification with the *given* chain Ω_γ. Therefore the skeleton of (W, Ω) is the given $(\Gamma, \{(G_\gamma, \Omega_\gamma) \mid \gamma \in \Gamma\})$.

Incidentally, this tells us that *every* chain Γ occurs as the chain of positive orbits of G_α for some transitive ℓ-permutation group (G, Ω). Simply take

$$(G, \Omega) = Wr\{(A(\mathbf{R}), \mathbf{R})_\gamma \mid \gamma \in \Gamma\},$$

the wreath product of Γ copies of $(A(\mathbf{R}), \mathbf{R})$.

Theorem 6.1 (Holland and McCleary [20]) *Let (G, Ω) be a transitive ℓ-permutation group with skeleton $(\Gamma, \{(G_\gamma, \Omega_\gamma) \mid \gamma \in \Gamma\})$. Then (G, Ω) can be embedded in a natural way in its wreath product $Wr\{(G_\gamma, \Omega_\gamma) \mid \gamma \in \Gamma\}$.*

The embedding of the given chain Ω into the wreath chain is not in general onto when Γ is infinite. We return to this matter at the end of the next section.

7 The classification of o-primitive transitive ℓ-permutation groups

The o-Primitive Classification Theorem says that every o-primitive transitive ℓ-permutation group looks very much like one of the first three examples of Section 2. Recall that a transitive ℓ-permutation group is *o-primitive* if it has no proper convex congruences; or equivalently, if it has no proper o-blocks; or again equivalently, if the point stabilizers G_α are maximal convex (prime ℓ-) subgroups of G. (See Theorems 5.1 and 5.2.)

To understand o-primitive groups, and even to give the precise *statement* of the Classification Theorem, we need to look at a slight enlargement of our chain Ω. For an arbitrary Ω, we denote by $\bar{\Omega}$ the completion of by Dedekind cuts of Ω (without endpoints). That is, $\bar{\Omega}$ is formed from Ω by filling in each proper Dedekind cut (each "hole") of Ω with a point. Since Ω is dense in $\bar{\Omega}$, each $g \in A(\Omega)$ can be uniquely extended to an order-preserving permutation of $\bar{\Omega}$ (which we also denote by g). Moreover, $\bar{\omega}(f \vee g) = \max\{\bar{\omega}f, \bar{\omega}g\}$ and dually, even for $\bar{\omega} \in \bar{\Omega} \backslash \Omega$. Thus we can regard $A(\Omega)$ to be acting as an ℓ-permutation group on $\bar{\Omega}$.

Example 6 *$(A(\mathbf{Q}), \mathbf{Q})$ is o-2-transitive, but of course $(A(\mathbf{Q}), \bar{\mathbf{Q}}) = (A(\mathbf{Q}), \mathbf{R})$ is not, having $\mathbf{Q}$ as one orbit and the chain of irrationals as the other.*

For transitive (G, Ω), a non-singleton o-block Δ cannot have an endpoint in Ω (either in Δ or in $\Omega \backslash \Delta$). Necessarily its supremum, denoted by $\sup \Delta$, lies in $\bar{\Omega} \backslash \Omega$ (unless $\Delta = \Omega$). Thus every transitive ℓ-permutation group acting on $\Omega = \mathbf{R}$ is o-primitive.

It is easy to see that the first three examples in Section 2 are o-primitive, and that every o-2-transitive ℓ-permutation group is actually *primitive* (in the unordered sense).

We now offer a redescription of Example 3, the periodic example with $G = \{g \in A(\mathbf{R}) \mid (\alpha + 1)g = \alpha g + 1 \ \forall \alpha \in \Omega\}$. Let $z \in A(\mathbf{R})$ be translation by 1. Then $G = \{g \in A(\mathbf{R}) \mid zg = gz\}$, the centralizer $Z_{A(\mathbf{R})}z$ in $A(\mathbf{R})$ of z. Vice versa, z generates $Z_{A(\mathbf{R})}G$ since for any $f \in Z_{A(\mathbf{R})}G$ and $\alpha \in \mathbf{R}$, αf is fixed by G_α, whence it follows easily that f is a power of z.

Now we need to consider a slightly more general kind of period z, lying in $A(\bar{\Omega})$ but not necessarily in $A(\Omega)$. A *period* of a transitive ℓ-permutation group (G, Ω) is an element $z \in A(\bar{\Omega})^+$, one (hence each) of whose orbits $\{\alpha z^n \mid n \in \mathbf{Z}\}$ is coterminal in $\bar{\Omega}$, and which generates (as a group) $Z_{A(\bar{\Omega})}G$. An o-primitive transitive ℓ-permutation group (G, Ω) is called *periodically o-primitive* if it has a period z (except that $(\mathbf{Z}, \mathbf{Z})$ is not counted as periodically o-primitive). The effect of any $g \in G$ on any interval $[\alpha, \alpha z)$ determines its effect "one period up" on $[\alpha z, \alpha z^2)$, and similarly throughout all of Ω.

We now prove that every o-primitive transitive ℓ-permutation group is of one of the three types we have discussed.

Theorem 7.1 (o-Primitive Classification Theorem, McCleary [30]) *Every o-primitive transitive ℓ-permutation group is of one of the following three mutually exclusive types, and all transitive ℓ-permutation groups of these types are o-primitive:*

(1) *o-2-transitive.*

(2) *The regular representation of a subgroup of the additive reals.*

(3) *Periodically o-primitive.*

To prove the Classification Theorem we need a lemma:

Lemma 7.2 *Let (G, Ω) be an o-primitive transitive ℓ-permutation group, and let $\bar{\omega} \in \bar{\Omega}$. Then the orbit $\bar{\omega}G$ is dense in $\bar{\Omega}$. Consequently the action of G on $\bar{\omega}G$ is faithful and $(G, \bar{\omega}G)$ is o-primitive, making $G_{\bar{\omega}}$ a maximal prime convex ℓ-subgroup of G.*

Proof. Suppose $\bar{\omega}G$ were not dense in $\bar{\Omega}$. Necessarily $\bar{\omega} \in \bar{\Omega} \backslash \Omega$. Set $\sigma C \tau$ if and only if there is no $\bar{\omega}g$ ($g \in G$) between σ and τ. Then C would be a proper convex congruence of (G, Ω).

Proof of the Classification Theorem. Let (G, Ω) be an o-primitive transitive ℓ-permutation group not satisfying (1) or (2).

Suppose by way of contradiction that for one (hence every) $\alpha \in \Omega$, G_α fixes every $\beta < \alpha$. Then $G_\alpha \subseteq G_\beta$, making $G_\alpha = G_\beta$ by the maximality of G_α, so that *all* the stabilizer subgroups G_δ coincide, whence it follows that all are $\{e\}$. Thus (G, Ω) is regular (i.e., uniquely transitive). Then G is totally ordered since if $\alpha g \leq \alpha h$ then $\delta g \leq \delta h$ for all δ, else $gh^{-1} \vee e$ would move δ while fixing α, contradicting regularity. The o-group G has no proper convex subgroups since $G_\alpha = \{e\}$ is a maximal convex subgroup. This makes G Archimedean since if (in additive notation) $0 < a < b$ and $na < b$ for all $n \in \mathbf{N}$, then $\operatorname{Conv}\{ma \mid m \in \mathbf{Z}\}$ would be a proper convex subgroup of G. By Hölder's Theorem (see Example 2 ff.), we have (2).

Therefore we may pick $\beta < \alpha$ which is moved by G_α (see Figure 7). Our immediate goal is to show that G_α has a first positive orbit $\Delta_1 = (\Delta_1)_\alpha$. Let $\Delta = \operatorname{Conv}(\beta G_\alpha)$ be the orbit of G_α containing β. Since Δ is bounded above, we may set $\bar{\tau} = \sup \Delta \in \bar{\Omega}$. Pick $g \in G$ such that $\beta g = \alpha$, so that $\alpha < \bar{\tau}g$. Let $\Delta_1 = \operatorname{Conv}((\bar{\tau}g)G_\alpha)$, the orbit of G_α containing $\bar{\tau}g$ (in its Dedekind completion). To show that Δ_1 is the first positive orbit of G_α, we need to show that for any λ such that $\alpha < \lambda < \bar{\tau}g$, there exists $f \in G_\alpha$ such that $(\bar{\tau}g)f \leq \lambda$. By the lemma we may pick $h \in G$ such that $\alpha < \bar{\tau}h \leq \lambda$. Since $\alpha h^{-1} < \bar{\tau} = \sup \Delta$ and $\alpha g^{-1} = \beta \in \Delta$, we may pick $k \in G_\alpha$ such that $(\alpha g^{-1})k \geq \alpha h^{-1}$. Now $\alpha g^{-1}kh \geq \alpha$ and $(\bar{\tau}g)g^{-1}kh = \bar{\tau}kh = \bar{\tau}h \leq \lambda$ (since $\bar{\tau}$ is fixed by $k \in G_\alpha$). Taking $f = g^{-1}kh \wedge e$, we have $f \in G_\alpha$ and $(\bar{\tau}g)f \leq \lambda$, making Δ_1 the first positive orbit of G_α. Δ_1 must be a long orbit, else Ω would be discrete and by o-primitivity would in fact be a copy of $\mathbf{Z}$, yielding (2).

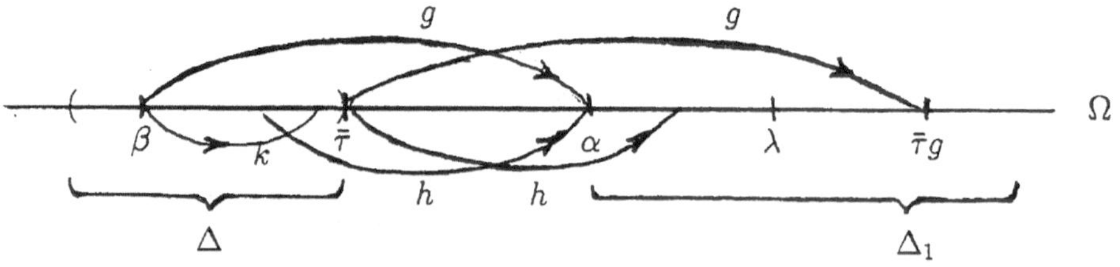

Figure 7: Proof that G_α has a first positive orbit Δ_1

Let $\bar{\omega}_\alpha = \sup(\Delta_1)_\alpha \in \bar{\Omega}$. (Were $(\Delta_1)_\alpha$ unbounded above, putting all points above α in the same G_α-orbit, (G, Ω) would be o-2-transitive.) Then $\bar{\omega}_{\alpha g} = \bar{\omega}_\alpha g$ for all $g \in G$ since $(\Delta_1)_{\alpha g} = (\Delta_1)_\alpha g$. $G_\alpha \subseteq G_{\bar{\omega}_\alpha}$, and by the maximality of G_α we have $G_\alpha = G_{\bar{\omega}_\alpha}$ Then $\alpha < \mu$ implies $\bar{\omega}_\alpha < \bar{\omega}_\mu$; for if $\alpha g = \mu$ for some $g \in G^+ \backslash G_\alpha$, then $\bar{\omega}_\alpha g = \bar{\omega}_{\alpha g} = \overline{\omega}_\mu$ and $g \in G^+ \backslash G_{\overline{\omega}_\alpha}$.

Define $z : \Omega \to \bar{\Omega}$ by $\alpha z = \bar{\omega}_\alpha$, a one-to-one order-preserving map from Ω into $\bar{\Omega}$. For any $\alpha \in \Omega$ and $g \in G$ we have

$$\bar{\omega}_{\alpha g} = \bar{\omega}_\alpha g, \text{ i.e., } (\alpha g)z = (\alpha z)g.$$

Hence z maps Ω onto a dense subset of $\bar{\Omega}$ (see the lemma) and thus has a unique extension to an order-preserving permutation of $\bar{\Omega}$ which we also denote by z; and $z \in Z_{A(\bar{\Omega})}G$.

$\operatorname{Conv}\{\alpha z^n \mid n \in \mathbf{Z}\}$ is an o-block Π of (G, Ω). For if

$$\alpha z^n \leq \alpha z^m g \leq \alpha z^{n+1} \; (m \in \mathbf{Z},\, g \in G)$$

then for all $p \in \mathbf{Z}$ we have

$$\alpha z^{m+p} g = (\alpha z^m g) z^p,.$$

which lies between αz^{n+p} and αz^{n+1+p}. By o-primitivity, $\Pi = \Omega$ and the z-orbit $\{\alpha z^n \mid n \in \mathbf{Z}\}$ is coterminal in Ω.

Because z centralizes G, every αz^n ($n \in \mathbf{Z}$) must be fixed by G_α, and no element of $\bar{\Omega}$ between αz^n and αz^{n+1} can be fixed by G_α (since this holds for $n = 0$ because $(\alpha, \alpha z) = \Delta_1$). Therefore the Ω-interval $(\alpha z^n, \alpha z^{n+1})$ constitutes a single G_α-orbit Δ_{n+1}. Now let $w \in Z_{A(\bar{\Omega})}G$. Since αw must be fixed by G_α, $\alpha w = \alpha z^{n_\alpha}$ for some n_α. In fact n_α is independent of α since for all $g \in G$ we have

$$(\alpha g)w = \alpha w g = \alpha z^{n_\alpha} g = (\alpha g) z^{n_\alpha}.$$

Hence w is a power of z. This shows that z is a period of (G, Ω), making (G, Ω) periodically o-primitive. This concludes the proof of the Classification Theorem.

Corollary 7.3 *In Lemma 7.2, $(G, \bar{\omega}G)$ is of the same o-primitivity type and subtype (o-2-transitive (pathological/non-pathological), regular, or periodic) as (G, Ω).*

Proof. Now that we have the Classification Theorem, a quick scan of the several cases makes it clear that the o-primitivity type of $(G, \bar{\omega}G)$ cannot differ from that of (G, Ω).

Corollary 7.4 (Lloyd [25] and Holland [15]) *If $(A(\Omega), \Omega)$ is transitive and o-primitive, it must be non-pathologically o-2-transitive or regular.*

Proof. $(A(\Omega), \Omega)$ cannot be periodically o-primitive. For pick $\alpha \in \Omega$ and $e \neq g \in G_\alpha$. Let $\hat{g}$ be the element of $A(\Omega)$ obtained from g by "depressing" g (i.e., making it agree with e) off the first positive long orbit Δ_1 of G_α. Then $\hat{g}$ doesn't commute with the period z.

Now suppose $(A(\Omega), \Omega)$ is o-2-transitive (and thus o-3-transitive by Proposition 2.2). Pick $\alpha_1 < \alpha_2 < \alpha_3 \in \Omega$, and pick $h \in G$ which moves α_2 strictly up but fixes α_1 and α_3. Now to obtain a non-identity element of bounded support, let $k = h \vee e$ and depress k off the supporting interval of k which contains α_2.

It is easy to use this notion of "depressing" to check that any regular $(A(\Omega), \Omega)$ is automatically o-primitive, and thus is the regular representation (G, G) of some subgroup of the additive reals. One obvious example of a G giving rise to a regular $(A(\Omega), \Omega)$ is $G = \mathbf{Z}$. It is far from clear that there are any *dense* G's which do so. However $2^{2^{\aleph_0}}$ such groups were discovered by T. Ohkuma [38] These groups are known as Ohkuma groups.

Corollary 7.5 (Holland [16] and McCleary [30]) *Let (G, Ω) be an o-primitive transitive ℓ-permutation group. Then G is ℓ-simple (i.e., has no proper ℓ-ideals) unless (G, Ω) is o-2-transitive and has elements of unbounded support.*

Proof. Consider the case in which (G,Ω) is periodically o-primitive with period z. Let $H \neq \{e\}$ be an ℓ-ideal of G, and pick $e \neq h \in H$. Replacing h by $h \vee h^{-1}$, we may suppose that $e < h$. For every $\bar{\omega} \in \bar{\Omega}$, the density of $\bar{\omega}G$ in $\bar{\Omega}$ guarantees that $\bar{\omega}$ is moved by some conjugate $h_{\bar{\omega}}$ of h. Pick $\alpha \in \Omega$. The supports of the $h_{\bar{\omega}}$'s for which $\alpha \leq \bar{\omega} \leq \alpha z$ provide an open cover of the compact $\bar{\Omega}$-interval $[\alpha, \alpha z]$, so there exists a finite collection $h_{\bar{\omega}_1}, \ldots, h_{\bar{\omega}_n}$ of these $h_{\bar{\omega}}$'s having no common fixed point $\bar{\omega}$ in $[\alpha, \alpha z]$. Let $k = h_{\bar{\omega}_1} \vee \ldots \vee h_{\bar{\omega}_n} \in H^+$. By periodicity $\bar{\tau}k > \bar{\tau}$ for all $\bar{\tau} \in \bar{\Omega}$. Now let $g \in G^+$. Then $\alpha k^m \geq \alpha z g$ for some m, else $\bar{\tau} = \sup\{\alpha k^m \mid m \in \mathbf{N}\}$ would be fixed by k. Now for all $\beta \in [\alpha, \alpha z]$ we have $\beta k^m \geq \alpha k^m \geq \alpha z g \geq \beta g$, and thus $\beta k^m \geq \beta g$ for all $\beta \in \Omega$ by periodicity. By the convexity of H, we have $g \in H$. Since G is generated by its positive elements, this establishes that $H = G$.

The o-2-transitive case (when all elements of G have bounded support) and the regular case are considerably easier, and we won't bother to treat them.

Later we shall need the following theorem, whose proof we omit:

Theorem 7.6 (McCleary [33]) *Let (G,Ω) be periodically o-primitive. Then the (faithful) action (G_α, Δ_n) on any long orbit Δ_n of G_α is non-pathologically o-2-transitive.*

The proof of the Classification Theorem showed that the chain of long orbits Δ_m of G_α is order-isomorphic to $\mathbf{Z}$. Between Δ_m and Δ_{m+1} lies exactly one element $\bar{\omega}_m$ of $\bar{\Omega}$, which may or may not lie in Ω. It is easy to extend the proof to show that for any periodically o-primitive (G,Ω) either

(a) There exists $n \in \mathbf{N}$ such that $\bar{\omega}_m \in \Omega$ iff $n|m$ (the Config$\,(n)$ case), or

(b) $\bar{\omega}_m \in \Omega$ only when $m = 0$ (the Config$\,(\infty)$ case).

Of course this configuration is independent of the choice of α. The periodic group in Example 3 has Config$\,(1)$. In fact, all of these configurations can occur with $\Omega = \mathbf{Q}$.

It is not hard to show that groups of Config$\,(\infty)$ are actually primitive (in the unordered sense), whereas those of finite Config$\,(n)$ are not primitive since $\{\alpha z^{nk} \mid k \in \mathbf{Z}\}$ forms a block. Moreover, when $G = Z_{A(\Omega)}z$, the full centralizer in $A(\Omega)$ of the period z, G is actually simple (as a group) when G has Config(∞), and for finite Config (n) the proper normal subgroups of G are precisely the subgroups contained in the center $\langle z^n \rangle$ of G.

We are now in a position to discriminate among the elements of $\bar{\Omega}$, some of which are more like elements of Ω than others. Let (G,Ω) be a transitive ℓ-permutation group, not necessarily o-primitive. We define a chain Ω^0 intermediate between Ω and $\bar{\Omega}$ (and depending on G as well as Ω). There are two cases:

(1) (G,Ω) may have a smallest non-trivial convex congruence $\mathcal{C}$, i.e., $\Gamma(G,\Omega)$ may have a smallest element $\gamma_0 = (\mathcal{E}, \mathcal{C}) = (\mathcal{C}_{\gamma_0}, \mathcal{C}^{\gamma_0})$. In this case (G,Ω) is said to be *locally o-primitive* and the $\mathcal{C}^{\gamma_0}$-classes are its *primitive segments.* We let Ω^0 be

the union of the completions $\bar{\Delta}$ of the primitive segments Δ.

(2) If (G, Ω) has no smallest C, let Ω^0 consist of those $\bar{\omega} \in \bar{\Omega}$ such that $\bar{\omega} \in \bar{\Delta}$ for o-blocks Δ from arbitrarily small non-trivial convex congruences C, i.e., for which $\bar{\omega}$ is the intersection of (the completions of) a tower of o-blocks coming from arbitrarily small non-trivial convex congruences.

In both cases, $\Omega \subseteq \Omega^0 \subseteq \bar{\Omega}$, and in view of Lemma 7.2, $\Omega^0 = \{\bar{\omega} \in \bar{\Omega} \,|\, \bar{\omega}G$ is dense in $\bar{\Omega}\}$.

We can now expand on the remark after Theorem 6.1 about the embedding of Ω in the wreath product chain. The chain embedding fills in certain holes of Ω with single points and others with non-singleton segments which turn out to be o-blocks of the wreath product. When (G, Ω) is locally o-primitive, no holes are filled in by single points; and when it is not, the holes filled in by single points are precisely the elements of $\Omega^0 \backslash \Omega$. In both cases, the (unpleasant) holes filled in by non-singleton segments are those holes $\bar{\omega}$ for which $\bar{\omega}$ is the intersection of (the completions of) a tower of o-blocks *not* coming from arbitrarily small non-trivial convex congruences. When $\Gamma(G, \Omega)$ is well ordered, no holes at all are filled in, and the wreath chain is just Ω itself.

8 Arbitrary ℓ-permutation groups: The structure theory

Without transitivity, there seems at first to be little hope of developing anything much like the transitive structure theory. To consider an extreme case, if $G = \{e\}$ then *every* convex subset $\Delta \subseteq \Omega$ is an o-block and *every* convex equivalence relation is a convex congruence of (G, Ω). Nevertheless, it does turn out to be possible to extend the transitive structure theory to arbitrary ℓ-permutation groups in a manner which parallels the transitive theory rather closely [34]. Here we first give a suggestive sketch of how the general theory relates to the transitive theory. Despite the lack of precision in that overview, we then manage to give a precise statement (Theorem 8.1) of the information crucial in applications of the theory, which combines analogues of part (2) of Theorem 5.3 (the reduction to the o-primitive case) and of the o-Primitive Classification Theorem .

The way out of the overabundance of o-blocks is to select certain "natural" ones, and ignore the others. We omit here the precise definition of "natural", but mention that in the transitive case (and in the following example) all o-blocks are natural.

Example 7 *Form (G, Ω) as follows: Begin with $(A(\mathbf{Q}), \bar{\mathbf{Q}})$. Form Ω by modifying $\bar{\mathbf{Q}}$, replacing each element of $\mathbf{Q}$ by a copy of $\mathbf{R}$ and each element of $\bar{\mathbf{Q}} \backslash \mathbf{Q}$ by a copy of $\mathbf{Z}$. Take $G = A(\Omega)$, which respects the set of local copies of $\mathbf{R}$ and the set of local copies of $\mathbf{Z}$. Then $\Gamma(G, \Omega)$ consists of three elements: a greatest γ (with $(G_\gamma, \Omega_\gamma) = (A(\mathbf{Q}), \bar{\mathbf{Q}})$), and under it two incomparable γ's, one with $(G_\gamma, \Omega_\gamma) = (A(\mathbf{R}), \mathbf{R})$ and one with $(G_\gamma, \Omega_\gamma) = (\mathbf{Z}, \mathbf{Z})$.*

As in the example, the collection of natural o-blocks containing any given $\alpha \in \Omega$ again forms a tower. The collection $\Gamma(G, \Omega)$ of covering pairs of natural convex congruences on appropriate subsets of Ω forms not a tower but a *root system*, meaning that the collection of covering pairs above any given one is totally ordered. Each pair $\alpha \neq \beta \in \Omega$ has a value $\mathrm{Val}(\alpha, \beta) = \gamma \in \Gamma(G, \Omega)$, associated with which is a (not necessarily transitive) *o-primitive component* $(G_\gamma, \Omega_\gamma)$. Here *o-primitive* means that $(G_\gamma, \Omega_\gamma)$ has no proper natural convex congruences, or equivalently, no proper natural o-blocks.

In the transitive case, a combination of the Classification Theorem and part (2) of Theorem 5.3 tells us that for each pair $\alpha \neq \beta \in \Omega$, there exists an o-block Δ (namely αC^γ) containing α and β, and a convex congruence C of the action (G_Δ, Δ) such that $\alpha C \neq \beta C$, and such that the transitive ℓ-permutation group $(G_\gamma, \Omega_\gamma)$ arising from the action of G_Δ on Δ/C comes from a short list of well understood types. This turns out to hold even in the general case. The general o-primitivity types are those listed in Theorem 8.1, and are hardly any different from the transitive types. The precision of Theorem 8.1 (which makes no reference to "natural" o-blocks) is made possible by this listing.

An ℓ-permutation group (G, Ω) is called *transitively derived* if it is obtained from an o-primitive transitive ℓ-permutation group (G, Σ), with Σ dense in itself (i.e., with (G, Σ) not isomorphic to $(\mathbf{Z}, \mathbf{Z})$), by selecting any set of orbits $\bar{\sigma}G$ of holes $\bar{\sigma}$ of Σ and augmenting Σ by filling in each hole in each selected orbit with a single point. (The action of any $g \in G$ is unambiguously determined on $\bar{\Sigma}$.) For example, $(A(\mathbf{Q}), \bar{\mathbf{Q}})$ is transitively derived, with $\mathbf{Q}$ and $\bar{\mathbf{Q}} \backslash \mathbf{Q}$ as the orbits.

The transitively derived ℓ-permutation groups themselves split into three types: o-2-transitively derived (pathological/non-pathological), regularly derived, and periodically derived. A transitively derived ℓ-permutation group can be derived from any one of its orbits, and the type is independent of which orbit of (G, Ω) we take to be Σ (see Lemma 7.2 and Corollary 7.3). Clearly a transitively derived ℓ-permutation group has no proper o-blocks, natural or otherwise.

(G, Ω) is called *integral* if (up to isomorphism) it is obtained from $(\mathbf{Z}, \mathbf{Z})$ by choosing a chain Π and taking Ω to be $\overleftarrow{\Pi \times \mathbf{Z}}$ (i.e., each point in $\mathbf{Z}$ is replaced by a copy of Π), with the action on Π of each integer translation t given simply by $(\pi, z)t = (\pi, zt)$. For example, if Π is a 2-point chain, then (G, Ω) is isomorphic to the ℓ-permutation group consisting of the chain $\mathbf{Z}$ with the group of even translations. Note that the regular transitive ℓ-permutation group $(\mathbf{Z}, \mathbf{Z})$ is classified as integral rather than as regularly derived.

Finally, (G, Ω) is called *static* if $G = \{e\}$.

Theorem 8.1 (McCleary [34]) *Let (G, Ω) be an arbitrary ℓ-permutation group. Let $\alpha \neq \beta \in \Omega$. Then there exists an o-block Δ containing α and β, and a convex congruence C of (G_Δ, Δ) such that $\alpha C \neq \beta C$, and such that the ℓ-permutation group arising from the action of G_Δ on the chain Δ/C of C-classes is of one of the following* *five mutually exclusive types:*

(1) *Transitively derived:*

(a) *o-2-Transitively derived.*

(b) *Regularly derived.*

(c) *Periodically derived.*

(2) *Integral.*

(3) *Static.*

When α and β lie in the same G-orbit, Δ and C are uniquely determined.

Now we relate Theorem 8.1 to our discussion of o-primitive components. When α and β lie in the same G-orbit, the uniquely determined Δ and C are both "natural", and the ℓ-permutation group arising from the action of G_Δ on Δ/C is an o-primitive component $(G_\gamma, \Omega_\gamma)$ of (G, Ω). Moreover, all non-static o-primitive components arise in this way.

9 Ordered permutation groups

As promised in the introduction, we pause to indicate the main structure theoretic results which apply more generally to *ordered* permutation groups (G, Ω), with G not required to be a sublattice of $A(\Omega)$, or which at least apply to *coherent o-*permutation groups. (Recall that (G, Ω) is *coherent* if whenever $\alpha < \beta$ for some $g \in G$ then $\alpha h = \beta$ for some $h \in G^+$.) The orbits of G_α need not be convex. Accordingly, the focus is instead on their convexifications $\mathrm{Conv}(\omega G_\alpha)$, which are known as *orbitals.* If $\omega_2 \in \mathrm{Conv}(\omega_1 G_\alpha)$ then $\mathrm{Conv}(\omega_2 G_\alpha) = \mathrm{Conv}(\omega_1 G_\alpha)$, so the set of orbitals inherits a total order from Ω.

Theorem 5.1 holds for ordered permutation groups, and Theorem 5.2 (for convex subgroups which contain G_α) holds for coherent permutation groups. Theorems 5.3 and 6.1 hold for ordered permutation groups. The o-Primitive Classification Theorem holds for coherent permutation groups provided that o-2-transitivity is relaxed to require not that G_α have only one positive orbit (a requirement equivalent to o-2-transitivity), but that it have only one positive *orbital.* Finally, Theorem 8.1 also holds with a similar modification.

10 Application to normal-valued ℓ-groups

In the study of abstract ℓ-groups, the class $\mathcal{N}$ of ℓ-groups satisfying

$$fg \leq g^2 f^2 \quad \forall f, g \in G^+ \qquad (*)$$

plays a very important role. $\mathcal{N}$ is a variety (equational class) of ℓ-groups since $(*)$ can be rewritten as a universally quantified equation "$w = e$" for some ℓ-group word w (constructed from the lattice operations as well as the group operations), specifically as

$$(g \vee e)^2(f \vee e)^2(g \vee e)^{-1}(f \vee e)^{-1} \wedge e = e \quad \forall f, g \in G.$$

An ℓ-group lying in $\mathcal{N}$ is called *normal-valued.* (This term stems from another characterization of $\mathcal{N}$ which makes reference to the notion of a "value" of an abstract ℓ-group, a different usage of the word "value" from that in ℓ-permutation groups.)

Obviously all abelian ℓ-groups are normal-valued. $A(\mathbf{R})$ is not, as can be seen by choosing $f, g \in A(\mathbf{R})^+$ in accord with Figure 8. The same holds for every o-2-transitive (G, Ω) because of Proposition 2.2. Our goal here is to illustrate how the structure theory is used by seeing how membership of G in $\mathcal{N}$ is reflected in the structure of (G, Ω). We deal first with the transitive case.

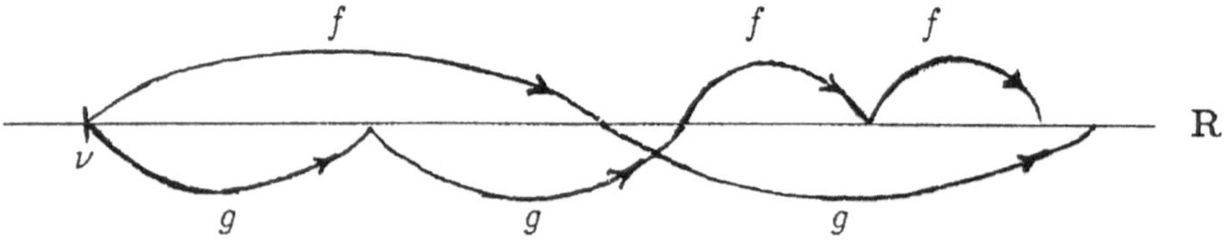

Figure 8: Failure of the identity $(*)$ in $A(\mathbf{R})$

Theorem 10.1 (J. A. Read [39]) *Let (G, Ω) be a transitive ℓ-permutation group. Then G is normal-valued if and only if all o-primitive components of (G, Ω) are regular.*

Proof$(\Rightarrow)$. Let $(G_\gamma, \Omega_\gamma)$ be an o-primitive component of (G, Ω), with $\Omega_\gamma = \alpha C^\gamma / C_\gamma$ (see Figure 6).

If $(G_\gamma, \Omega_\gamma)$ is o-2-transitive, then there exist elements $\hat{f}, \hat{g} \in G_\gamma^+$ re-creating Figure 8 (with $\mathbf{R}$ replaced by Ω_γ and ν by αC_γ). Moreover, $\hat{f}$ and $\hat{g}$ must be induced by some f and g in $G_{\alpha C^\gamma}$, and by replacing f by $f \vee e$ and g by $g \vee e$ we may assume that $f, g \in G_{\alpha C^\gamma}^+ \subseteq G^+$. Since

$$(\alpha C_\gamma) fg = (\alpha C_\gamma) \hat{f} \hat{g} > (\alpha C_\gamma) \hat{g}^2 \hat{f}^2 = (\alpha C_\gamma) g^2 f^2,$$

then certainly

$$\alpha fg > \alpha g^2 f^2.$$

Hence $G \notin \mathcal{N}$.

If $(G_\gamma, \Omega_\gamma)$ is periodically o-primitive, then Theorem 7.6 guarantees that Figure 8 can be re-created within $(G_\gamma, \Omega_\gamma)$, specifically within the action of a stabilizer subgroup $(G_\gamma)_{\alpha C_\gamma}$ on any one of its long orbits. Then just as in the o-2-transitive case, $G \notin \mathcal{N}$.

Proof ($\Leftarrow$). Suppose all o-primitive components of (G, Ω) are regular. Suppose by way of contradiction that for some $\alpha \in \Omega$ and $f, g \in G^+$ we have

$$\alpha fg > \alpha g^2 f^2 \ (\geq \alpha).$$

Let $\gamma = \mathrm{Val}(\alpha fg, \alpha)$. Let $\hat{f}$ and $\hat{g}$ be the elements of G_γ^+ induced on $\Omega_\gamma = \alpha C^\gamma / C_\gamma$ by f and g. Then

$$(\alpha C_\gamma)\hat{f}\hat{g} > (\alpha C_\gamma)\hat{g}^2 \hat{f}^2.$$

But this is impossible since $(G_\gamma, \Omega_\gamma)$ is the regular representation of a subgroup of the additive reals, and $\hat{f}$ and $\hat{g}$ are non-negative translations of Ω_γ.

Thanks to Theorem 8.1, the preceding theorem and proof carry over with hardly any change to intransitive ℓ-permutation groups (since $\mathrm{Val}(\alpha fg, \alpha)$ cannot be static):

Theorem 10.2 *Let (G, Ω) be an arbitrary ℓ-permutation group. Then G is normal-valued if and only if all o-primitive components of (G, Ω) are regularly derived, integral, or static.*

From the point of view of Theorems 10.1 and 10.2, relatively few ℓ-groups are normal-valued. Thus it comes as a surprise to discover that

Theorem 10.3 (Holland [18]) *$\mathcal{N}$ is the largest proper variety of ℓ-groups.*

Proof. Let $w(\mathbf{x}) = w(x_1, \ldots, x_n)$ be an ℓ-group word. Suppose $w(\mathbf{h}) \neq e$ for some substitution in *some* ℓ-group H. We must show that this happens in *every* ℓ-group G which is not normal-valued.

Always $w(\mathbf{x})$ can be written (though far from uniquely) in the standard form

$$w(\mathbf{x}) = \bigvee_i \bigwedge_j w_{ij}(\mathbf{x}) = \bigvee_i \bigwedge_j \prod_k x_{ijk},$$

a finite sup of finite infs of reduced group words $w_{ij}(\mathbf{x})$, with $x_{ijk} = x_m^{\pm 1}$. This is due to the distributivity of the lattices underlying ℓ-groups and a few other related ℓ-group properties.

Use the Holland Representation Theorem to view H as an ℓ-permutation group on some chain Σ. We are given that there exist $\sigma \in \Sigma$ and $h_1, \ldots, h_n \in H$ such that

$$\sigma w(\mathbf{h}) \neq \sigma, \text{ i.e., } \max_i \min_j \sigma w_{ij}(\mathbf{h}) \neq \sigma.$$

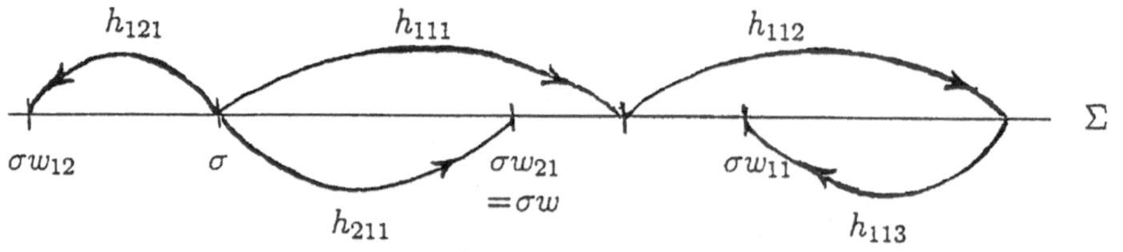

Figure 9: $\sigma w = \max\{\min\{\sigma w_{11}, \sigma w_{12}\}, \sigma w_{21}\} \neq \sigma$

We envision a picture showing how σ is moved successively by the h_{ijk}'s in w_{ij}, with a single picture showing this for all the finitely many w_{ij}'s (see Figure 9).

Now use the Holland Representation Theorem to represent G as an ℓ-permutation group (G, Ω). We assume temporarily that (G, Ω) is transitive. Since $G \notin \mathcal{N}$, Theorem 10.1 tells us that (G, Ω) has an o-primitive component $(G_\gamma, \Omega_\gamma)$ which is o-2-transitive or periodic. Proposition 2.2 and Theorem 7.6 guarantee that the above picture can be re-created in $(G_\gamma, \Omega_\gamma)$. (This is because for each h_m the picture stipulates the effect of h_m at only finitely many points, namely those points that are tails of h_{ijk}-arrows (when $x_{ijk} = x_m$) or heads of h_{ijk}-arrows (when $x_{ijk} = x_m^{-1}$)). Therefore there exists $\alpha \in \Omega$ and $g_1, \ldots, g_n \in G$ such that for the permutations $\hat{g}_1, \ldots, \hat{g}_n$ induced on $\Omega_\gamma = \alpha C^\gamma / C_\gamma$ by $g_1, \ldots, g_n$, we have

$$(\alpha C_\gamma) w(\hat{\mathbf{g}}) = \max_i \min_j (\alpha C_\gamma) w_{ij}(\hat{\mathbf{g}}) \neq \alpha C_\gamma.$$

That is to say,

$$(\alpha C_\gamma) w(\mathbf{g}) = \max_i \min_j (\alpha C_\gamma) w_{ij}(\mathbf{g}) \neq \alpha C_\gamma.$$

Then certainly

$$\alpha w(\mathbf{g}) \neq \alpha.$$

Therefore $w(\mathbf{g}) \neq e$, as desired.

The above assumption of transitivity for (G, Ω) was only for convenience—because of Theorem 8.1 the proof is essentially the same in the intransitive case. Alternately, the remarks at the end of Section 3 show that consideration of the transitive case is enough.

11 ℓ-group laws

In the proof of Theorem 10.3 it was mentioned that the standard form

$$w(\mathbf{x}) = \bigvee_i \bigwedge_j w_{ij}(\mathbf{x}) = \bigvee_i \bigwedge_j \prod_k x_{ijk}$$

is not unique. For example,

$$(x \wedge e) \vee (x^{-1} \wedge e) = e$$

in every ℓ-group (as can be seen by considering ℓ-permutation groups). How can one determine whether or not this is true of a given w? Equivalently, how can one determine in the free ℓ-group on $\{x_1, ..., x_n\}$ whether or not the element $w(\mathbf{x})$ is e? (The usual universal algebraic considerations establish that there is a unique free ℓ-group of any given rank.)

Theorem 11.1 (Holland and McCleary [21]) *Let w be an ℓ-group word. Then there is an algorithm to decide whether or not "$w = e$" is a law of ℓ-groups. That is, the word problem is solvable in free ℓ-groups.*

Proof (semi-formal). First put w in the standard form

$$w(\mathbf{x}) = \bigvee_i \bigwedge_j w_{ij}(\mathbf{x}) = \bigvee_i \bigwedge_j \prod_k x_{ijk}.$$

Draw a copy of $\mathbf{R}$, and mark the point 0. Draw a diagram like Figure 9 (with 0 playing the role of σ, and with an arrow for each x_{ijk}, labelling the arrow with x_m if $x_{ijk} = x_m$ and with x_m^{-1} if $x_{ijk} = x_m^{-1}$. Do this by proceeding in order through the x_{ijk}'s in any one w_{ij}, then (restarting at 0) in another w_{ij}, etc. When a given x_{ijk}-arrow is drawn, only finitely many points (ends of arrows) have so far appeared in the diagram. Choose the head of the new x_{ijk}-arrow to be any one of those finitely many points, or choose it to be (any point) in the interval between two consecutive ones of those points, subject to the following consistency requirement (where $x_{ijk} = x_m^{\pm 1}$):

> For each previously drawn x_m-arrow, the head of the new x_m-arrow must bear the same relation (left of, equal to, right of) the head of the previous x_m-arrow as the tail of the new x_m-arrow bears to the tail of the previous x_m-arrow. (Here the reverse of an $x_m^{\pm 1}$-arrow is considered to be an $x_m^{\mp 1}$-arrow.)

There are only finitely many possible diagrams, each of which determines a point $\pi = \max_i \min_j 0wij$ (see Figure 9). If some $\pi \neq 0$, then "$w = e$" is not a law of ℓ-groups—it fails in $A(\mathbf{R})$. Otherwise "$w = e$" *is* a law of ℓ-groups because as in the proof of Theorem 10.3 its failure in an ℓ-permutation group (H, Σ) would lead to a diagram illustrating that failure.

Despite the impression that the preceding proof may have given, it is fairly practical to use this algorithm by actually drawing the diagrams by hand, at least when a few refinements are added to cut down on the number of choices to be made.

12 Big subgroups of doubly transitive $A(\Omega)$

Our first interpretation of "big" is "maximal". When $G = A(\Omega)$ is o-2-transitive (and thus primitive in the unordered sense), the stabilizer subgroups $G_{\bar{\alpha}}$ $(\alpha \in \bar{\Omega})$ are maximal subgroups of G. It seems safe to say that there are no other obvious maximal subgroups. Any new ones would have to lack fixed points even in $\bar{\Omega}$, and it can be shown that they would have to contain $B(\Omega)$, the group of elements of bounded support. Examples have in fact been found by H. D. Macpherson [27] and by Glass and McCleary [10]. It is not known whether every proper subgroup of an o-2-transitive $A(\Omega)$ must be contained in a maximal subgroup.

Of course the $G_{\bar{\alpha}}$'s are also maximal prime convex ℓ-subgroups of G. Ball [1] had earlier produced examples of other maximal prime convex ℓ-subgroups of $A(\mathbf{R})$, using the Continuum Hypothesis. A curious connection was discovered in [10]:

Theorem 12.1 *Let $A(\Omega)$ be o-2-transitive. Then every maximal prime convex ℓ-subgroup of $A(\Omega)$ is in fact maximal as a subgroup.*

Another interpretation of "big" is "small index". For $G = A(\mathbf{Q})$, the stabilizer subgroups G_α of *points* $\alpha \in \mathbf{Q}$ have countable index, as do the stabilizer subgroups of finite sets of points. (Here the pointwise and setwise stabilizer subgroups coincide.)

Theorem 12.2 (J. K. Truss [41]) *The only proper subgroups of $A(\mathbf{Q})$ having index less than $2^{\aleph_0}$ are the stabilizer subgroups of finite subsets of $\mathbf{Q}$.*

Another nice result about subgroups of $A(\mathbf{Q})$ is

Theorem 12.3 (C. Gourion [11]) *$A(\mathbf{Q})$ is not the union of any countable tower of proper subgroups.*

13 Closed stabilizer subgroups

Given an ℓ-permutation group (G, Ω), one would like to be able to characterize the stabilizer subgroups G_α $(\alpha \in \Omega)$ in ℓ-group language. (In the next section, for example, we consider the problem of reconstructing the chain Ω from the ℓ-group G.) We mentioned in Section 3 that the G_α's are prime convex ℓ-subgroups of G, and that no more can be said in complete generality. Here we discuss an ℓ-group property which the stabilizer subgroups satisfy in many important cases, and look at some other closely related properties. These properties all involve suprema of arbitrary (infinite) subsets which may happen to exist in G; they are automatically valid for finite suprema. All are equivalent to their duals. The presence of these properties makes an ℓ-permutation group much easier to deal with.

An ℓ-subgroup H of an ℓ-group G is *closed in* G if whenever $g = \bigvee_{i\in I} h_i$ ($g \in G, h_i \in H$), then $g \in H$. We prove below (Theorem 13.2) that the stabilizer subgroups $A(\Omega)_\alpha$ of any $(A(\Omega), \Omega)$ are closed in $A(\Omega)$.

For any (G, Ω) each of the following three conditions implies the next (McCleary [31]):

(1) *Sups in* (G, Ω) *are pointwise*, i.e., if in G, $g = \bigvee_{i\in I} g_i$, then for each $\beta \in \Omega$, βg is the sup in Ω of $\{\beta g_i \mid i \in I\}$.

(2) (G, Ω) *has closed stabilizer subgroups*, i.e., G_α is closed in G for all $\alpha \in \Omega$. (Of course if G_α is closed so is G_β for all $\beta \in \alpha G$, so that when (G, Ω) is transitive all stabilizer subgroups are closed if any one of them is.)

(3) *The ℓ-group G is completely distributive*, i.e.,

$$\bigwedge_{i\in I}\bigvee_{j\in J} g_{ij} = \bigvee_{f\in J^I}\bigwedge_{i\in I} g_{if(i)}$$

for any collection $\{g_{ij} \mid i \in I,\ j \in J\}$ of elements of G for which the indicated sups and infs exist. (This is a statement about the ℓ-group G, independent of the particular representation (G, Ω).)

(1) implies (2) for the same reason that stabilizer subgroups G_α are ℓ-subgroups of G, and (1) implies (3) since (1) causes the (conditional) complete distributivity of the chain Ω to carry over to the ℓ-group G. We omit the proof that (2) implies (3). We mention a few other pleasant facts. Condition (1) for (G, Ω) is equivalent to (1) for $(G, \bar{\Omega})$. This is also valid for (2), so that if the stabilizer subgroups of the points α are closed, so are those of all the cuts $\bar{\alpha}$. (1) and (2) are essentially equivalent (and *are* equivalent when (G, Ω) is transitive); and when (G, Ω) is o-primitive (transitive or not), then (1), (2), and (3) are all equivalent. Every ℓ-group G satisfying (3) has a representation satisfying (1) and (2).

The following theorem is due to Lloyd [26]. Lloyd's proof required a great deal of ingenuity; the present proof from [34] shows how the structure theory enables us to obtain the theorem by a routine argument. The lemma helps explain why the least well understood o-primitive transitive ℓ-permutation groups are those that are pathologically o-2-transitive.

Lemma 13.1 (McCleary [32]) *Let (G, Ω) be an o-primitive transitive ℓ-permutation group. Then (G, Ω) has closed stabilizer subgroups if and only if (G, Ω) is not pathologically o-2-transitive.*

Proof ($\Leftarrow$). Suppose $\bigvee_{i\in I} h_i = g$ ($h_i \in G_\alpha$, $g \in G$), but that $g \notin G_\alpha$, making $\alpha < \alpha g$ and thus $\alpha g^{-1} < \alpha$. Suppose first that (G, Ω) is non-pathologically o-2-transitive. Pick $e < f \in G$ such that $\mathrm{supp}(f) \subseteq (\alpha g^{-1}, \alpha)$. Then $g > f^{-1}g \geq h_i$ for each i (since when $\omega \in \mathrm{supp}(f^{-1})$ we have $\omega f^{-1} g \geq (\alpha g^{-1})g = \alpha = \alpha h_i \geq \omega h_i$), a contradiction. The periodic case is treated similarly, using Theorem 7.6 and the periodicity itself. The regular case is trivial since then the stabilizer subgroups are $\{e\}$.

Proof($\Rightarrow$). Omitted.

Because of the close relation of arbitrary o-primitive ℓ-permutation groups to the transitive ones, Corollary 7.3 makes it clear that Lemma 13.1 (with the exceptional case now being "non-pathologically o-2-transitively derived") doesn't actually require transitivity.

Theorem 13.2 (Lloyd [26]) *$A(\Omega)$ has closed stabilizer subgroups.*

Proof. First we consider the transitive case, phrasing the argument so it will readily carry over to the intransitive case. Let $G = A(\Omega)$, and let $\alpha \in \Omega$. Suppose $\bigvee_{i \in I} h_i = g$ ($h_i \in G_\alpha$, $g \in G$), but that $\alpha < \alpha g$. Let $\gamma = \mathrm{Val}(\alpha, \alpha g)$. We may assume that g and the h_i's are the identity on $\Omega \backslash \alpha C^\gamma$ since modifying them to make them the identity outside αC^γ leaves them within $A(\Omega)$ and leaves the sup still valid. Let Δ be any C_γ-class within αC^γ. Then $A(\Delta) = \{e\}$ (which in the transitive case forces Δ to be a singleton). For if not, then by Lemma 7.2 there exists $k \in G$ such that Δk meets the C_γ-interval $(\alpha g^{-1} C_\gamma, \alpha g C_\gamma]$, and then there exists $e < f \in A(\Delta k)$, yielding $g > f_1^{-1} g \geq g_i$, where f_1 is the extension of f to Ω which is the identity off Δk. Thus there is a *unique* order-isomorphism from Δ onto Δg. To obtain a contradiction it would be enough to show that the o-primitive component $(G_\gamma, \Omega_\gamma)$ has closed stabilizer subgroups. *That* we already know because of Corollary 7.4 and Lemma 7.2.

In view of the remark after Lemma 13.1, the argument works even without transitivity once we observe that $(G_\gamma, \Omega_\gamma)$ is obviously not static and that the existence of the desired k is clear in the integral case.

Lemma 13.1 helps to shed light on free ℓ-groups. The free ℓ-group F_η of rank $\eta > 1$ has a representation as a pathologically o-2-transitive ℓ-permutation group. Such representations are the main tool for studying free ℓ-groups ([7],[24],[35],[36],[5],[4]). Indeed, in view of Lemma 13.1 the mere existence of such a representation tells us that F_η is not completely distributive (as it "shouldn't" be).

Lemma 13.1 can be generalized. (G, Ω) is said to have the *support property* if every non-singleton o-block Δ supports some $e \neq g \in G$.

Theorem 13.3 (McCleary [31]) *Let (G, Ω) be a transitive ℓ-permutation group having the support property. Then (G, Ω) has closed stabilizer subgroups if and only if (G, Ω) is not locally pathologically o-2-transitive.*

Corollary 13.4 *A wreath product $Wr\{(G_\gamma, \Omega_\gamma) \mid \gamma \in \Gamma\}$ of o-primitive transitive*

ℓ-permutation groups has closed stabilizer subgroups unless Γ has a smallest element γ_0 and $(G_{\gamma_0}, \Omega_{\gamma_0})$ is pathologically o-2-transitive.

14 Reconstruction

To what extent does the ℓ-group $A = A(\Omega)$ determine the chain Ω? We have seen that the stabilizer subgroups $A_{\bar\alpha}$ $(\bar\alpha \in \bar\Omega)$ are closed prime convex ℓ-subgroups of A. Conversely, *every* closed prime convex ℓ-subgroup $(\neq A)$ of A is a stabilizer subgroup $A_{\bar\alpha}$ (McCleary [29]). This encourages the hope that we might be able to reconstruct the chain Ω, or at least its completion $\bar\Omega$, from the collection of closed prime convex ℓ-subgroups of $A(\Omega)$.

We cannot hope to do this in complete generality since $A(\Omega) = \{e\}$ for every well ordered Ω. Thus we restrict to the transitive case. Even then we cannot hope to reconstruct Ω itself (as opposed to $\bar\Omega$) since clearly $A(\mathbf{Q})$ and $A(\bar{\mathbf{Q}}\backslash\mathbf{Q})$ are isomorphic as ℓ-groups via extension to $\bar{\mathbf{Q}} = \mathbf{R}$ and restriction to the irrationals $\bar{\mathbf{Q}}\backslash\mathbf{Q}$. But the reconstruction of $\bar\Omega$ works! In fact we can sharpen the conclusion a bit by cutting $\bar\Omega$ down to Ω^0 (defined at the end of Section 7 and not equal to $\bar\Omega$ unless (G, Ω) is o-primitive) because for transitive $A(\Omega)$

$$\Omega^0 = \{\bar\omega \in \bar\Omega \mid A_{\bar\omega} \text{ is a representing subgroup of } A\}.$$

Various parts of the next two theorems are due to Lloyd [25], Holland [15], and McCleary [29].

Theorem 14.1 *Let $A(\Omega_1)$ and $A(\Omega_2)$ be transitive. Then every ℓ-group isomorphism $\psi : A(\Omega_1) \to A(\Omega_2)$ is induced by some order-isomorphism $\phi : \Omega_1^0 \to \Omega_2^0$, meaning that $g\psi = \phi^{-1}g\phi$ for all $g \in A(\Omega_1)$. Hence the pair (ψ, ϕ) gives an ℓ-permutation group isomorphism from $(A(\Omega_1), \Omega_1)$ onto $(A(\Omega_2), \Omega_1\phi)$, and $\Omega_1\phi$ is an orbit of $(A(\Omega_2), \Omega_2^0)$. In particular, every ℓ-group automorphism of $A(\Omega_1)$ is just conjugation by some $\phi \in A(\Omega_1^0)$.*

Since $(A(\mathbf{R}), \bar{\mathbf{R}})$ has only one orbit and since the two orbits of $(A(\mathbf{Q}), \bar{\mathbf{Q}})$ are not order-isomorphic, we get the next two corollaries:

Corollary 14.2 *Every ℓ-group automorphism of $A(\mathbf{R})$ is inner, and similarly for $A(\mathbf{Q})$.*

Corollary 14.3 *Let $(A(\Omega), \Omega)$ be transitive. If $A(\Omega)$ is isomorphic as an ℓ-group to $A(\mathbf{R})$, then Ω is order-isomorphic to $\mathbf{R}$; and if $A(\Omega)$ is isomorphic to $A(\mathbf{Q})$, then Ω is order-isomorphic to either $\mathbf{Q}$ or $\mathbf{R}\backslash\mathbf{Q}$.*

For an o-2-transitive $A(\Omega)$ to have an outer ℓ-group automorphism, it is necessary and sufficient that there be an orbit $\neq \Omega$ of $(A(\Omega), \bar{\Omega})$ which is order-isomorphic to Ω. It is not at all obvious that any such Ω's exist, but in fact they do—an example has been constructed by Holland [17].

For the locally o-primitive case, we can relax these hypotheses considerably. The next theorem says roughly that if an ℓ-group G has two faithful representations as transitive ℓ-permutation groups, with one of them fairly decent and the other almost arbitrary, the two representations must be exceedingly much alike.

Theorem 14.4 *Let (G_1, Ω_1) be a transitive ℓ-permutation group which is locally o-primitive but not locally pathologically o-2-transitive and which has the support property. Let (G_2, Ω_2) be any transitive ℓ-permutation group such that G_2 does not act faithfully on any non-trivial convex congruence of (G_2, Ω_2). Then every ℓ-group isomorphism $\psi : G_1 \to G_2$ is induced by some order-isomorphism $\phi : \Omega_1^0 \to \Omega_2^0$. Hence the pair (ψ, ϕ) gives an ℓ-permutation group isomorphism from (G_1, Ω_1) onto $(G_2, \Omega_1\phi)$, and $\Omega_1\phi$ is an orbit of (G_2, Ω_2^0).*

We mention that the slightly cleaner statement of the preceding theorem, in comparison with [8, Theorem 7A], is enabled by the fact that if (G, Ω) is a transitive ℓ-permutation group and $\bar{\omega} \in \Omega^0$, then $(\bar{\omega}G)^0$, formed from $(G, \bar{\omega}G)$, coincides with the original Ω^0.

So far we have been dealing with isomorphisms that preserve both the group structure and the lattice structure. Sometimes preservation of just one of these is enough!

The history of *group* isomorphisms involves obtaining conclusions like that of Theorem 14.1 under ever weaker hypotheses. For example, Corollaries 14.2 and 14.3 hold for group isomorphisms except that in the former the conjugator may be an order-*reversing* permutation of $\mathbf{R}$ or $\mathbf{Q}$ respectively. As it turns out, such conclusions sometimes hold for subgroups G of $A(\Omega)$ which are not necessarily sublattices and do not necessarily act transitively on Ω:

(G, Ω) is said to be *approximately o-2-transitive* if for all $\alpha_1 < \alpha_2$ and $\beta_1 < \beta_2 \in \Omega$ and any open intervals Δ_1 containing β_1 and Δ_2 containing β_2, there exists $g \in G$ such that $\alpha_1 g \in \Delta_1$ and $\alpha_2 g \in \Delta_2$.

Theorem 14.5 (McCleary and M. Rubin [37]) *Let (G_1, Ω_1) and (G_2, Ω_2) be approximately o-2-transitive ordered permutation groups having non-identity elements of bounded support. Then every* group *isomorphism $\psi : G_1 \to G_2$ is induced by some order-isomorphism or anti-order-isomorphism $\phi : \bar{\Omega}_1 \to \bar{\Omega}_2$.*

Here is another remarkable theorem involving only the group and not the lattice (cf. Corollaries 14.2 and 14.3):

Theorem 14.6 (Yu. Gurevich and Holland [12]) *There is a sentence in the elementary language of groups which characterizes $A(\mathbf{Q})$ in the sense that it holds in a transitive $A(\Omega)$ if and only if Ω is order-isomorphic to $\mathbf{Q}$ or to $\mathbf{R}\backslash\mathbf{Q}$; and similarly for $A(\mathbf{R})$.*

The other half of the story is due to M. Giraudet [22]:

Theorem 14.7 (Giraudet) *Let $(A(\Omega_1), \Omega_1)$ and $(A(\Omega_2), \Omega_2)$ be o-2-transitive. Then every lattice-with-e isomorphism $\psi : A(\Omega_1) \to A(\Omega_2)$ is induced by some order-isomorphism or anti-order-isomorphism $\phi : \bar{\Omega}_1 \to \bar{\Omega}_2$ (where in the latter case this is applied to g^{-1}, i.e., $g\psi = \phi^{-1}g^{-1}\phi$ for all $g \in G$).*

Giraudet also established a lattice-with-e analogue of Theorem 14.6.

We have been dealing with the ℓ-group obtained by equipping $A(\Omega)$ with the pointwise order. How much choice did we really have? That is, how else could the abstract group $A(\Omega)$ have been lattice-ordered? This leads to a different type of reconstruction problem, this time of the order (the exact order, not just up to isomorphism) from the group operation!

Theorem 14.8 (Holland [19]) *Let $A(\Omega)$ be o-2-transitive. Then the only lattice-orderings of the* group *$A(\Omega)$ yielding an ℓ-group are the pointwise order and its dual.*

And vice versa!

Theorem 14.9 (M. Darnel, Giraudet, and McCleary [3]) *The only group operations on the* lattice *$A(\mathbf{R})$ having the usual identity e are composition and reverse composition. This holds also for $A(\mathbf{Q})$.*

References

[1] R. N. Ball, *Full convex ℓ-subgroups of a lattice-ordered group,* Ph.D. Thesis, University of Wisconsin, Madison, Wis., USA, 1974.

[2] R. N. Ball and M. Droste, *Normal subgroups of doubly transitive automorphism groups of chains,* Trans. Amer. Math. Soc. **290**(1985), 647-664.

[3] M. R. Darnel, M. Giraudet, and S. H. McCleary, *Uniqueness of the group operation on the lattice of order-automorphisms of the real line,* Algebra Universalis **33**(1995), 419-427.

[4] M. Droste, *Representations of free lattice-ordered groups,* Order **10**(1993), 375-381.

[5] M. Droste and S. H. McCleary, *The root system of prime subgroups of a free lattice-ordered group (without GCH),* Order **6**(1989), 305-309.

[6] M. Droste and S. Shelah, *A construction of all normal subgroup lattices of 2-transitive automorphism groups of linearly ordered sets,* Israel J. Math. **51**(1985), 223-261.

[7] A. M. W. Glass, *ℓ-simple lattice-ordered groups,* Proc. Edinburgh Math. Soc. **19**(1974), 133-138.

[8] ----------, *Ordered Permutation Groups,* London Math. Soc. Lecture Notes Series No. **55**, Cambridge University Press, Cambridge, England, 1981.

[9] A. M. W. Glass and W. C. Holland (editors), *Lattice-ordered groups: Advances and Techniques,* Kluwer Academic Pub., Dordrecht, The Netherlands, 1989.

[10] A. M. W. Glass and S. H. McCleary, *Big subgroups of automorphism groups of doubly homogeneous chains,* Ordered Algebraic Structures: The 1991 Conrad Conference, ed. J. Martinez and C. Holland, Kluwer Academic Publishers, Dordrecht, The Netherlands, 1993, 51-71.

[11] C. Gourion, *A propos du groupe des automorphismes de* $(\mathbf{Q}; \leq)$, C. R. Acad. Sci. Paris Sér. I Math. 315(1992), 1329-1331..

[12] Yu. Gurevich and W. C. Holland, *Recognizing the real line,* Trans. Amer. Math. Soc. **265**(1981), 527-534.

[13] G. Higman, *On infinite simple groups,* Publ. Math. Debrecen **3**(1954), 221-226.

[14] W. C. Holland, *The lattice-ordered group of automorphisms of an ordered set,* Michigan Math. J. **10**(1963), 399-408.

[15] ----------, *Transitive lattice-ordered permutation groups,* Math. Zeit. **87**(1965), 420-433.

[16] ----------, *A class of simple lattice-ordered permutation groups,* Proc. Amer. Math. Soc. **16**(1965), 326-329.

[17] ----------, *Outer automorphisms of ordered permutation groups,* Proc. Edin-

burgh Math. Soc. **19**(1975), 331-344.

[18] ———, *The largest proper variety of lattice-ordered groups,* Proc. Amer. Math. Soc. **57**(1976), 25-28.

[19] ———, *Partial orders of the group of automorphisms of the real line,* Contemporary Math. **131**(1992), 197-207.

[20] W. C. Holland and S. H. McCleary, *Wreath products of ordered permutation groups,* Pacific J. Math. **31**(1969), 703-716.

[21] ———, *Solvability of the word problem in free lattice-ordered groups,* Houston J. Math. **5**(1979), 99-105.

[22] M. (Jambu-)Giraudet, *Bi-interpretable groups and lattices,* Trans. Amer. Math. Soc. **278**(1983), 253-269.

[23] H. Kneser, *Kurvenscharen auf Ringflächen,* Math. Ann. **91**(1924), 135-154.

[24] V. M. Kopytov, *Free lattice-ordered groups,* Algebra and Logic **18**(1979), 259-270 (English translation).

[25] J. T. Lloyd, *Lattice-ordered groups and o-permutation groups,* Ph.D. Thesis, Tulane University, New Orleans, La., USA, 1964.

[26] ———, *Complete distributivity in certain infinite permutation groups,* Michigan Math. J. **14**(1967), 393-400.

[27] H. D. Macpherson, *Large subgroups of infinite symmetric groups,* Proc. NATO AST Conf. on Finite and Infinite Combinatorics, Banff, Alberta, Canada, 1991.

[28] S. H. McCleary, *Orbit configurations of ordered permutation groups,* Ph.D. Thesis, University of Wisconsin, Madison, Wis., USA, 1967.

[29] ———, *The closed prime subgroups of certain ordered permutation groups,* Pacific J. Math. **31**(1969), 745-753.

[30] ———, *O-primitive ordered permutation groups,* Pacific J. Math. **40**(1972), 349-372.

[31] ———, *Closed subgroups of lattice-ordered permutation groups,* Trans. Amer. Math. Soc. **173**(1972), 303-314.

[32] ———, *O-2-transitive ordered permutation groups,* Pacific J. Math. **49**(1973), 425-429.

[33] ———, *O-primitive ordered permutation groups II,* Pacific J. Math. **49**(1973), 431-443.

[34] ———, *The structure of intransitive ordered permutation groups,* Algebra Universalis **6**(1976), 229-255.

[35] ———, *Free lattice-ordered groups represented as o-2-transitive ℓ-permutation groups,* Trans. Amer. Math. Soc. **290**(1985), 69-79.

[36] ———, *An even better representation for free lattice-ordered groups,* Trans. Amer. Math. Soc. **290**(1985), 81-100.

[37] S. H. McCleary and M. Rubin, *Locally moving groups and the reconstruction problem for chains and circles,* in preparation.

[38] T. Ohkuma, *Sur quelques ensembles ordonnés linéairement,* Fund. Math. **43**(1955), 326-337.

[39] J. A. Read, *Wreath products of nonoverlapping lattice-ordered groups,* Canad. Math. Bull. **17**(1975), 713-722.

[40] J. Schreier and S. Ulam, *Eine Bemerkung über die Gruppe der topologischen Abbildungen der Kreislinie auf sich selbst,* Studia Math. **5**(1935), 155-159.

[41] J. K. Truss, *Infinite permutation groups: subgroups of small index,* J. Algebra **120**(1989), 495-515.

On Recovering Structures from Quotients of their Automorphism Groups

J. K. Truss

Department of Pure Mathematics
University of Leeds
Leeds LS2 9JT
England

1 Introduction

The question as to what information about a structure is provided by knowledge of its automorphism group has been considered in many cases. In those which I am principally interested in here, the relevant work was carried out in [10] and [17] (for symmetric groups), and [7] (for ordered permutation groups). In addition I consider the rather harder matter of how much we can find out if we are only given suitable *quotients* of the automorphism group. In the first of these two cases this is related to De Bruijn's question on embeddings of quotients of infinite permutation groups.

If μ is an cardinal (finite or infinite), I write $S(\mu)$ for the full symmetric group on μ, that is to say, the group of all permutations of μ ($1-1$ mappings from μ onto μ under function composition). It is well known (see [16] for instance) that, apart from the alternating group $A(\mu)$ of even permutations, the non-trivial normal subgroups of $S(\mu)$ all have the form $S_\kappa(\mu) = \{g \in S(\mu) : |supp\ g| < \kappa\}$ for some infinite cardinal κ, where the *support* of $g \in S(\mu), supp\ g$, is the set of elements moved by g. Subsets of μ of cardinality less than κ are called *small*.

Two issues which we may address in this connection are the following. In the first place, can we distinguish the quotient groups $S_\lambda(\mu)/S_\kappa(\mu)$ for different values of κ, λ, and μ? This question was posed for instance in [16]. Two obvious senses of 'distinguish' here are 'up to isomorphism', and 'up to elementary equivalence'. The sense used in [16] was the former. (In [4] it was initially stated that this problem was still open, but then it was remarked in proof that the groups can be distinguished up to isomorphism if we assume the GCH.) As it turns out, a solution to this problem in a large number of cases had already been included in [11] Theorem 4.3, which gives a suitable first order bi-interpretability result in the case where $\lambda = \mu^+$ and $cf(\kappa) > 2^{\aleph_0}$, covering both the isomorphism and elementary equivalence problems. In [3] De Bruijn asked as a related question for which κ and

W. C. Holland (ed.), Ordered Groups and Infinite Permutation Groups, 63–95.

μ, $S(\mu)/S_\kappa(\mu)$ is embeddable in $S(\mu)$, and results about this are presented in [4].

The second one concerns outer automorphisms. In [1] it is shown that the outer automorphism group of $S(\omega)/S_\omega(\omega)$ is infinite cyclic, where a typical outer automorphism is induced by the map $n \mapsto n+1$. The method used there incorporates a second order interpretation of the relevant ring of sets in the quotient group. One of the by-products of [19] is another proof (in addition to [11]) that this can actually be done in a first order fashion, and indeed this had already been shown in [11] Theorem 4.2. A similar result was proved for quotients of order-preserving permutation groups in [5]. It was suggested by M. Droste and M. Rubin that a better way to view these results was as reconstructions of the *tail* of a suitable structure from the quotient group, which would also imply the outer automorphism result, but which would cast it in a more natural setting; and that is the viewpoint adopted here.

The key idea in all the work described in this paper is to try to interpret (in a precise sense) as much as possible of the underlying structure on which a group acts in the group itself. To set everything in context I begin by surveying how this goes for symmetric groups (before passing to quotients), and ordered permutation groups. For symmetric groups this is essentially classical, at any rate in the finite case, and the technique is described in many standard textbooks ([8] for instance). But I shall emphasize the first order nature of the interpretation, as described for example in [10]. For ordered permutation groups, an elegant result of Gurevich and Holland [7] tells us that there is a first order sentence of the language of group theory which holds in the (full) automorphism group of a doubly homogeneous chain if and only if that chain is isomorphic to the real numbers $\mathbf{R}$. The idea of the proof employs an interpretation of $\mathbf{R}$ inside its automorphism group, and following on from this one can immediately read off what the outer automorphisms are of $\mathrm{Aut}(\mathbf{R}, <)$, as is done for instance in [6].

The notation used is fairly standard. I use κ, λ, and μ to stand for cardinals (usually infinite), and $|X|$ for the cardinality of the set X. If g and h are elements of the group G, g^h stands for the conjugate $h^{-1}gh$, and h is called a *conjugator* of g to g^h. Permutations (and sometimes other maps) are written on the right of their arguments.

Overlines such as $\overline{x}$ are used to stand for 'tuples' (finite sequences $(x_1, \ldots, x_n)$ of the appropriate length). If $\mathcal{A}$ and $\mathcal{B}$ are first order structures, we say that F is a (first order) *interpretation* of $\mathcal{B}$ in $\mathcal{A}$ if for some n it maps some 0-definable (meaning definable without parameters) subset D of A^n onto B, the equivalence relation E on D given by $\overline{a}E\overline{a}' \Leftrightarrow F(\overline{a}) = F(\overline{a}')$ is 0-definable, and is such that for every m-ary relation symbol $\underline{R}$ of the language of $\mathcal{B}$ there is a corresponding formula $\varphi_{\underline{R}}$ in the language of $\mathcal{A}$ with mn free variables such that for all $\overline{a}_1, \ldots, \overline{a}_m \in A^n$, $\mathcal{A} \models \varphi_{\underline{R}}(\overline{a}_1, \ldots, \overline{a}_m)$ if and only if $\mathcal{B} \models \underline{R}(F(\overline{a}_1), \ldots, F(\overline{a}_m))$, with similar conditions for any function or constant symbols of the language of $\mathcal{B}$. The structures $\mathcal{A}$ and $\mathcal{B}$ are *bi-interpretable* if there are interpretations F_1 of $\mathcal{B}$ in $\mathcal{A}$ and F_2 of $\mathcal{A}$ in $\mathcal{B}$ such that F_2F_1 is 0-definable in $\mathcal{A}$ and F_1F_2 is 0-definable in $\mathcal{B}$ (under a natural definition of 'composition' of interpretations). We may also consider interpretations with parameters, but do not need them here. Moreover these are not so useful in studying questions about outer automorphisms, since we would then need to restrict attention to automorphisms fixing the parameters. There is also a notion of *uniform interpretability* of one class of structures in another, meaning that all the interpretations are given by the same formulae. It will be clear that the interpretations we describe are uniform over large classes of structures, though we do not labour the point.

I would particularly like to thank Mati Rubin for introducing me to many of the ideas discussed here, and also Saharon Shelah, for patiently explaining his methods as described in §4 (which are more fully detailed in [19]). Some of the results I discuss are either consequences of, or generalized by, work of Rubin's [11, 12, 14], and some of these issues are also discussed in his paper in this volume. I give citations where appropriate. My main goal however has been to lead up to and elucidate the results on outer automorphisms, in the natural style mentioned above, and so I have not given all the possible intermediate results in full generality.

2 Interpretability Results for Symmetric Groups

In this section I review the results from [10] about what can be interpreted in the symmetric groups $S(\kappa)$ for cardinals κ. Essentially one can interpret the structure (here just a 'set') inside the group. It is an easy consequence that $S(\kappa)$ has no outer automorphisms. Or, according to the viewpoint adopted here, we should rather say that any isomorphism between two symmetric groups is induced by a bijection between their underlying sets. Since we wish the methods to apply even to the finite case, we have to stipulate that $\kappa \neq 6$. (It is well known that $S(6)$ is the only symmetric group admitting an outer automorphism.)

On the whole we represent the underlying structure by means of involutions, since these are abundant in symmetric groups, and generate the whole group (in a small number of steps). Although involutions will still be used for the case of the quotient groups, things do not work so well, and we have to work quite hard by an alternative approach (due to Shelah [19]) before they can again be invoked.

Consider a symmetric group $S(\kappa)$ (where κ may be finite). The idea is that we may represent the members of κ by families of pairs of transpositions. Specifically $\alpha \in \kappa$ will be represented by the family of all pairs (g, h) of transpositions for which $\{\alpha\} = supp\ g \cap supp\ h$ (so that in the definition of 'interpretation' we shall take $n = 2$ and $D = \{(g,h) : g, h \text{ transpositions and } |supp\ g \cap supp\ h| = 1\}$). The steps required are to show how we can define (i) g is a transposition, (ii) g_1 and g_2 are transpositions for which $|supp\ g_1 \cap supp\ g_2| = 1$, (iii) g_1, g_2, h_1 and h_2 are transpositions for which $|supp\ g_1 \cap supp\ g_2| = |supp\ h_1 \cap supp\ h_2| = 1$ and $supp\ g_1 \cap supp\ g_2 = supp\ h_1 \cap supp\ h_2$.

For (i) we use the formula

$trans_1(x) : x \neq 1 \wedge x^2 = 1 \wedge (\forall y)((xx^y)^2 = 1 \vee (xx^y)^3 = 1).$

Lemma 2.1 *If $\kappa \geq 3 \wedge \kappa \neq 4, 6$ then for any $g \in S(\kappa), S(\kappa) \models trans_1(g)$ if and only if g is a transposition.*

Proof. If g is a transposition $(\alpha\beta)$, then so is g^h, $(\gamma\delta)$ say. If $\{\alpha,\beta\} = \{\gamma,\delta\}$ then $gg^h = 1$. If $\{\alpha,\beta\} \cap \{\gamma,\delta\} = \emptyset$ then gg^h has order 2. Otherwise $|\{\alpha,\beta\} \cap \{\gamma,\delta\}| = 1$. Suppose that $\alpha = \delta$. Then $gg^h = (\alpha\beta\gamma)$, which has order 3.

Conversely suppose that $S(\kappa) \models trans_1(g)$. Then g has order 2, so is a product of one or more 2-cycles. Suppose for a contradiction that there are at least two, $(\alpha\beta)$ and $(\gamma\delta)$. Since $\kappa \neq 4$, there is some $\epsilon \notin \{\alpha,\beta,\gamma,\delta\}$. Letting $h = (\alpha\ \beta\ \gamma\ \delta\ \epsilon)$ we find that gg^h maps α to γ to ϵ, so it must have order 3, and hence g moves ϵ. Suppose g interchanges ϵ and ζ. Since $\kappa \neq 6$, there is some other element η say. This time letting $h = (\alpha\ \beta\ \gamma\ \delta\ \epsilon\ \zeta\ \eta)$ we find that gg^h maps α to γ to ϵ to η, contrary to gg^h of order 2 or 3.

Note that $trans_1$ is true of (01)(23) in $S(4)$ even though it is not a transposition. In this case we can use an *ad hoc* method to distinguish the transpositions:
$trans(x) : trans_1(x) \wedge (\exists y)(x^y \neq x \wedge (xx^y)^3 = 1)$.

This is still clearly true of all transpositions (for $\kappa \geq 3$), but is not true for the products of two 2-cycles in $S(4)$. The group $S(6)$ however remains as an exception, and this is because there is an automorphism in this case which interchanges (01) and (01)(23)(45). It follows that there can be no way of defining the class of transpositions here, for any definition would have to be preserved by this automorphism.

To characterize when $|supp(g) \cap supp(h)| = 1$ we use the formula
$overlap(x,y) : trans(x) \wedge trans(y)\ \wedge\ x \neq y \wedge (xy)^3 = 1$.

Lemma 2.2 *If $\kappa \geq 3$ and $\kappa \neq 6$ then for any $g, h \in S(\kappa)$,*

$$S(\kappa) \models overlap(g,h)$$

if and only if

$$g, h \text{ are transpositions for which } |supp(g) \cap supp(h)| = 1.$$

Proof. Let $g = (\alpha\ \beta)$ and $h = (\gamma\ \delta)$. If $\{\alpha,\beta\}$ and $\{\gamma,\delta\}$ intersect in a single point, $\alpha = \delta$ say, then $gh = (\alpha\ \beta)(\gamma\ \alpha) = (\alpha\ \beta\ \gamma)$, which has order 3. If they are disjoint, gh has order 2.

Since a pair of overlapping transpositions is meant to 'code' the point of intersection, we also need to express when two such intersections are equal. To ease this we use as an auxiliary formula
$triple(x,y,z) : overlap(x,y) \wedge overlap(x,z) \wedge overlap(y,z) \wedge x^y \neq z$.

In terms of this we can introduce the formula
$equal(x_1,y_1,x_2,y_2) : overlap(x_1,y_1) \wedge overlap(x_2,y_2) \wedge [(x_1 = x_2) \vee (x_1 = y_2) \vee triple(x_1,x_2,y_2)] \wedge [(y_1 = x_2) \vee (y_1 = y_2) \vee triple(y_1,x_2,y_2)]$.

Lemma 2.3 *If $\kappa \geq 3$ and $\kappa \neq 6$ then for any $g, h, k \in S(\kappa)$,*

$$S(\kappa) \models triple(g, h, k)$$

if and only if there are distinct $\alpha, \beta, \gamma, \delta \in \kappa$ such that $g = (\alpha\ \beta), h = (\alpha\ \gamma)$, and $k = (\alpha\ \delta)$.

Proof. Suppose that $S(\kappa) \models triple(g, h, k)$. By Lemma 2.2 we may write $g = (\alpha\ \beta), h = (\alpha\ \gamma)$. If $k = (\beta\ \gamma)$ then $g^h = k$, contrary to hypothesis. So $supp\ g \cap supp\ k = supp\ h \cap supp\ k = \{\alpha\}$ as required.

Lemma 2.4 *If $\kappa \geq 3$ and $\kappa \neq 6$ then for any $g_1, h_1, g_2, h_2 \in S(\kappa)$,*

$$S(\kappa) \models equal(g_1, h_1, g_2, h_2)$$

if and only if g_i, h_i are all transpositions, and there is $\alpha \in \kappa$ such that $supp\ g_1 \cap supp\ h_1 = supp\ g_2 \cap supp\ h_2 = \{\alpha\}$.

Proof. If $S(\kappa) \models equal(g_1, h_1, g_2, h_2)$ we let $g_1 = (\alpha\beta), h_1 = (\alpha\gamma), g_2 = (\delta\epsilon)$, and $h_2 = (\delta\zeta)$, and suppose for a contradiction that $\alpha \neq \delta$. If $triple(g_1, g_2, h_2)$ then $\delta \in \{\alpha, \beta\}$ and so $\delta = \beta$. Then $triple(h_1, g_2, h_2)$ must fail (as otherwise $\delta = \gamma$) and so $h_1 = g_2$ or h_2. But each of these is impossible since $\beta \notin \{\alpha, \gamma\}$. It follows in the same way that $\neg triple(h_1, g_2, h_2)$. Since $g_1 \neq h_1$ and $g_2 \neq h_2$ we have either $g_1 = g_2 \wedge h_1 = h_2$ or $g_1 = h_2\ \wedge\ h_1 = g_2$, each of which implies $\alpha = \delta$ again.

Conversely suppose that $g_1 = (\alpha\beta), h_1 = (\alpha\gamma), g_2 = (\alpha\delta), h_2 = (\alpha\epsilon)$ where $\beta \neq \gamma$ and $\delta \neq \epsilon$. Then $overlap(g_1, h_1)$ and $overlap(g_2, h_2)$. The cases $\beta = \delta, \epsilon$ and $\beta \notin \{\delta, \epsilon\}$ correspond to $g_1 = g_2, h_2$, and $triple(g_1, g_2, h_2)$ respectively, and similarly for the final clause.

Corollary 2.5 *If $\kappa \geq 3$ and $\kappa \neq 6$, then $(\kappa, =)$ is interpretable in $S(\kappa)$.*

Proof. We let $n = 2$ and $D = \{(g, h) : S(\kappa) \models overlap(g, h)\}$. Then the map F is given by $F((g, h)) = \alpha$ provided $\{\alpha\} = supp\ g \cap supp\ h$. The only relation on the structure is equality, and the formula *equal* introduced above serves as in the definition of 'interpretation' to express when two pairs encode the same point.

Theorem 2.6 *Let X and Y be sets of cardinality $\neq 6$, and let θ be an isomorphism $S(X) \rightarrow S(Y)$. Then there is a bijection $f : X \rightarrow Y$ such that for every $g \in S(X), g^\theta = f^{-1}gf$.*

Proof. If X or Y is finite then so is the other, and as $|S(X)| = |S(Y)|, |X| = |Y|$. This means that we can check the cases $|X| = 1, 2$ directly. Now assuming $|X|, |Y| \geq 3$ we may use the interpretation given above. Specifically we let $af = b$ if there are distinct $a_1, a_2 \in X - \{a\}$ and distinct $b_1, b_2 \in Y - \{b\}$ such that $(a\ a_i)^\theta = (b\ b_i)$ for $i = 1, 2$. This is well-defined since if a_i', b_i' are other choices, then by Lemma 2.4, $S(X) \models equal((a,a_1),(a,a_2),(a,a_1'),(a,a_2'))$, and as θ is an isomorphism, $S(Y) \models equal((b,b_1),(b,b_2),(a,a_1')^\theta,(a,a_2')^\theta)$. By Lemma 2.4 again, $(a\ a_i')^\theta = (b\ b_i')$. Similarly one checks that f is $1-1$ and onto.

For the last part we start by noting that for distinct $a, b \in X, (a\ b)^\theta = (af\ bf)$. For by Lemma 2.1 and the remark after it the class of transpositions is definable, so that $(a\ b)^\theta = (c\ d)$ for some $c, d \in Y$. It suffices to show that $af \in \{c,d\}$. Choose $b' \in X - \{a,b\}$. By Lemma 2.2, $S(X) \models overlap((a\ b),(a\ b'))$ and so $S(Y) \models overlap((a\ b)^\theta,(a\ b')^\theta)$. Hence $supp((a\ b)^\theta) \cap supp((a\ b')^\theta)$ has just one member, and by definition of f, this element, which lies in $\{c,d\}$, is equal to af.

Therefore if $g \in S(X)$ and $a \neq b, (agf\ bgf) = (ag\ bg)^\theta = (g^{-1}(a\ b)g)^\theta = (g^{-1})^\theta(a\ b)^\theta g^\theta = (g^\theta)^{-1}(af\ bf)(g^\theta) = (afg^\theta\ bfg^\theta)$. Letting a, b, c be any three distinct members of X, agf is represented in $S(Y)$ by the pair $((agf\ bgf),(agf\ cgf))$ and so also by $((afg^\theta\ bfg^\theta),(afg^\theta\ cfg^\theta))$, and it follows that $agf = afg^\theta$. As a was arbitrary, $gf = fg^\theta$ so that $g^\theta = f^{-1}gf$ as required.

3 Automorphism Groups of Chains

In this section I consider automorphism groups of chains, that is to say, linearly ordered sets. If $(\Omega,<)$ is a chain, we write $A(\Omega)$ for the group of all order-automorphisms of Ω, as in [6], and we write $\overline{\Omega}$ for the order-completion of Ω. Then $A(\Omega)$ may be regarded as a subgroup of $A(\overline{\Omega})$ in the natural way. This is useful for various purposes; for instance it enables us to refer to arbitrary 'convex' subsets of Ω as 'intervals', even though their endpoints may not lie in Ω itself; but we do have to keep track of where a member of $A(\Omega)$ is being viewed as acting.

It is usually easiest to restrict attention to *doubly homogeneous* chains, which by definition are those such that for any $x_1 < x_2$ and $y_1 < y_2$ there is $g \in A(\Omega)$ such that $x_1g = y_1$ and $x_2g = y_2$. Much of the reconstruction work can be carried out more generally, but the details are considerably more complicated, and from now on I shall therefore assume Ω is doubly homogeneous. Provided this holds, there is an elegant characterization due to Holland of when two elements of $A(\Omega)$ are conjugate (see [6]), in terms of the 'orbitals' of these elements. If $g \in A(\Omega)$ and $x \in \Omega$ let us say that the *orbital* of g containing x is the subset $\{y : (\exists m,n \in \mathbf{Z})(xg^m \leq y \leq xg^n)\}$. Thus an orbital is a convex subset 'spanned' by an orbit. For any orbital X of g, g is the identity on X (in which case $|X| = 1$), or it is strictly increasing, $x < xg$, for all $x \in X$, or is strictly decreasing, $xg < x$ for all $x \in X$. We say that the *parity* of the orbital is $0, +1, -1$ in these three cases respectively, and the orbitals of parity ± 1 are called *non-trivial*. It is clear that the union of all non-trivial orbitals of g is equal to its support. The family of all orbitals of g is itself linearly ordered in the obvious way, and the *orbital pattern* of g is the ordered set of orbitals together with the function assigning to each orbital its parity. Holland's criterion then says that (for doubly homogeneous Ω) two elements of $A(\Omega)$ are conjugate if and only if they have isomorphic orbital patterns.

We let the *cofinality* (*coinitiality*) of Ω be the least cardinality of a cofinal (respectively coinitial) subset, and its *coterminality* is the greater of the two.

It is rather easy to see that in all cases $A(\Omega)$ has three normal subgroups, usually written $L(\Omega)$, $R(\Omega)$, and $B(\Omega)$, and defined by

$$L(\Omega) = \{g \in A(\Omega) : supp\ g \text{ is bounded above}\},$$

$$R(\Omega) = \{g \in A(\Omega) : supp\ g \text{ is bounded below}\},$$

and

$$B(\Omega) = L(\Omega) \cap R(\Omega) = \{g \in A(\Omega) : supp\ g \text{ is bounded above and below}\}.$$

In the most familiar examples, ($\Omega = \mathbf{Q}$ or $\mathbf{R}$) these are the only proper non-trivial normal subgroups, but for general Ω (for instance the 'long line') there will be many others. See [2] for instance.

For certain chains, very strong interpretability results are known. Specifically, in [7] it is shown that there is are sentences φ and ψ such that if Ω is a doubly homogeneous chain, then $A(\Omega) \models \varphi$ if and only if $\Omega \cong \mathbf{R}$ and $A(\Omega) \models \psi$ if and only if $\Omega \cong \mathbf{Q}$ or Ir (the irrationals). The proof given in [7] goes via the elegant theory of *lattice-ordered* groups. For my purposes it is easier to stick to the pure language of group theory, and indicate how we can interpret many properties inside $A(\Omega)$, sufficiently well to be able to recover $\mathbf{R}$ and $\mathbf{Q}$ or Ir.

The idea of the method presented here is to try to represent Ω inside $A(\Omega)$ by means of carefully chosen elements. The first attempt is to use elements which have just one non-trivial orbital, which seem the easiest to handle. These are referred to in [6] as *bumps*, in view of the shape of their graphs (which deviate from the identity function in just one interval). In [5] and [20] we used bumps whose support takes the form (a, ∞) for some $a \in \overline{\Omega}$. Indeed in [5] it was essential to do so, since in the quotient group $A(\Omega)/L(\Omega)$ all bumps with bounded support become trivial. The formula which enables us to capture elements of this type (at any rate for suitable Ω), and which was introduced in [20] is

$$conj1(x) : x \neq 1 \wedge (\forall y)(\exists z)(xx^y = x^z).$$

This asserts that x is not 1, and the product of x with any conjugate is again conjugate to x. More generally one can consider for $n \geq 1$ a formula $conjn(x)$ which may be needed in order to make finer distinctions, and which asserts that x is not 1 and the products of x with a conjugate of x fall into exactly n conjugacy classes. Study of the formula $conj1$ works well in the case where Ω has countable cofinality, and it was used in [20] to give an easy way of distinguishing the elementary theories of $A(\mathbf{Q})$ and $A(\mathbf{R})$. In this case it is rather easy to check that $g \in A(\Omega)$ satisfies $conj1$ if and only if it has exactly one non-trivial orbital, which is unbounded (above, below, or both). Again in [5] the same formula was used to investigate properties of the corresponding quotient groups, but almost entirely for Ω having countable coterminality. Since in this section I do not yet want to restrict attention to the case of countable coterminality, I focus rather on bumps with bounded support. Even these are really too restrictive, since any non-trivial orbital clearly has coterminality $\aleph_0$, and so any particular bump will only suffice to 'encode' points a of $\overline{\Omega}$ which happen to be limits of countable sequences (either increasing or decreasing). For the interpretation we instead use (equivalence classes of) elements, not necessarily bumps, which can 'encode' arbitrary cuts in Ω.

I now introduce one by one various formulae of the language of group theory which will express certain properties of elements of $A(\Omega)$ (always under the assumption of double homogeneity). The method is a modification of the one described in [6]. The following is one of the key points (in partially recovering the lattice ordering of $A(\Omega)$ inside the pure group).

$comp(x) : (\exists y)(\exists z)(y \neq 1 \wedge z \neq 1 \wedge (\forall t)(yz^{x^t} = z^{x^t}y))$.

Lemma 3.1 *For any $p \in A(\Omega), A(\Omega) \models comp(p)$ if and only if either p has no orbitals of parity $+1$, or it has no orbitals of parity -1.*

Proof. Here *comp* stands for 'comparability' of p with the identity 1, for in the natural lattice ordering on $A(\Omega)$, an element is comparable with 1 if and only if it satisfies the given condition (it is ≤ 1 if it has no orbitals of parity $+1$ and it is ≥ 1 if it has none of parity -1).

Suppose first that p has no orbitals of parity -1 (so it is 'non-decreasing'). We choose bumps f and g with disjoint supports (a, b) and (c, d) (where as remarked above, the endpoints lie in $\overline{\Omega}$), and such that $b < c$. This is possible by double homogeneity. Then (by Holland's criterion, for instance), any conjugate p^h of p is also non-decreasing, and so $(c, d)p^h$ is disjoint from (a, b). But $supp(g^{p^h}) = (c, d)p^h$ and it follows that f and g^{p^h} commute. Similarly if p has no orbitals of parity $+1$ (this time taking $d < a$).

Conversely, suppose that p has orbitals of parity $+1$ and -1, and let $f, g \neq 1$. We have to show that g has a conjugate g^h, where h is conjugate to p, which does *not* commute with f. Let (a, b) be non-trivial orbital of f, and (c, d) a non-trivial orbital of g. By replacing f and/or g by their inverses if necessary we may suppose that these orbitals have parity $+1$. Pick if possible $x \in (a, b)$ and $y \in (c, d)$ such that $yg^2 < x$. Let $q \in A(\Omega)$ map y to x, yg to xf, and yg^2 to xf^3. Since p has an orbital of parity $+1, q$ may be taken to be a conjugate p^h of p. Then $xfg^{p^h} = xfq^{-1}gq = xf^3$ and $xg^{p^h}f = xq^{-1}gqf = xf^2$, so $fg^{p^h} \neq g^{p^h}f$. If no choice of x, y is possible, then $b \leq c$, and a similar argument using an orbital of p of parity -1 is applied.

$apart(x, y) : (\exists z)(comp(z) \wedge z \neq 1 \wedge (\forall t)(xy^{z^t} = y^{z^t}x))$.

Lemma 3.2 *For any* $f, g \in A(\Omega), A(\Omega) \models apart(f, g)$ *if and only if either* $supp(f) < supp(g)$ *or* $supp(f) > supp(g)$ *(where* $supp(f) < supp(g)$ *means* $x \in supp(f) \wedge y \in supp(g) \Rightarrow x < y$*).*

Proof. If $supp(f) < supp(g)$ take any 'positive' h, (meaning that it is not equal to 1, and it has no orbitals of parity -1). Then any conjugate h^k of h is also positive, and so $supp(f) < supp(g^{h^k})$. Thus f and g^{h^k} have disjoint supports so commute. Similarly if $supp(f) > supp(g)$ (but now using negative h).

Conversely suppose neither $supp(f) < supp(g)$ nor $supp(f) > supp(g)$. Take any $h \neq 1$ and suppose it has an orbital of parity $+1$. Since $supp(f) < supp(g)$ is false there are $x \in supp(f)$ and $y \in supp(g)$ with $x \geq y$. As in the previous proof we assume for ease that $x < xf$ and $y < yg$. Then there is a conjugate h^k of h taking yg^{-2} to x, yg^{-1} to xf, and y to xf^3, and f does not commute with g^{h^k} since $xfg^{h^k} = xf^3$ but $xg^{h^k}f = xf^2$. So $A(\Omega) \models \neg apart(f, g)$.

$bump(x) : x \neq 1 \wedge (\forall y)(\forall z)(apart(y, z) \wedge x = yz \rightarrow (y = 1 \vee z = 1))$.

Lemma 3.3 *For any* $f \in A(\Omega), A(\Omega) \models bump(f)$ *if and only if* f *is a bump.*

Proof. If g and h are elements of $A(\Omega)$ with disjoint supports then the family of non-trivial orbitals of gh is clearly equal to the unions of the corresponding families for g and h separately. Hence if f is a bump in $A(\Omega)$ it cannot be written as gh where g, h are non-trivial elements with disjoint supports, and so $A(\Omega) \models bump(f)$.

Conversely, if $f \neq 1$ and f is not a bump, it has at least 2 non-trivial orbitals, and so there is $x \in \overline{\Omega}$ fixed by f such that each of $(-\infty, x) \cap supp(f)$ and $(x, \infty) \cap supp(f)$ is non-empty. We let $yg = yf$ if $y < x$ and $yh = yf$ if $y > x$, and g and h fix all other points. Then g and h witness $\neg bump(f)$.

$bbump(x) : bump(x) \wedge (\exists y)(comp(y) \wedge apart(x, x^y) \wedge apart(x, x^{y^{-1}}))$.

Lemma 3.4 *For any* $f \in A(\Omega), A(\Omega) \models bbump(f)$ *if and only if* f *is a bump with bounded support.*

Proof. If f is a bump with $supp(f) \subseteq (a, b)$, let $a_1 < b_1 < a < b < a_2 < b_2$, and let p be a positive element taking a_1 to a to a_2 and b_1 to b to b_2. This p is a witness for y, and so $A(\Omega) \models bbump(f)$.

Conversely, if $bbump(f)$ and p is a witness for y, we either have $supp(f^{p^{-1}}) < supp(f) < supp(f^p)$ (if p is positive), or $supp(f^{p^{-1}}) > supp(f) > supp(f^p)$ (if p is negative).

$orbital(x, y) : bump(x) \wedge (\exists y_1)(\exists y_2)(apart(x, y_1) \wedge apart(x, y_2) \wedge y = xy_1y_2)$.

Lemma 3.5 *For any* $g, h \in A(\Omega), A(\Omega) \models orbital(g, h)$ *if and only if* g *is the restriction of* h *to one of its non-trivial orbitals.*

Proof. Here by saying that g is a *restriction* of h to X we mean that for every $x \in X, xg = xh$, and for all $x \notin X, xg = x$. If g *is* a restriction of h to an orbital (a, b) of h we let h_1, h_2 be the restrictions of h to $(-\infty, a)$ and (b, ∞) respectively. Then h_1, h_2 are witnesses for y_1 and y_2.

Conversely, if $orbital(g, h)$ is true, and h_1, h_2 are witnesses for y_1 and y_2, then $supp(g)$ must be disjoint from $supp(h_1) \cup supp(h_2)$, and g must be equal to the restriction of h to its support. Since g is a bump, it is the restriction of h to one of its orbitals.

We introduce some further formulae, and state their properties without proof in a lemma.

$disj(x, y) : (\forall z)(\forall t)(orbital(z, x) \wedge orbital(y, t) \rightarrow apart(z, t))$.
$dense(x) : (\forall y)(disj(x, y) \rightarrow y = 1)$.
$half(x) : (\exists y)(y \neq 1 \wedge apart(x, y) \wedge dense(xy))$.
$end(x, y) : (\exists z)(z \neq 1 \wedge apart(x, z) \wedge apart(y, z) \wedge dense(xz) \wedge dense(yz))$.

Lemma 3.6 *(i) For any $g, h \in A(\Omega), A(\Omega) \models disj(g, h)$ if and only if g and h have disjoint supports.*

(ii) For any $g \in A(\Omega), A(\Omega) \models dense(g)$ if and only if $supp(g)$ is dense in Ω.

(iii) For any $g \in A(\Omega), A(\Omega) \models half(g)$ if and only if for some $a \in \overline{\Omega}, supp(g)$ is a dense subset of (a, ∞) or $(-\infty, a)$.

(iv) For any $g, h \in A(\Omega), A(\Omega) \models end(g, h)$ if and only if for some $a \in \overline{\Omega}$, $supp(g)$ and $supp(h)$ are both dense subsets of (a, ∞) or are both dense subsets of $(-\infty, a)$

Lemma 3.7 *For any $a \in \overline{\Omega}$ there are $g, h \in A(\Omega)$ such that $supp(g)$ is a dense subset of (a, ∞) and $supp(h)$ is a dense subset of $(-\infty, a)$. Moreover g, h may be chosen positive (or negative).*

Proof. For any $x \in (b, c) \subseteq (a, \infty)$ with $b < c$ there is a bump with support a subset of (b, c) and containing x, as follows by double homogeneity. By Zorn's Lemma we choose a maximal pairwise disjoint family of open intervals which are supports of bumps of this form, and let g be chosen having the members of this family as non-trivial orbitals. Similarly for h.

These results have served to indicate how we may *attempt* to interpret Ω, or more accurately, $\overline{\Omega}$, inside $A(\Omega)$. Since $A(\Omega)$ has an outer automorphism, at any rate in the cases where Ω is isomorphic to its reversed chain Ω^* (such as $\mathbf{Q}$ and $\mathbf{R}$), we cannot hope for this as it stands. This is one reason why in [6] and [7] the lattice-ordered group language is used. However we may find an interpretation with one parameter (that is by expanding the language to 'tell left from right').

Theorem 3.8 *For any doubly homogeneous chain $\Omega, \overline{\Omega}$ is interpretable in $(A(\Omega), g)$ where g is a suitable member of $A(\Omega)$.*

Proof. Choose $a \in \Omega$ and let g be a member of $A(\Omega)$ whose support is dense in (a, ∞). In the definition of 'interpretation' we take $n = 1$ and D to be the set of elements of $A(\Omega)$ whose support is dense in (b, ∞) for some $b \in \overline{\Omega}$. Then D is definable (without parameters) in $(A(\Omega), g)$ since $h \in D$ if and only if $A(\Omega) \models half(h) \wedge \exists k_1 \exists k_2 (subset(g^{k_1}, h) \wedge subset(h^{k_2}, g))$ where $subset(x, y)$ is a formula expressing '$supp(x) \subseteq supp(y)$' (for instance $(\forall z)(disj(y, z) \rightarrow disj(x, z))$). If we let $F(h) = b$ provided $supp(h)$ is dense in (b, ∞), the fact that F is surjective follows from Lemma 3.7, and the definability of 'equality' and $<$ uses the formulae *end* and *subset*.

One would really like to interpret Ω itself in $(A(\Omega), g)$, but this is not possible unless Ω is order-complete (when it is trivial), since then $A(\Omega) = A(\overline{\Omega} - \Omega)$ (under the natural identification). This is the reason for the difference in the results of [7] for **R** and **Q**. In the latter case we can only 'recognize' **Q** up to complementation in **R**. We remark that the results described here for doubly homogeneous chains have been extensively generalized in (at least) two directions by Rubin, in [13] for topological spaces (where the results are phrased in terms of 'locally movable groups'; see Item 11 in the list of faithful classes; also see his paper [15] in this volume), and for 'trees' in [14].

Here I just state the consequences of the development so far for outer automorphisms.

Theorem 3.9 *Suppose that $(\Omega_1, <)$ and $(\Omega_2, <)$ are doubly homogeneous chains and that θ is an isomorphism from $A(\Omega_1)$ to $A(\Omega_2)$. Then there is a bijection f from $\overline{\Omega}_1$ to $\overline{\Omega}_2$, which is either order-preserving or order-reversing, and such that for every $g \in A(\Omega_1), g^\theta = f^{-1}gf$. Moreover f takes Ω_1 to an orbit of $A(\Omega_2)$ in $\overline{\Omega}_2$.*

Proof. We let $af = b$ for $a \in \overline{\Omega}_1, b \in \overline{\Omega}_2$, if there are $g \in A(\Omega_1), h \in A(\Omega_2)$ such that $supp(g)$ is dense in (a, ∞) and h is dense in (b, ∞) or $(-\infty, b)$, and $h = g^\theta$. The fact that this definition is independent of the choices of g and h follows from the previous lemmas. Moreover it is easy to check that if $supp(h) \subseteq (b, \infty)$ $((-\infty, b))$ for some b and choices of g and h, then this holds in all cases, and then f is order-preserving (order-reversing, respectively), and in either instance f is $1-1$.

Finally we have to show that for each $g \in A(\Omega_1), g^\theta = f^{-1}gf$. For ease we suppose that f is order-preserving. Pick any $a \in \overline{\Omega}_1$ and pick g_1 with support dense in (a, ∞). Then by definition of f, g_1^θ has support dense in (af, ∞), and as $supp(g_1^g) = (supp(g_1))g$, which is dense in (ag, ∞), similarly $supp((g_1^g)^\theta)$ is dense in (agf, ∞). But $(g_1^g)^\theta = (g_1^\theta)^{(g^\theta)}$, and as $supp(g_1^\theta)$ is dense in $(af, \infty), supp((g_1^\theta)^{(g^\theta)})$ is dense in (afg^θ, ∞). We deduce that $agf = afg^\theta$. As a was arbitrary, $gf = fg^\theta$ and so $g^\theta = f^{-1}gf$.

For the last part, let $a \in \Omega_1$ and $g \in A(\Omega_1)$. Then $(ag)f = (af)g^f = (af)g^\theta$, so as $g^\theta \in A(\Omega_2), af$ and $(ag)f$ are in the same $A(\Omega_2)$–orbit of $\overline{\Omega}_2$. Since the argument can be reversed it follows that the image of Ω_1 under f is equal to an orbit of $\overline{\Omega}_2$ under the action of $A(\Omega_2)$.

Corollary 3.10 *(i) If Ω is doubly homogeneous, order-complete, and $\Omega \cong \Omega^*$ (the reversed chain) then the outer-automorphism group of $A(\Omega)$ has order 2.*

(ii) If Ω is order-complete and $\Omega \not\cong \Omega^*$ then $A(\Omega)$ has no outer automorphisms.

This gives as an example of a doubly homogeneous chain with no outer automorphisms the 'right long line' $\mathbf{L}^+$ (see [5]). To obtain corresponding results for Ω which are not complete, it appears that we would need to strengthen the interpretability result. In cases where we can show that Ω itself is interpretable in $A(\Omega)$, the analogue of this corollary will apply. For $\Omega = \mathbf{Q}$ this is given in [6].

4 Interpretability in $S_\lambda(\mu)/S_\kappa(\mu)$

In this section I show how it is possible to interpret many 'set-theoretical' properties inside $S_\lambda(\mu)/S_\kappa(\mu)$, by representing subsets of μ via supports of suitably chosen elements (always up to less than 'κ mistakes'), and consequently to interpret the ring of subsets of μ of cardinality less than λ modulo those of cardinality less than κ. This work was originally carried out in a rather more general setting in [11]. The method presented here is due to Shelah, where the key idea is to use sequences whose entries are transitive representations of a specific finite *non-commutative* group to represent the subsets, which enables us to capture disjointness of their supports via a commutativity condition. To make things work more smoothly, it helps if the group used is simple, as this ensures that its action will be either faithful or trivial, and the number of cases to be considered is reduced. Any non-abelian finite simple group would probably do, but we fix on the alternating group $A(5)$ for definiteness. (In fact at one point in the argument we use the fact that the outer automorphism group of $A(5)$ is known (it is just $S(5)$). We introduce the necessary formulae one by one, and outline why they represent what is required. Fuller details are given in [19].

If $\overline{g}$ is a tuple of elements of a group, we let $\langle\overline{g}\rangle$ be the subgroup generated by the set of its entries, and $supp(\overline{g})$ be the union of their supports. If $\overline{f}$ and $\overline{g}$ are tuples of the same length n then $\overline{f} \cdot \overline{g}$ is the pointwise product $(f_1g_1, \ldots, f_ng_n)$. Let G be a fixed finite group, and let $\overline{G} = (a_1, a_2, \ldots, a_n)$ be an enumeration of G (with a_1 = the identity, for ease). Then we let $diag(\overline{G}, \overline{x})$ be the conjunction over all i, j, k between 1 and n for which $a_ia_j = a_k$ of the formulae $x_ix_j = x_k$. This is intended to say that $(x_1, x_2, \ldots, x_n)$ is a 'copy' of G (in the specified enumeration), but actually just says that it is a homomorphic image.

Lemma 4.1 *Suppose $\overline{f} \in S_\lambda(\mu)$ is such that $S_\lambda(\mu)/S_\kappa(\mu) \models$*

$diag(\overline{G}, S_\kappa(\mu).\overline{f})$. Then there is a small union X of $\langle\overline{f}\rangle$-orbits such that if $\alpha \in \mu - X$, and $a_i a_j = a_k$, then $\alpha f_i f_j = \alpha f_k$.

This is plausible, but not quite obvious in the case where $\kappa = \omega$, since it might appear that we have to close up the set of exceptional points under the action of $\langle\overline{f}\rangle$. The proof is given in [19].

Lemma 4.2 *For any $\overline{f}, S_\lambda(\mu)/S_\kappa(\mu) \models diag(\overline{A(5)}, S_\kappa(\mu).\overline{f})$ if and only if there is a small union X of orbits of $\langle\overline{f}\rangle$ on μ such that for every orbit Y of $\langle\overline{f}\rangle$ on $\mu - X$, the action of $\langle\overline{f}\rangle$ on Y is isomorphic to some transitive action of $A(5)$ (so that $|Y| = 1, 5, 6, 10, 12, 15, 20, 30$, or 60, and then we say that $\overline{f}$ acts as $A(5)$ with this degree).*

Proof. This is immediate from Lemma 4.1 on remarking that for orbits Y outside $\mu - X$, the action of $\langle\overline{f}\rangle$ on Y precisely corresponds to some transitive action of $A(5)$. The fact that the possible values of $|Y|$ are as stated follows from the fact that any transitive action of $A(5)$ is isomorphic to its action on a coset space $[A(5) : H]$ for some subgroup H of $A(5)$, and the possible orders of subgroups of $A(5)$ are $1, 2, 3, 4, 5, 6, 10, 12$, and 60.

With this lemma in mind we may define for any $\overline{f} \in S_\lambda(\mu)$ of length 60 the cardinals $\nu_m(\overline{f})$ for $m \in \{1, 5, 6, 10, 12, 15, 20, 30, 60\}$ by $\nu_m(\overline{f})$ = the number of $\langle\overline{f}\rangle$-orbits on μ on which $\overline{f}$ acts as $A(5)$ with degree m. The significant values (that is, those which are preserved on passing to the coset $S_\lambda(\mu).\overline{f}$) are those $\nu_m(\overline{f})$ which are $\geq \kappa$, and these provide a 'profile' of $S_\kappa(\mu).f$ characterizing it up to conjugacy.

Ideally we would like to restrict attention to those $\overline{f}$ for which $\nu_m(\overline{f}) < \kappa$ for $m \notin \{1, 5\}$ Thus the action of $\langle\overline{f}\rangle$ on all but a small set of its orbits would be either trivial or 'natural' (that is, the intended faithful action of degree 5). And if we are viewing expressing disjointness as the main goal at this stage then the natural way to express $S_\kappa(\mu).\overline{f}, S_\kappa(\mu).\overline{g}$ disjoint would be to say that every entry of $S_\kappa(\mu).\overline{f}$ commutes with every entry of $S_\kappa(\mu).\overline{g}$.

Unfortunately neither of these ideas works so straightforwardly, and we have to go by a longer route. In the first place the formula $diag(\overline{A(5)}, \overline{x})$ does not rule out the other representations. And there are several ways in which $S_\kappa(\mu).\overline{f}, S_\kappa(\mu).\overline{g}$ can commute entrywise without $|supp(\overline{f}) \cap supp(\overline{g})| < \kappa$ holding. To analyse the possibilities for this let us focus attention on the commuting actions of $\overline{f}$ and $\overline{g}$ on subsets X of κ such that $\langle\overline{f}, \overline{g}\rangle$ acts transitively on X and $\langle\overline{f}\rangle, \langle\overline{g}\rangle$ each act on all their orbits in X as $A(5)$.

The case we really wish to capture is where either $\overline{f}$ or $\overline{g}$ acts trivially on X. Other possibilities however are these: it may be possible to write $X = \{x_{ij} : i < m, j < n\}$ so that the orbits of $\overline{f}$ on X are $X_i = \{x_{ij} : j < n\}$ for $i < m$ and those of $\overline{g}$ are $Y_j = \{x_{ij} : i < m\}$ for $j < n$, and the action of $\overline{f}, \overline{g}$ on X is entirely determined from that on $\overline{f}$ on X_0 and of $\overline{g}$ on Y_0 by $x_{ij} f_k g_l = x_{i'j'} \Leftrightarrow x_{0j} f_k = x_{0j'} \;\&\; x_{i0} g_l = x_{i'0}$ (so that $\overline{f}, \overline{g}$ essentially act as a direct product). Or, for another example, we may have $|X| = 60$ and each of $\overline{f}, \overline{g}$ acting regularly on X; this possibility arises in view of the fact that the left and right regular actions of $A(5)$ on itself commute.

The key to devising a suitable way of expressing disjointness is to look at those faithful actions which are *indecomposable*, in the sense that they cannot be non-trivially expressed as $\overline{f} \cdot \overline{g}$ where $\overline{f}, \overline{g}$ are two transitive actions. Now it is rather clear that the faithful actions of degrees 60 and 30 are not indecomposable. This is because if we take $m = 12, n = 5$ and $m = 6, n = 5$ respectively, allow $\overline{f}, \overline{g}$ to act as a direct product, and consider $\overline{f} \cdot \overline{g}$, we obtain faithful actions of degrees 60, 30 (since m and n are coprime). It is equally clear on arithmetical grounds that the faithful transitive actions of smaller degree *are* indecomposable, but this is not good enough. We have to consider faithful, not necessarily transitive actions, where the orbits are of these lengths, and see whether they can be built up as $\overline{f} \cdot \overline{g}$ where $\overline{f}, \overline{g}$ are commuting faithful actions.

Having sketched the ideas behind what follows, I state the technical lemmas we need (proved in [19]).

Lemma 4.3 *Suppose that H and K are proper subgroups of $A(5)$. Then for some $a \in A(5), |H \cap a^{-1}Ka| \leq 3$. Moreover, if there is a such that $|H \cap a^{-1}Ka| = 3$ but no b such that $|H \cap b^{-1}Kb| < 3$, then $|H| = |K| = 12$.*

The purpose of this lemma is to show that in product-like actions derived as in the above sketch from non-trivial $\overline{f}$ and $\overline{g}, \overline{f} \cdot \overline{g}$ has orbits of length at least 20.

Lemma 4.4 *Let D be the diagonal subgroup $\{(a_i, a_i) : 1 \leq i \leq 60\}$ of $A(5) \times A(5)$. Then for any subgroup H of $A(5) \times A(5)$ of order 12 or 36, there is a such that $|a^{-1}Ha \cap D| \neq 3$.*

This lemma further analyses the final clause of Lemma 4.3. Note that the diagonal subgroup corresponds to the product $\overline{f} \cdot \overline{g}$ of the two associated coset actions of $A(5)$.

Lemma 4.5 *Suppose that $\overline{f}, \overline{g}$ are subgroups of* Sym*(X) isomorphic to $A(5)$ (in the specified listings) which centralize each other, and such that $\langle \overline{f}, \overline{g} \rangle$ is transitive on X. Then $\overline{f} \cdot \overline{g}$ has an orbit of length at least 20. Moreover, if $\overline{f} \cdot \overline{g}$ has an orbit of length 20 then it also has an orbit of some other length greater than 1.*

This lemma reveals explicitly the purpose of the two previous lemmas. Its proof involves the consideration of three cases: the first where $\overline{f}, \overline{g}$ each act transitively on X, in which case their actions will be regular; the second where one action is transitive and the other is not, which turns out to be impossible; and the third where neither action on its own is transitive, and this is where Lemmas 4.3 and 4.4 are used.

We now move towards the construction of a formula which is intended to say that $\overline{x}$ acts as $A(5)$ on all but a small set of its orbits, and that each such orbit has length 1 or 5. Actually we stop short of doing this (even though it can be done) and just find a formula restricting the range of representations possible, as this provides a quicker route to our goal. We require the following auxiliary formulae:

$comm(\overline{x}, \overline{y})$: $\bigwedge_{1\leq i\leq m, 1\leq j\leq n} x_i y_j = y_j x_i$,
where m and n are the lengths of $\overline{x}$ and $\overline{y}$. This asserts that each entry of $\overline{x}$ commutes with each entry of $\overline{y}$.

$conj(\overline{x}, \overline{y})$: $(\exists z)(\overline{x}^z = \overline{y})$.

$indec(\overline{x})$: $diag(\overline{A(5)}, \overline{x}) \wedge (\forall \overline{y})(\forall \overline{z})(comm(\overline{y}, \overline{z}) \wedge diag(\overline{A(5)}, \overline{y}) \wedge diag(\overline{A(5)}, \overline{z}) \wedge \overline{x} = \overline{y}\cdot\overline{z} \rightarrow (conj(\overline{x}, \overline{y}) \vee conj(\overline{x}, \overline{z}))$.

Lemma 4.6 *For any sequence $\overline{f}$ of elements of $S_\lambda(\mu)$ of length* 60, *$S_\lambda(\mu)/S_\kappa(\mu) \models indec(S_\kappa(\mu).\overline{f})$ if and only if $\langle\overline{f}\rangle$ acts as $A(5)$ on all orbits outside a small subset of μ, $\nu_{30}(\overline{f}), \nu_{60}(\overline{f}) < \kappa$, and there is at most one $m \in \{5, 6, 10, 12, 15, 20\}$ for which $\nu_m(\overline{f}) \geq \kappa$.*

Proof. Let us say that $\overline{f} \notin S_\kappa(\mu)$ is *indecomposable* if

$$S_\lambda(\mu)/S_\kappa(\mu) \models indec(S_\kappa(\mu).\overline{f}).$$

If $\nu_m(\overline{f}) \geq \kappa$ for two values of $m > 1$ then $\overline{f}$ may be expressed in the form $\overline{g}\cdot\overline{h}$ where $\overline{g}$ is the restriction of $\overline{f}$ to the union of its orbits of length m for one such value of m, and $\overline{h}$ is its restriction to the complement. Since $S_\kappa(\mu).\overline{g}$ and $S_\kappa(\mu).\overline{h}$ clearly cannot be conjugate to $S_\kappa(\mu).\overline{f}$, this violates $indec(S_\kappa(\mu).\overline{f})$. If $\nu_{30}(\overline{f}) \geq \kappa$ then we may express $\overline{f}$ as a commuting product $\overline{g}\cdot\overline{h}$ of $\overline{g}, \overline{h}$ for which $\nu_6(\overline{g}) \geq \kappa, \nu_5(\overline{h}) \geq \kappa$, contrary to *indec*, and similarly if $\nu_{60}(\overline{f}) \geq \kappa$, with 12, 5 instead of 6, 5.

Conversely if $m \in \{5, 6, 10, 12, 15, 20\}$ is the unique value for which $\nu_m(\overline{f}) \geq \kappa$, the fact that $indec(S_\kappa(\mu).\overline{f})$ holds is verified by appeal to Lemma 4.5.

We are now able to express 'disjointness', which (as we saw in studying $A(\Omega)$ in §3) is the key to recovering the appropriate ring of sets. We shall refine this in a moment so that we can apply it to involutions, and then to arbitrary members of $S_\lambda(\mu)/S_\kappa(\mu)$. But for now we use elements satisfying *indec*. Let us say that two indecomposable elements $S_\kappa(\mu).\overline{f}$ and $S_\kappa(\mu).\overline{g}$ have the *same action* if $\nu_m(\overline{f}) \geq \kappa$ and $\nu_m(\overline{g}) \geq \kappa$ for some $m > 1$.

$disj_1(\overline{x}, \overline{y}) : indec(\overline{x}) \wedge indec(\overline{y}) \wedge comm(\overline{x}, \overline{y}) \wedge indec(\overline{x}\cdot\overline{y})$.

Lemma 4.7 *For any sequences $\overline{f}$ and $\overline{g}$ of elements of $S_\lambda(\mu) - S_\kappa(\mu)$ of length* 60, *$S_\lambda(\mu)/S_\kappa(\mu) \models disj_1(S_\kappa(\mu)\overline{f}, S_\kappa(\mu)\overline{g})$ if and only if $\overline{f}$ and $\overline{g}$ are indecomposable with the same action, and $|supp\ \overline{f} \cap supp\ \overline{g}| < \kappa$.*

Proof. It is clear that if two indecomposable elements of $S_\lambda(\mu)/S_\kappa(\mu)$ have the same action, and almost disjoint supports (meaning that the intersection of their supports has cardinality less than κ), then they commute, and their pointwise product is also indecomposable. Conversely suppose that $S_\lambda(\mu)/S_\kappa(\mu) \models$

$disj_1(S_\kappa(\mu)\overline{f}, S_\kappa(\mu)\overline{g})$. Then as in the previous proof, if the supports of $\overline{f}$ and $\overline{g}$ are not almost disjoint, then indecomposability of $\overline{f}\cdot\overline{g}$ is violated. It also follows that $\overline{f}, \overline{g}$, and $\overline{f}\cdot\overline{g}$ must all have the same action.

We actually employ involutions to encode subsets of μ. The use of $A(5)$ *en route* was for technical reasons, but since $A(5)$ has elements of order 2 the fact that we can express disjointness of indecomposables having the same action means that we can also express disjointness of involutions. In fact it is easy to check that in its transitive action of degree 12, $A(5)$ has fixed-point free elements of order 2. Let a_i be one such. Then for $g \in S_\lambda(\mu), S_\kappa(\mu).g$ has order 2 if and only if there is some indecomposable $\overline{g}$ with $\nu_{12}(\overline{g}) \geq \kappa$ such that $S_\kappa(\mu).g = S_\kappa(\mu).g_i$. (If a_i had fixed points and $\lambda = \mu^+$ then we might have to 'piece together' such a g from two moieties, which would only slightly complicate the argument).

$disj_2(x,y) : x^2 = y^2 = 1 \wedge \exists\overline{z}\exists\overline{t}(z_i = x \wedge t_i = y \wedge disj_1(\overline{z},\overline{t}))$.
$disj(x,y) : \exists x_1\exists x_2\exists y_1\exists y_2(x = x_1x_2 \wedge y = y_1y_2 \wedge \bigwedge_{1\leq i,j\leq 2} disj_2(x_i,y_j))$.

Lemma 4.8 *(i) For any* $f,g \in S_\lambda(\mu), S_\lambda(\mu)/S_\kappa(\mu) \models disj_2(S_\kappa(\mu).f, S_\kappa(\mu).g)$ *if and only if* $S_\kappa(\mu).f$ *and* $S_\kappa(\mu).g$ *are involutions (or the identity) such that* $|supp(f) \cap supp(g)| < \kappa$.

(ii) For any $f,g \in S_\lambda(\mu), S_\lambda(\mu)/S_\kappa(\mu) \models disj(S_\kappa(\mu).f, S_\kappa(\mu).g)$ *if and only if* $|supp(f) \cap supp(g)| < \kappa$

Proof. (ii) follows from the fact that any permutation may be written as a product of two involutions.

In terms of 'disjointness' we can now move quickly to express the other properties required to interpret the suitable (quotient) ring of sets in $S_\lambda(\mu)/S_\kappa(\mu)$. We state the the following two formulae and what they express without proof.

$subset(x,y) : \forall z(disj(y,z) \rightarrow disj(x,z))$.
$restr(x,y) : \exists z(disj(x,z) \wedge xz = y)$.

Then $S_\lambda(\mu)/S_\kappa(\mu) \models subset(S_\kappa(\mu).f, S_\kappa(\mu).g) \Leftrightarrow |supp(f) - supp(g)| < \kappa$, and $S_\lambda(\mu)/S_\kappa(\mu) \models restr(S_\kappa(\mu).f, S_\kappa(\mu).g) \Leftrightarrow$ for some $f' \in S_\kappa(\mu).f, g' \in S_\kappa(\mu).g, f'$ is a restriction of g'.

The following theorem is (essentially) the same as Theorem 4.2 of [11], though Rubin's methods are somewhat different (and in particular do not make use of the 'trick' involving $A(5)$).

Theorem 4.9 *Let* $\mathcal{P}_\lambda(\mu), \mathcal{P}_\kappa(\mu)$ *be the rings of subsets of* μ *of cardinality less than* λ, κ *respectively (in the language* $\{=,\cup,\cap,\subseteq\}$*). Then* $\mathcal{P}_\lambda(\mu)/\mathcal{P}_\kappa(\mu)$ *is interpretable in* $S_\lambda(\mu)/S_\kappa(\mu)$.

Proof. Let D be the family of involutions in $S_\lambda(\mu)/S_\kappa(\mu)$ and for $S_\kappa(\mu).g \in D$ let $F(S_\kappa(\mu).g)$ be the equivalence class in $\mathcal{P}_\lambda(\mu)/\mathcal{P}_\kappa(\mu)$ of $supp(g)$. Then $F(S_\kappa(\mu).f) = F(S_\kappa(\mu).g) \Leftrightarrow S_\lambda(\mu)/S_\kappa(\mu) \models subset(S_\kappa(\mu).f, S_\kappa(\mu).g) \wedge subset(S_\kappa(\mu).g, S_\kappa(\mu).f)$, and 'subset' also enables the boolean operations (except for complementation) to be expressed. For instance unions may be expressed by the formula (needed later) $union(x,y,z) : \forall t(subset(x,t) \wedge subset(y,t) \leftrightarrow subset(z,t))$.

Now we move on to attempt to get information about outer automorphisms. In particular we show how the methods presented here can be used to prove the result of [1]. Unfortunately, unlike in §§2,3, the interpretability of $\mathcal{P}_\lambda(\mu)/\mathcal{P}_\kappa(\mu)$ does not appear to be sufficient to derive this without extra work, and the same applies in §5. What is required rather is to be able to interpret the family of *sequences* of a suitable length (as ever modulo less than κ 'mistakes') inside $S_\lambda(\mu)/S_\kappa(\mu)$. To do this we use very similar ideas to [1]; the main difference is that the reasoning there is mostly second order (a study of centralizers being one of the key points). The methods at present only work for $\kappa = \omega$, which is assumed from now on, and I list various formulae which are needed.

$minimal(x) : x \neq 1 \wedge \forall y(restr(y,x) \rightarrow (y = 1 \vee y = x))$.
$fincycle(x) : \forall y(restr(y,x) \rightarrow \neg minimal(y)) \wedge \exists y(y \neq 1 \wedge disj(x,y))$.
$norepeat(x) : fincycle(x) \wedge \forall y \forall z(y \neq 1 \wedge disj(y,y^z) \rightarrow \neg restr(yy^z, x))$.
$allfincycle(x) : norepeat(x) \wedge \forall y \exists z(norepeat(y) \rightarrow restr(y^z, x))$.
$infcycle(x) : \forall y(restr(y,x) \wedge y \neq 1 \rightarrow \neg fincycle(y))$.
$cut(x,y) : \forall z(z \neq 1 \wedge restr(z,x) \rightarrow (\neg disj(z,y) \wedge \neg subset(z,y)) \wedge$
$\forall z(z \neq 1 \wedge restr(z,y) \rightarrow (\neg disj(z,x) \wedge \neg subset(z,x)))$.
$seq(x,y) : x \neq 1 \wedge x^2 = 1 \wedge allfincycle(y) \wedge cut(x,y) \wedge \exists z(z \neq 1 \wedge$
$disj(x,z) \wedge disj(y,z))$.
$sameseq_1(x_1,y_1,x_2,y_2) : seq(x_1,y_1) \wedge seq(x_2,y_2) \wedge \exists z(y_1^z = y_2 \wedge disj(x_1,z) \wedge$
$disj(x_2,z))$.
$sameseq(x_1,y_1,x_2,y_2) : \exists x_1' \exists y_1' \exists x_2' \exists y_2'(sameseq_1(x_1,y_1,x_1',y_1') \wedge$
$sameseq_1(x_1',y_1',x_2',y_2') \wedge sameseq_1(x_2',y_2',x_2,y_2))$.

The intended meanings of these should be fairly clear, and I sum up their properties in the next lemma.

Lemma 4.10 *(i)* $S_\lambda(\mu)/S_\omega(\mu) \models minimal(x)$ *if and only if there is g having a single infinite cycle and fixing all other points such that* $x = S_\omega(\mu).g$.

(ii) $S_\lambda(\mu)/S_\omega(\mu) \models fincycle(S_\omega(\mu).g)$ *if and only if all cycles of g are finite and g has infinitely many fixed points.*

(iii) $S_\lambda(\mu)/S_\omega(\mu) \models norepeat(x)$ *if and only if there is g having all cycles finite, infinitely many fixed points, and no two cycles of the same length greater than 1 such that* $x = S_\omega(\mu).g$.

(iv) $S_\lambda(\mu)/S_\omega(\mu) \models allfincycle(x)$ *if and only if there is g as in (iii) having a unique cycle of every possible finite length greater than* 1.

(v) $S_\lambda(\mu)/S_\omega(\mu) \models infcycle(x)$ *if and only if for some* g *with* $S_\omega(\mu).g = x$, g *has no non-trivial finite cycles.*

(vi) $S_\lambda(\mu)/S_\omega(\mu) \models cut(x,y)$ *if and only if there are* g, h *with* $S_\omega(\mu).g = x$ *and* $S_\omega(\mu).h = y$ *such that every non-trivial cycle of* g *intersects some non-trivial cycle of* h *and contains a fixed point of* h, *and vice versa.*

(vii) $S_\lambda(\mu)/S_\omega(\mu) \models seq(x,y)$ *if and only if there are* g *of order* 2 *and* h *all of whose cycles are finite and having a unique cycle of each length greater than* 1, *with* $S_\omega(\mu).g = x$ *and* $S_\omega(\mu).h = y$ *and* $\mu - (supp(g) \cup supp(h))$ *infinite, and the non-trivial cycles of* g *and* h *can be put into* $1-1$ *correspondence* φ *by letting* $C^\varphi = C' \Leftrightarrow |C \cap C'| = 1$.

(viii) $S_\lambda(\mu)/S_\omega(\mu) \models sameseq(x_1, y_1, x_2, y_2)$ *if and only if there are* g_1, h_1, g_2, h_2 *as for seq such that for each* $n > 1$ *the intersections of the cycles of* h_1 *and* h_2 *of length* n *with the supports of* g_1, g_2 *respectively, are equal.*

The idea here is that we are now able to encode (countable) sequences of members of μ by pairs (g,h) where $S_\lambda(\mu)/S_\omega(\mu) \models seq(S_\omega(\mu).g, S_\omega(\mu).h)$. For $n > 1$ the nth entry of the sequence so encoded is to be the unique member of the cycle of h of length n which intersects $supp(g)$. Since this appears to depend on the choices of g and h, we have to be able to express the fact that two choices encode the same sequence, and this is what *sameseq* is for.

In order to indicate what is now required, let us suppose that X_1 and X_2 are sets and that θ is an isomorphism from $S_\lambda(X_1)/S_\omega(X_1)$ to $S_\lambda(X_2)/S_\omega(X_2)$. The aim is to show that this is induced by a bijection f from a cofinite subset of X_1 to a cofinite subset of X_2. For any countable sequence $\mathbf{a} = (a_0, a_1, \ldots)$ of members of X_1 we find a partial $1-1$ map $p_\mathbf{a}$ from its set of entries into X_2, via the interpretation of sequences in $S_\lambda(X_1)/S_\omega(X_1)$ as given in the previous paragraph. Let (g,h) be chosen as described, and let $S_\omega(X_2).g', S_\omega(X_2).h'$ be the images of $S_\omega(X_1).g$ and $S_\omega(X_1).h$ under θ. Since $S_\lambda(X_1)/S_\omega(X_1) \models seq(S_\omega(X_1).g, S_\omega(X_1).h)$, and θ is an isomorphism, $S_\lambda(X_2)/S_\omega(X_2) \models seq(S_\omega(X_2).g', S_\omega(X_2).h')$. By Lemma 4.10(vii), there is a sequence $\mathbf{a}'$ encoded by $(S_\omega(X_2).g', S_\omega(X_2).h')$, by Lemma 4.10(viii) this depends only on $\mathbf{a}$ (up to finitely many mistakes) and not on g and h, and we let $p_\mathbf{a}$ be given by $a_i p_\mathbf{a} = a'_i$.

The main extra piece of work which is needed is to show that these 'approximations' to the map f are compatible. Here what is required is that if $\mathbf{b}$ is a subsequence of $\mathbf{a}$, then $\{i : b_i p_\mathbf{a} \neq b_i p_\mathbf{b}\}$ is finite. For this we devise a method of 'storing' the behaviour of a subsequence of a sequence, and the following formulae are introduced with this in mind.

$store_n(x_1, \ldots, x_n, y_1, \ldots, y_n, z) : infcycle(z) \wedge \bigwedge_{1 \le i,j \le n} disj(x_i, y_j) \wedge$
$\bigwedge_{1 \le i \le n} ((x_i y_i)^z = y_i) \wedge \forall t (restr(t,z) \wedge t \neq z \rightarrow \bigvee_{1 \le i \le n} disj(t, x_i)),$ $(n = 1, 2, 3, \ldots)$.

$equalfincycle(x) : fincycle(x) \wedge \forall y \forall z (disj(y,z) \wedge x = yz \rightarrow (conj(x,y) \vee conj(x,z)).$

$cycles_n(x) : equalfincycle(x) \wedge x^n = 1 \wedge \bigwedge_{1 \le m < n} (x^m \neq 1),$ $(n = 2, 3, \ldots)$.

$init_{ij}(x, y, z) : cycles_2(x) \wedge cycles_i(y) \wedge cycles_j(z) \wedge cut(x, y) \wedge cut(x, z) \wedge$
$cycles_{(j-i+1)}(y^{-1}z)$, $(2 \leq i < j)$.

Lemma 4.11 *(i)* $S_\lambda(\mu)/S_\omega(\mu) \models store_n(x_1, \ldots, x_n, y_1, \ldots, y_n, z)$ *if and only if there are* g_i, h_i, k *such that each* g_i *and* h_j *have disjoint supports, the non-trivial cycles of* k *are all infinite,* $\bigcup_{1 \leq i \leq n} supp(g_i)$ *forms a transversal for these infinite cycles of* k*, each* h_i *is the product of the conjugates of* g_i *by positive powers of* k *(where the product of an infinite sequence of pairwise disjoint permutations is defined in the natural way), and* $S_\omega(\mu).g_i = x_i, S_\omega(\mu).h_i = y_i, S_\omega(\mu).k = z$.

(ii) $S_\lambda(\mu)/S_\omega(\mu) \models equalfincycle(x)$ *if and only if there is* g *having all non-trivial cycles of the same finite length such that* $x = S_\omega(\mu).g$.

(iii) $S_\lambda(\mu)/S_\omega(\mu) \models cycles_n(x)$ *if and only if there is* g *having all non-trivial cycles of length* n *such that* $x = S_\omega(\mu).g$.

(iv) $S_\lambda(\mu)/S_\omega(\mu) \models init_{ij}(x, y, z)$ *if and only if there are* g, h, k *such that* $x = S_\omega(\mu).g, y = S_\omega(\mu).h, z = S_\omega(\mu).k$*, all non-trivial cycles of* g, h, k *have lengths* $2, i, j$ *respectively, the cycles of* g *and* h *are in* $1-1$ *correspondence* φ *given by* $C' = C^\varphi \Leftrightarrow |C \cap C'| = 1$*, and similarly for the cycles of* g *and* k*, and if* $(\alpha_0\ \beta)$, $(\alpha_0\ \alpha_1 \ldots \alpha_{i-1})$*, and* $(\alpha_0\ \alpha'_1 \ldots \alpha'_{j-1})$ *are cycles of* g, h, k *intersecting in* $\{\alpha_0\}$ *then* $(\alpha_0, \alpha_1, \ldots, \alpha_{i-1})$ *is an initial segment of* $(\alpha_0, \alpha'_1, \ldots, \alpha'_{j-1})$.

Lemma 4.12 *Let* $\mathbf{a}$ *be a countable sequence of members of* X_1*, and let* $\mathbf{b}$ *be an infinite subsequence. Let* $p_\mathbf{a}, p_\mathbf{b}$ *be the* $1-1$ *maps from* $\{a_i : i \in \omega\}$ *and* $\{b_j : j \in \omega\}$ *into* X_2 *respectively, determined as above (by choice of suitable pairs satisfying seq). Then* $\{j : b_j p_\mathbf{a} \neq b_j p_\mathbf{b}\}$ *is finite.*

Proof. Find g, h as in the definition of $p_\mathbf{a}$ with $X_1 - (supp(g) \cup supp(h))$ infinite, to allow 'spare space for storage'. Let us write $g = \prod_{i=2}^{\infty}(a_i\ c_i)$ and $h = \prod_{i=2}^{\infty}(a_i\ a_{i1}\ a_{i2} \ldots a_{i\ i-1})$ where all the a_i, c_i, a_{ij} are distinct. Let

$$B = \{i : \exists j(b_j = a_i)\}$$

and define g_1 and h_1 by $g_1 = \prod_{i \in B, i \geq 2}(a_i\ c_i), h_1 = \prod_{i \in B, i \geq 2}(a_i\ a_{i1}\ a_{i2} \ldots a_{i\ j_i-1})$ where for $i \in B, a_i = b_{j_i}$. Then (g_1, h_1) is suitable for the definition of $p_\mathbf{b}$. We have to 'copy' (g, h, g_1, h_1) infinitely often to ensure that its pattern is not lost even under the distortion caused by passing to the quotient by $S_\omega(X_1)$.

Find k_1, k_2, k_3, k_4, l which 'store' (g, h, g_1, h_1). For this we first choose l having just infinite cycles (and fixed points) and so that each infinite cycle of l contains just one member of $supp(g) \cup supp(h)$. Then we let $ak_1 = c \Leftrightarrow (\exists n > 0)(al^{-n}g = cl^{-n})$, $ak_2 = c \Leftrightarrow (\exists n > 0)(al^{-n}h = cl^{-n})$, and similarly for k_3 and k_4. Thus $S_\lambda(X_1)/S_\omega(X_1) \models store(S_\omega(X_1).g, S_\omega(X_1).h, S_\omega(X_1).g_1,$
$S_\omega(X_1).h_1, S_\omega(X_1).k_1, S_\omega(X_1).k_2, S_\omega(X_1).k_3, S_\omega(X_1).k_4, S_\omega(X_1).l)$. In fact we could say that (k_1, k_2, k_3, k_4, l) 'stores' (g, h, g_1, h_1) 'accurately' even in $S_\lambda(X_1)$. The images of the nine cosets under θ also satisfy *store* in $S_\lambda(X_2)/S_\omega(X_2)$, and this

means that there must be $g', h', g'_1, h'_1, k'_1, k'_2, k'_3, k'_4, l'$ such that $(S_\omega(X_1).g)^\theta = S_\omega(X_2).g'$ and so on, and where for suitable choices of distinct $a'_i, c'_i, a'_{ij} \in X_2, g' = \prod_{i=2}^{\infty}(a'_i\ c'_i), h' = \prod_{i=2}^{\infty}(a'_i\ a'_{i1}\ a'_{i2} \ldots a'_{i\ i-1})$ and k'_1, k'_2, l' are determined from g' and h' as k_1, k_2, l were from g and h.

We show that (up to finitely many mistakes) $g'_1 = \prod_{i \in B, i \geq 2}(a'_i\ b'_i)$ and $h' = \prod_{i \in B, i \geq 2}(a'_i\ a'_{i1}\ a'_{i2} \ldots a'_{i\ j_i-1})$. Let $i \in B$. Then if r_1, r_2, r_4 are the restrictions of k_1, k_2, k_4 to the union of the cycles of l intersecting $\{a_i, c_i, a_{i0}, \ldots, a_{i\ j_i-1}\}$, then $S_\lambda(\mu)/S_\omega(\mu) \models restr(S_\omega(\mu).r_1, S_\omega(\mu).k_1) \wedge restr(S_\omega(\mu).r_2, S_\omega(\mu).k_2) \wedge restr(S_\omega(\mu).r_4, S_\omega(\mu).k_4) \wedge init_{j_i i}(S_\omega(\mu).k_1, S_\omega(\mu).k_4, S_\omega(\mu).k_2)$. This formula is preserved under θ, and it follows that for all but finitely many cycles of r'_1, r'_2, r'_4 (representatives of the corresponding images), the non-trivial cycles of r'_2, r'_4 can be put into $1-1$ correspondence with those of r'_1 by virtue of having non-empty intersection, they have lengths $2, j_i, i$, and they are restrictions of k'_1, k'_2, k'_4 respectively. It follows that $\prod_{i \in B, i \geq 2}(a'_i\ c'_i)$ and $\prod_{i \in B, i \geq 2}(a'_i\ a'_{i1}\ a'_{i2} \ldots a'_{i\ j_i-1})$ serve for g'_1 and h'_1 as in the definition of $p_{\mathbf{b}}$, and so must actually be equal to g'_1 and h'_1 apart from finitely many mistakes. Therefore $\{i \in B : b_{j_i} p_{\mathbf{b}} \neq a'_i\}$ is finite, and since $\{i : a_i p_{\mathbf{a}} \neq a'_i\}$ is finite, we deduce that $\{j : b_j p_{\mathbf{b}} \neq b_j p_{\mathbf{a}}\}$ is also finite, as required.

Theorem 4.13 *Let X_1 and X_2 be countable sets, and let θ be an isomorphism from $S(X_1)/S_\omega(X_1)$ to $S(X_2)/S_\omega(X_2)$. Then there are finite sets $A \subseteq X_1$ and $B \subseteq X_2$ and a bijection $f : (X_1 - A) \to (X_2 - B)$ such that for every $g \in S(X_1), (S_\omega(X_1).g)^\theta = S_\omega(X_2).f^{-1}gf$.*

Proof. We remark that strictly speaking, $f^{-1}gf$ does not lie in $S(X_2)$, which means that we cannot form the coset $S_\omega(X_2).f^{-1}gf$. But $f^{-1}gf$ maps $X_2 - (B \cup (Ag^{-1} - A)f)$ $1-1$ onto $X_2 - (B \cup (Ag - A)f)$, and as $|B \cup (Ag^{-1} - A)f| = |(B \cup (Ag - A)f|$, there is $h \in S(X_2)$ extending $f^{-1}gf$, and so we may unambiguously define $S_\omega(X_2).f^{-1}gf$ to equal $S_\omega(X_2).h$ for any such h.

Let $X_1 = \{a_i : i \in \omega\}$. Then if $\mathbf{a}_0 = (a_0, a_2, a_4, \ldots)$ and $\mathbf{a}_1 = (a_1, a_3, a_5, \ldots), p_{\mathbf{a}_0}$ and $p_{\mathbf{a}_1}$ are both defined, and we let $f' = p_{\mathbf{a}_0} \cup p_{\mathbf{a}_1}$. We show that f' is a $1-1$ map from X_1 into X_2, apart from finitely many exceptions, meaning as in the statement of the theorem that there are finite subsets $A \subseteq X_1$ and $B \subseteq X_2$ such that if f is the restriction of f' to $X_1 - A$, then f is a bijection from $X_1 - A$ to $X_2 - B$ (and also has the property required, in that 'conjugation' by f induces θ on the quotients).

First let Y be the intersection of the ranges of $p_{\mathbf{a}_0}$ and $p_{\mathbf{a}_1}$. Let $\mathbf{a}'$ be a sequence having $\mathbf{a}_0$ as a subsequence, and having the members of $Yp_{\mathbf{a}_1}^{-1}$ as its other entries. By Lemma 4.12, $C = \{a_{2i} : a_{2i}p_{\mathbf{a}_0} \neq a_{2i}p_{\mathbf{a}'}\}$ and $D = \{a_{2j+1} : a_{2j+1}p_{\mathbf{a}_1} \neq a_{2j+1}p_{\mathbf{a}'}\}$ are finite. We show that Y is contained in $Cp_{\mathbf{a}_0} \cup Dp_{\mathbf{a}_1}$, and hence is finite. If not there is $a_{2i}p_{\mathbf{a}_0} = a_{2j+1}p_{\mathbf{a}_1} \in Y - (Cp_{\mathbf{a}_0} \cup Dp_{\mathbf{a}_1})$. Thus $a_{2i} \notin C$ and $a_{2j+1} \notin D$ and so $a_{2i}p_{\mathbf{a}'} = a_{2i}p_{\mathbf{a}_0} = a_{2j+1}p_{\mathbf{a}_1} = a_{2j+1}p_{\mathbf{a}'}$, contrary to $p_{\mathbf{a}'}$ $1-1$.

We deduce that $p_{\mathbf{a}_0} \cup p_{\mathbf{a}_1}$ is $1-1$ on $X_1 - A$ where $A = \{a_{2i} : \exists j(a_{2i}p_{\mathbf{a}_0} = a_{2j+1}p_{\mathbf{a}_1})\}$. Let $f = f'|(X_1 - A)$. Since the whole argument may evidently be reversed, we deduce also that $B = X_2 - (X_1 - A)f$ is finite.

Finally we have to show that for each $g \in S(X_1), (S_\omega(X_1).g)^\theta = S_\omega(X_2).f^{-1}gf$ (bearing in mind the above-mentioned proviso). Let $\mathbf{a}$ be a fixed $1-1$ sequence with coinfinite range, and let (g_1, h_1) be a pair representing $\mathbf{a}$ as in the definition of $p_{\mathbf{a}}$. Then $((S_\omega(X_1).g_1)^\theta, (S_\omega(X_1).h_1)^\theta)$ represents $\mathbf{a}f = \mathbf{a}'$. Moreover, $\mathbf{a}g$ is represented by (g_1^g, h_1^g) and so $(\mathbf{a}g)f$ is represented by $(S_\omega(X_1).g_1^g, S_\omega(X_1).h_1^g)^\theta = ((S_\omega(X_1).g_1^g)^\theta, (S_\omega(X_1).h_1^g)^\theta) = ((S_\omega(X_1)g_1)^\theta, (S_\omega(X_1).h_1)^\theta)^{(S_\omega(X_1).g)^\theta}$. But this also represents $(\mathbf{a}f)^{(S_\omega(X_1).g)^\theta}$, and so we deduce that up to finitely many mistakes $(\mathbf{a}g)f = (\mathbf{a}f)g'$, where $S_\omega(X_2).g' = (S_\omega(X_1).g)^\theta$. We may now let $\mathbf{a}$ equal $\mathbf{a}_0$ and $\mathbf{a}_1$ in turn, and deduce that $\{i : a_igf \neq a_ifg'\}$ is finite, and therefore that $S_\omega(X_1).gf = S_\omega(X_1).fg'$, from which $(S_\omega(X_1).g)^\theta = S_\omega(X_2).f^{-1}gf$ follows.

Corollary 4.14 [1] *The outer automorphism group of $S(\omega)/S_\omega(\omega)$ is infinite cyclic, and is generated by θ given by $g^\theta = f^{-1}gf$ where $nf = n+1$ for all n.*

Now let us remark on possible extensions of these results to other values of μ. At present the restriction to $\kappa = \omega$ seems unavoidable, without use of stronger methods. Subject to this the general problem then boils down to showing that the various 'approximations' $p_{\mathbf{a}}$ can be consistently fitted together to make one big map f which agrees with each of them on all but finitely many points. The particular case analysed in Theorem 4.13 worked because the whole of the set could be covered by *two* countable coinfinite sets. The best we can do at present uses a similar idea, but with $\lambda = \mu^+$ (in other words where the top group is the full symmetric group).

Theorem 4.15 *Let X_1 and X_2 be sets of cardinality μ, and let θ be an isomorphism from $S(X_1)/S_\omega(X_1)$ to $S(X_2)/S_\omega(X_2)$. Then there are finite sets $A \subseteq X_1$ and $B \subseteq X_2$ and a bijection $f : (X_1 - A) \to (X_2 - B)$ such that for every $g \in S(X_1), (S_\omega(X_1).g)^\theta = S_\omega(X_2).f^{-1}gf$. Therefore the outer automorphism group of $S(\mu)/S_\omega(\mu)$ is infinite cyclic and is generated by θ given by $g^\theta = f^{-1}gf$ where $nf = n+1$ for all $n \in \omega$, and $\alpha f = \alpha$ for infinite $\alpha < \mu$.*

Proof. We just indicate the extra trick required. Instead of using *single* countable sequences $\mathbf{a}$, we use sets $A = \{\mathbf{a}_\alpha : \alpha \in \mu\}$ of pairwise disjoint countable sequences of elements of μ, the complement of whose union has cardinality μ (so that the range is a 'moiety'). We require modifications of the formulae previously used to encapsulate this new situation. In place of *seq* we shall have a formula *setofseq* with three variables. The third will satisfy *infcycle* and each of its (infinite) cycles will be the union of those cycles of the first two elements which encode $\mathbf{a}_\alpha$ for a fixed α. There will also be a formula *samesetofseq* saying that two of these triples encode the same sets of sequences.

Given that this can be expressed, we are able to define a $1-1$ map p_A for any such set A of sequences which takes the union of the ranges of its elements into X_2.

Moreover, for any countable subset of A, the situation is as before, since everything can be encoded into just one sequence. This means for instance that we do not need to store the whole of such a set (though we could). For the analogue of Lemma 4.12 actually follows from that case. This is because if $\{b : bp_B \neq bp_A\}$ is infinite, then this fact may be established by looking at just a countable subset of A, which will then contradict Lemma 4.12.

The proof of the theorem is completed as for Theorem 4.13 using the fact that X_1 may be covered by two moieties.

5 Quotients of Automorphism Groups of Chains

In this section I carry out a similar exercise to the one described for quotients of symmetric groups in §4, but here for quotients of automorphism groups of chains. This work was described in [5], and so I shall just highlight the most significant points, and derive the modified statement (about recovery of the 'tail' of a structure), rather than concentrating on the outer automorphism problem. Moreover, although a certain amount of generality was achieved in [5] at some points, the main goal was to show that $A(\mathbf{R})/L(\mathbf{R})$ has no outer automorphisms, and so I shall focus on that case here. This has the effect of simplifying some of the lemmas.

The notation for automorphism groups of chains Ω was all introduced in §3, so we can begin by listing the key properties of $A(\Omega)/L(\Omega)$ which we intend to express in the first order language of groups. One major difference from §3 is that this time the (lattice-) ordering on $A(\mathbf{R})/L(\mathbf{R})$ is definable from the abstract group. The 'trick' is reminiscent of the one used in formulating the property *comp* (in §3). Another point is that for the case of $A(\Omega)$ we were able to capture points of $\overline{\Omega}$ by use of the formula *half*. In the quotient group we instead have to aim to interpret unbounded subsets or sequences of Ω (or $\overline{\Omega}$). If as in §4 we try to use countable sequences then these will only exist in the case of countable cofinality, and things work most easily for $\mathbf{R}$ (and $\mathbf{Q}$), since in addition all points (even of the completion) are limits of countable increasing or decreasing sequences.

In view of this, we shall use the formulae *conjn* mentioned earlier, which (for $n = 1, 2$) work well in this case.

$conj1(x) : x \neq 1 \wedge \forall y \exists z (x x^y = x^z)$.

$conj2(x) : x \neq 1 \wedge \neg conj1(x) \wedge \exists u \exists v \forall y \exists z (x x^y = (x x^u)^z \vee x x^y = (x x^v)^z)$.

Lemma 5.1 *For Ω a doubly homogeneous chain having countable cofinality, $A(\Omega)/L(\Omega) \models conj1(L(\Omega).g)$ if and only if g has an unbounded orbital.*

Proof. If g has an unbounded orbital, then so does any conjugate, and these orbitals have the same parity (± 1). The product also then has an unbounded orbital which is again of the same parity (and which contains the intersection of the individual orbitals given), and so, by applying Holland's criterion (*mod* $L(\Omega)$) is again conjugate to g.

Conversely, if g does not have an unbounded orbital, then as it is not 1 (in the quotient group), it has a cofinal collection of non-trivial orbitals (a_n, b_n) say with $a_n < b_n \leq a_{n+1}$ for each n. Moreover by passing to an infinite subset we may suppose that these all have the same parity. By taking the product of $L(\Omega).g$ with a suitable conjugate (whose corresponding orbitals contain $[b_n, a_{n+1}]$) we obtain an element which *does* have an unbounded orbital, and which cannot therefore be conjugate to $L(\Omega).g$.

It is shown in [5] that there are (at least) two possible kinds of elements in $A(\Omega)/L(\Omega)$ which may satisfy $conj2$, called there 'contiguous' and 'separated'. If Ω is order-complete then no elements can be separated, but if Ω is not order-complete, then both possibilities arise, and how many orbits they fall into depends on the number of orbits of $A(\Omega)$ on $\overline{\Omega}$. We say that g is *contiguous* if there is an increasing divergent sequence $(a_0, a_1, \ldots)$ in $\overline{\Omega}$, all entries of which lie in the same $A(\Omega)$-orbit, such that (beyond some point) all (a_n, a_{n+1}) are non-trivial orbitals of g of the same parity. It is said to be *separated* if there is such a sequence in $\overline{\Omega}$ such that all (a_{2n}, a_{2n+1}) are non-trivial orbitals of the same parity, g fixes all points of each (a_{2n+1}, a_{2n+2}) pointwise, and there are distinct orbits X and Y of the action of $A(\Omega)$ on $\overline{\Omega}$ such that each a_{2n} lies in X and each a_{2n+1} lies in Y.

Lemma 5.2 *For any $g \in A(\mathbf{R}), A(\mathbf{R})/L(\mathbf{R}) \models conj2(L(\mathbf{R}).g)$ if and only if g is contiguous.*

Proof. If g is contiguous, then the proof that $A(\mathbf{R})/L(\mathbf{R}) \models conj2(L(\mathbf{R}).g)$ is rather similar to that of the previous lemma. On taking the product of g with a conjugate, either all but finitely many of the fixed points are deleted, in which case the resulting element has an unbounded orbital, of parity the same as the eventual common parity of the non-trivial orbitals of g, or infinitely many fixed points remain, in which case the resulting product is still conjugate ($mod\ L(\Omega)$) to g. (The 'eventual' fixed point set of the product of g with its conjugate is just the intersection of the two sets of fixed points). Thus two conjugacy classes result.

Conversely suppose that $A(\mathbf{R})/L(\mathbf{R}) \models conj2(L(\mathbf{R}).g)$. Since $\neg conj1(L(\mathbf{R}).g)$, g does not have an unbounded orbital, and since $g \notin L(\mathbf{R})$, it has a cofinal set of non-trivial orbitals ordered in type ω. By passing to an infinite subset we may suppose that these all have the same parity, $+1$ say. By taking products of g with suitable conjugates, we can build (a) a contiguous element, and (b) an element with an unbounded orbital. Since we are assuming that there are only two such elements in $A(\mathbf{R})/L(\mathbf{R})$ up to conjugacy, it follows that g itself must be contiguous.

In terms of these special elements we now give the formulae used to recapture the lattice ordering.

$pos1(x) : conj1(x) \wedge \forall y \forall z (conj2(y) \wedge (\exists u) yy^z = y^u \rightarrow (\exists u) yy^z = y^{x^u}).$
$pos(x) : (\exists y)(pos1(xx^y)) \wedge (\forall y) \neg pos1(x^{-1}(x^{-1})^y).$

Lemma 5.3 *For any $g \in A(\mathbf{R}), A(\mathbf{R})/L(\mathbf{R}) \models pos1(L(\mathbf{R}).g)$ if and only if g has an unbounded orbital of parity $+1$.*

Proof. Choose a contiguous element h having positive orbitals $(n, n+1)$ for $n \in \omega$. Then we know that for any k, $L(\mathbf{R}).hh^k$ either has an unbounded orbital, or is conjugate to $L(\mathbf{R}).h$. In the latter case, a conjugator must take (a tail of) ω to (a tail of) the fixed point set of hh^k. Now this fixed point set is a subset of ω, so if $L(\mathbf{R}).g$ has an unbounded orbital of parity $+1$, a conjugator of $L(\mathbf{R}).h$ to $L(\mathbf{R}).hh^k$ can be found which is itself conjugate to $L(\mathbf{R}).g$. If on the other hand, for any such k a suitable conjugate of $L(\mathbf{R}).g$ exists, then we can choose k so that the fixed point set of hh^k contains eventually just the even integers, and this means that the conjugator, and hence also $L(\mathbf{R}).g$, cannot be negative, and as g has an unbounded orbital, it must have parity $+1$.

Let us say that $L(\mathbf{R}).g$ is *positive* if $\{a : a < ag\}$ is unbounded above, but $\{a : ag < a\}$ is not. This notion of positivity corresponds to the natural ordering on $A(\mathbf{R})/L(\mathbf{R})$ induced from the pointwise ordering on $A(\mathbf{R})$.

Lemma 5.4 *For any $g \in A(\mathbf{R}), A(\mathbf{R})/L(\mathbf{R}) \models pos(L(\mathbf{R}).g)$ if and only if $L(\mathbf{R}).g$ is positive.*

Proof. If $\{a : a < ag\}$ is unbounded above, then there is h such that gg^h has an unbounded orbital of parity $+1$, and if $\{a : ag < a\}$ is unbounded above, there is h such that gg^h has an unbounded orbital of parity -1. The given condition therefore precisely says that $\{a : a < ag\}$ is unbounded above, but $\{a : ag < a\}$ is not.

Once one has shown how to define the positive elements inside the group, it is then easy to define the (lattice-) ordering: $g \geq h \Leftrightarrow gh^{-1} \geq 1$, and from this many other things can be defined, of course including the lattice operations $\wedge$ and $\vee$, 'absolute value' $|g| = g \vee g^{-1}$, and also in terms of these, disjointness of supports, containment, and restriction. All these will apply *eventually*; for instance, saying that $L(\mathbf{R}).g$ and $L(\mathbf{R}).h$ have 'disjoint' supports here is taken to mean that $supp(g) \cap supp(h)$ is bounded above. In addition to this proviso, saying that the support of $L(\mathbf{R}).g$ is contained in that of $L(\mathbf{R}).h$ means that $supp(g) - supp(h)$ is bounded above (compare the proof of Theorem 3.8).

$disj(x, y) : |x| \wedge |y| = 1.$
$subset(x, y) : (\forall z)(disj(y, z) \rightarrow disj(x, z)).$
$restr(x, y) : (\exists z)(disj(x, z) \wedge xz = y).$

The main technical point needed to prove that $A(\mathbf{R})/L(\mathbf{R})$ has no outer automorphisms is to show how to interpret the family of strictly increasing divergent sequences $\mathbf{a} = (a_n : n \in \omega)$ of reals in $A(\mathbf{R})/L(\mathbf{R})$ (up to finitely many mistakes). There are various ways to do this, but the one chosen, as for the case of $S_\lambda(\mu)/S_\omega(\mu)$, employs a pair of elements, the first to serve as a marker (or rather to provide a sequence of markers), and the other to encode the increasing sequence of positive integers. To build up to this, we first need a formula distinguishing those positive elements whose support (eventually) has the form $\bigcup\{(a_{2n}, a_{2n+1}) : n \in \omega\}$ (which are the first kind of element).

$moi1(x) : pos(x) \wedge \exists y(y \neq 1 \wedge disj(x, y) \wedge conj2(xy))$.
$moi(x) : moi1(x) \wedge (\forall y)(\exists z)(moi1(y) \rightarrow restr(x^z, y))$.

By a *moiety* we understand a positive element g of $A(\mathbf{R})$ whose non-trivial orbitals are of the form (a_{2n}, a_{2n+1}) for some strictly increasing divergent sequence $\mathbf{a}$ of reals.

Lemma 5.5 *For any $g \in A(\mathbf{R}), A(\mathbf{R})/L(\mathbf{R}) \models moi(x)$ if and only if $x = L(\mathbf{R}).g$ for some moiety g.*

Proof. Let $(a_n : n \in \omega)$ be a typical strictly increasing divergent sequence. Its entries thus form the (eventual) fixed point set of an element satisfying $conj2$. The formula $moi1$ thus says that an infinite coinfinite subset of the set of the intervals (a_n, a_{n+1}) is chosen as the (eventual) family of non-trivial (and positive) orbitals of the element in question. The extra condition imposed by moi is just that the element should be 'minimal' among such elements, or more precisely, it can be conjugated to a suitable restriction of any element satisfying $moi1$. This has the effect of ensuring that for all but finitely many $n, (a_n, a_{n+1})$ and (a_{n+1}, a_{n+2}) are not both selected, so by relabelling the a_ns, the support can be expressed in the desired form.

To encode strictly increasing sequences $(a_n : n \in \omega)$ we use pairs (g, h) of moieties such that the non-trivial orbitals of g have the form (a_n, b_n) for some $b_n \in (a_n, a_{n+1})$, and for each n there are exactly n non-trivial orbitals of h contained in (b_n, a_{n+1}).

$link(x, y) : disj(x, y) \wedge moi(xy) \wedge x, y \neq 1 \wedge \forall z[restr(z, x) \wedge \exists u \exists v(disj(u, y) \wedge$
$\quad disj(v, y) \wedge z = x^u x^v) \rightarrow z = x]$.
$inc(x, y) : link(x, y) \wedge \exists z \exists t[disj(z, t) \wedge y = zt \wedge link(t, x) \wedge link(x, t) \wedge$
$\quad \exists u(pos(u) \wedge disj(u, z) \wedge t^u = x) \wedge \exists u \exists v \exists w(pos(u) \wedge pos(v) \wedge pos(w) \wedge$
$\quad disj(u, x) \wedge disj(v, t) \wedge disj(w, y^u) \wedge y^{uv} = z \wedge x^w = t]$.

Lemma 5.6 *(i) For any p and $q, A(\mathbf{R})/L(\mathbf{R}) \models link(p, q)$ if and only if there are moieties g and h in $A(\mathbf{R})$ with disjoint supports and with no two positive orbitals of g or h sharing an endpoint, such that $p = L(\mathbf{R}).g$ and $q = L(\mathbf{R}).h$, and in between any two positive orbitals of g there is at least one positive orbital of h.*

(ii) For any p and $q, A(\mathbf{R})/L(\mathbf{R}) \models inc(p, q)$ if and only if there are moieties g and h in $A(\mathbf{R})$ with disjoint supports and with no two positive orbitals of g or h sharing an endpoint, such that $p = L(\mathbf{R}).g$ and $q = L(\mathbf{R}).h$, and such that if the positive orbitals of g are (a_n, b_n) for $n \in \omega$ with $a_0 < b_0 < a_1 < b_1 < a_2 \ldots$ then for each n there are exactly $n + 1$ positive orbitals of h contained in (b_n, a_{n+1}).

Proof. Suppose that g and h are as described in (i). Then gh is a moiety. Let k be any element which can be expressed as a product of two conjugates by conjugators disjoint from h. Then k must have at least one positive orbital in any interval determined by the orbitals of h which contains a positive orbital of g, so if it also is a restriction of g, must have *exactly* one. On passing to the quotient by $L(\mathbf{R})$, the conclusion is that k must agree with g on all but finitely many orbitals. This verifies $A(\mathbf{R})/L(\mathbf{R}) \models link(L(\mathbf{R}).g, L(\mathbf{R}).h)$.

Conversely if *link* holds for p and q then there are moieties g and h with $p = L(\mathbf{R}).g$ and $q = L(\mathbf{R}).h$ having disjoint supports and with the positive orbitals of g, h sharing no endpoints. Let (a_n, b_n) be the positive orbitals of g. Then it suffices to show that for all but finitely many n, there is at least one positive orbital of h in (b_n, a_{n+1}). In cases where this is not true, if for instance $(a_n, b_{n+1}) \cap supp(h) = \emptyset$, we can find witnesses k and l corresponding to u and v in *link*, with supports disjoint from h, so that forming the product $g^k g^l$ has the effect of 'coalescing' the two orbitals (a_n, b_n) and (a_{n+1}, b_{n+1}). By this method we may produce an element which infinitely often has fewer positive orbitals in between consecutive positive orbitals of h than g does, and g cannot then be a restriction of this element.

(ii) Rather than giving full details, I shall just explain the various parts of the formula, which should indicate how a proof would go. If the given condition holds for g and h, let (c_n, d_n) be the last orbital of h in (b_n, a_{n+1}) for each n, and take as witnesses to z and t in *inc* the elements $L(\mathbf{R}).k$ and $L(\mathbf{R}).l$ where l and k are the restrictions of $L(\mathbf{R}).h$ to the union of all the (c_n, d_n) and to the union of all the other positive orbitals of h respectively. The fact that $link(L(\mathbf{R}).l, L(\mathbf{R}).g)$ and $link(L(\mathbf{R}).g, L(\mathbf{R}).l)$ just corresponds to the alternation of the positive orbitals of g and l. The condition $\exists u(pos(u) \wedge disj(u, z) \wedge t^u = x)$ asserts that the positive orbital of l contained in any interval of positive orbitals of g is the last positive orbital of h in this interval, which agrees with our choice. Since k has one fewer positive orbital than h contained in each interval of positive orbitals of g, to conjugate h to k (even modulo $L(\mathbf{R})$), a conjugator has to carry the positive orbitals of h contained in (b_n, a_{n+1}) to the positive orbitals of k contained in (b_{n+1}, a_{n+2}) for each n. The final clause in *inc* tells us that this can be done: u moves these orbitals into (d_n, a_{n+1}), and v moves them into (b_{n+1}, a_{n+2}); w is needed to ensure that after conjugating by u, these orbitals have moved to the right of d_n.

Conversely, by (i) we know that on the basis of *link*, if *inc* holds for p and q we may find corresponding moieties g and h with disjoint supports such that if (a_n, b_n) for $n \in \omega$ are the positive orbitals of g, for each n there are m_n positive orbitals of h in (b_n, a_{n+1}) where $1 \leq m_n < \infty$. Running the sketch of the previous paragraph in reverse we find that for each $n, m_{n+1} = m_n + 1$, and this means that choices of g and h can be made to meet the stated conditions.

Now that we can encode strictly increasing divergent sequences, we can move towards the main result. For this we suppose that we are given an isomorphism θ from $A(\Omega_1)/L(\Omega_1)$ to $A(\Omega_2)/L(\Omega_2)$, where Ω_1 and Ω_2 are doubly homogeneous chains, each having a tail isomorphic to that of **R** (so that in fact Ω_1, Ω_2 are isomorphic to either **R** or the 'left long line' $\mathbf{L}^-$, see [5]). The aim is to show that θ is induced by some isomorphism between tails of Ω_1 and Ω_2. We can now begin to build up this desired isomorphism. We use the same notation inside each of Ω_1 and Ω_2 as we have been doing in **R**. In particular from now on we write **a**, **b** and so on to stand for strictly increasing divergent sequences $(a_0, a_1, a_2, \ldots), (b_0, b_1, b_2, \ldots)$ in Ω_1 or Ω_2. For any such **a** in Ω_1 choose moieties $g, h \in A(\Omega_1)$ whose positive orbitals share no endpoints, those of g are (a_n, b_n) for $a_n < b_n < a_{n+1}$, and h has exactly n positive orbitals in (b_n, a_{n+1}) for each n. Since θ carries the family of pairs in $A(\Omega_1)/L(\Omega_1)$ satisfying *inc* to those in $A(\Omega_2)/L(\Omega_2)$ satisfying *inc*, by Lemma 5.6 there must be g' and h' with these same properties such that $(L(\Omega_1).g)^\theta = L(\Omega_2).g'$ and $(L(\Omega_1).h)^\theta = L(\Omega_2).h'$. Let the positive orbitals of g' be (a'_n, b'_n). Then we define $p_{\mathbf{a}}$ to have domain $\{a_n : n \in \omega\}$ and to take a_n to a'_n for each n.

Of course we do not expect the whole of $p_{\mathbf{a}}$ to be accurately determined, but it is meant to agree with the final isomorphism (which will carry $L(\Omega_1).g$ to $(L(\Omega_2).g)^\theta$ for all g) on all but finitely many points. To achieve this we need to verify the following:
(i) the value of $p_{\mathbf{a}}$ is independent of the choice of g and h,
(ii) for any **a** and **b**, $p_{\mathbf{a}}$ and $p_{\mathbf{b}}$ are compatible,
(iii) there is some isomorphism k from a final segment ('tail') of Ω_1 onto a final segment of Ω_2 whose restriction to the domain of any $p_{\mathbf{a}}$ agrees with it on all but finitely many points,
(iv) if k is as given in (iii) (which is then clearly unique beyond some point), then $L(\Omega_2).g^k = (L(\Omega_1).g)^\theta$, (where g^k, which is strictly speaking an isomorphism from one final segment of Ω_2 to another, defines a unique coset $L(\Omega_2).g^k = L(\Omega_2).l$ where $l \in A(\Omega_2)$ is any extension of g^k to an automorphism).

Lemma 5.7 *Let (g_1, h_1) and (g_2, h_2) be two possible choices of (g, h) as used in defining $p_{\mathbf{a}}$, and let p_1, p_2 be the resulting partial maps. Then $\{n : a_n p_1 \neq a_n p_2\}$ is finite.*

Proof. We devise a formula *sameseq* analogous to the one used in §4.

$between(x, y, z) : inc(x, y) \wedge disj(xy, z) \wedge moi(xyz) \wedge \exists u \exists v[disj(z, u) \wedge link(x, zu) \wedge link(zu, x) \wedge v > 1 \wedge disj(v, y) \wedge (zu)^v = x]$.
$sameseq(x, y, z, u) : inc(x, y) \wedge inc(z, u) \wedge \exists v[subset(v, x) \wedge \exists w(disj(w, y) \wedge v^w = x) \wedge subset(v, z) \wedge \exists w(disj(w, u) \wedge v^w = z) \wedge \exists w(disj(v, w) \wedge y^w = u)] \wedge \forall v[(subset(v, x) \rightarrow \neg between(z, u, v)) \wedge (subset(v, z) \rightarrow \neg between(x, y, v))]$.

These two formulae have the following intended meanings:
between(x, y, z): z has at most one positive orbital between any two consecutive positive orbitals of x, and if there is such an orbital, it lies to the right of all non-trivial orbitals of y in that interval, and the positive orbitals of x, y, z share no endpoints.

sameseq(x, y, z, u): the sequences encoded by (x, y) and (z, u) are eventually equal.

In *between* the element u provides a positive orbital in those intervals between the positive orbitals of x where z has none. In *sameseq*, if the positive orbitals of x and z do have the same left hand endpoints, then we may take the positive orbitals of v to be given by their intersections, and the condition that $y^w = u$ ensures that the correct ones correspond. The final clause tells us that the left hand endpoints are indeed equal.

Now it is clear that with g_1, h_1, g_2, h_2 as given, $A(\Omega_1)/L(\Omega_1) \models$ $sameseq(L(\Omega_1).g_1, L(\Omega_1).h_1, L(\Omega_1).g_2, L(\Omega_1).h_2)$, and since θ is an isomorphism, also $A(\Omega_2)/L(\Omega_2) \models sameseq(L(\Omega_2).g_1', L(\Omega_2).h_1', L(\Omega_2).g_2', L(\Omega_2).h_2')$. Hence $\{n : a_n p_1 \neq a_n p_2\}$ is finite.

Lemma 5.8 *For any* $\mathbf{a}, \mathbf{b}$ *there is an isomorphism* k *from a final segment of* Ω_1 *to a final segment of* Ω_2 *such that* $\{i : a_i p_{\mathbf{a}} \neq a_i k\} \cup \{i : b_i p_{\mathbf{b}} \neq b_i k\}$ *is finite.*

Proof. First we reduce to the case where $\mathbf{b}$ is an infinite subsequence of $\mathbf{a}$ and its complement in $\mathbf{a}$ is also infinite, where we show that $\{i : b_i p_{\mathbf{a}} \neq b_i p_{\mathbf{b}}\}$ is finite. For let $\mathbf{c}$ be chosen so that each of $\mathbf{a}$ and $\mathbf{b}$ are subsequences of $\mathbf{c}$ and each has an infinite complement in $\mathbf{c}$, and let k be an isomorphism from a final segment of Ω_1 to a final segment of Ω_2 extending $p_{\mathbf{c}}$. Then by the special case, $\{i : a_i p_{\mathbf{a}} \neq a_i p_{\mathbf{c}}\}$, $\{i : b_i p_{\mathbf{b}} \neq b_i p_{\mathbf{c}}\}$ are finite, and hence so is $\{i : a_i p_{\mathbf{a}} \neq a_i k\} \cup \{i : b_i p_{\mathbf{b}} \neq b_i k\}$.

To prove the special case we will use a method precisely analogous to that of Lemma 4.12. Consider the following formula:

$store(x_1, x_2, \ldots, x_n, y, z) : \exists x(x = x_1 x_2 \ldots x_n \wedge inc(y, z) \wedge$
$\bigwedge_{1 \leq r < r' \leq n} disj(x_r, x_{r'}) \wedge link(x, yz) \wedge link(yz, x) \wedge \exists s \exists t[disj(s, t) \wedge x =$
$st \wedge link(s, z) \wedge link(z, s) \wedge \exists s \exists t[disj(s, t) \wedge x = st \wedge link(s, z) \wedge link(z, s) \wedge$
$\exists u(pos(u) \wedge disj(u, z) \wedge t^u = y) \wedge \exists u \exists v \exists w(pos(u) \wedge pos(v) \wedge pos(w) \wedge$
$disj(u, y) \wedge disj(v, t) \wedge disj(w, x^u) \wedge x^{uv} = s \wedge y^w = t \wedge$
$\bigwedge_{1 \leq r \leq n} x_r^{uv} = s \wedge x_r)]$.

This is intended to store disjoint elements g_i whose product is a moiety with positive orbitals $I_0 < I_1 < I_2 < \ldots$. Essentially the intuition is similar to that behind the construction of the formula *inc* (though more complicated). Given the g_i we form moieties h_i such that g_i is a restriction of h_i and for each n there are exactly $n+1$ positive orbitals of $\prod h_i$ between I_n and I_{n+1}, and they are distributed among the h_i in precisely the same order as the orbitals of $\prod g_i$ to the left of I_{n+1} are among the g_i. Thus $\prod h_i$ 'remembers' the pattern of the g_is cofinally often, and so this pattern is not destroyed on passing to the quotient by $L(\Omega_1)$. To control recovery of the stored pattern we insert additional markers in the shape of a pair satisfying *inc*.

In the formula, x_i stands for $L(\Omega_1).h_i$, and y, z for the additional markers; t stands for $L(\Omega_1).\prod g_i$ and positive elements are used to conjugate $\prod h_i$ to the right so as exactly to match s after passing through one of the positive orbitals of y.

Now to apply this formula we choose a pair (g,h) satisfying *inc* with a_i the left hand endpoints of the positive orbitals I_i of g, and with J_{ij} for $j \leq i$ the positive orbitals of h lying between I_i and I_{i+1}, to define $p_{\mathbf{a}}$. We use the restriction of g to $\bigcup\{I_i : a_i$ an entry of $\mathbf{b}\}$ and a corresponding restriction of h by taking the correct number of J_{ij}s for such i as required to define $p_{\mathbf{b}}$. We have four restrictions g_1, g_2, g_3, g_4 of gh therefore to remember, the restrictions of g, h just described, and their restrictions to the complement. What matters is the order these come in. We form elements k and l so that $A(\Omega_1)/L(\Omega_1) \models$ $store(L(\Omega_1).g_1, \ldots, L(\Omega_1).g_4, L(\Omega_1).k, L(\Omega_1).l)$. As θ preserves the truth of *store* there are g_i', k', l' such that $L(\Omega_1).g_i^\theta = L(\Omega_2).g_i'$ and so on, and $A(\Omega_2)/L(\Omega_2)$ satisfies the corresponding *store* statement. The outcome as in Lemma 4.12 is that the arrangements of the orbitals of g_1, g_2, g_3, g_4 and of g_1', g_2', g_3', g_4' are the same. In other words there is an isomorphism from a final segment of Ω_1 to a final segment of Ω_2 taking g_i to g_i' for each i. Hence $\{i : b_i p_{\mathbf{a}} \neq b_i p_{\mathbf{b}}\}$ is finite, as required.

Theorem 5.9 *Let Ω_1, Ω_2 be doubly homogeneous chains having final segments isomorphic to $\mathbf{R}$. Then for any isomorphism $\theta : A(\Omega_1)/L(\Omega_1) \to A(\Omega_2)/L(\Omega_2)$ there is an isomorphism from a final segment of Ω_1 onto a final segment of Ω_2 which induces θ.*

Proof. Given the development so far it follows that for any $\mathbf{a}$ there is a map $p_{\mathbf{a}}$ obtained naturally from θ, defined on the set of entries of $\mathbf{a}$ and taking $\mathbf{a}$ to some strictly increasing divergent sequence in Ω_2. The previous lemma showed that any two such $p_{\mathbf{a}}$s are compatible, and so the remaining task is to show how they can all be fitted together to form the desired k. In §4 we were able to handle this in cases where the whole set could be expressed in terms of finitely many (two in fact) of the approximants. To simulate the same situation here there are various options available. The most straightforward, as in [5], involves use of finitely many elements in $A(\Omega_1)$ such that there is an orbit of the group they generate which is dense on a final segment of Ω_1.

One of the tricks used in [7] involved consideration of translations g and h of $\mathbf{R}$ such that $\langle g,h\rangle$ has a dense orbit, for instance $xg = x + 1$ and $xh = x + \sqrt{2}$. We can envisage copying this behaviour on intervals arbitrarily far to the right in Ω_1, and hence forming a grid which will sufficiently well approximate all points of a final segment of Ω_1.

Choose points $\{a_n : n \in \mathbf{Z}\}$ in Ω_1 ordered in the natural way and unbounded above, and elements g_1, g_2, g_3, g_4 so that $supp(g_1) = supp(g_2) = \bigcup\{(a_{2n}, a_{2n+1}) : n \in \mathbf{Z}\}, supp(g_3) = supp(g_4) = \bigcup\{(a_{2n+1}, a_{2n+2}) : n \in \mathbf{Z}\}$, and for each n there is an isomorphism from $\mathbf{R}$ to (a_n, a_{n+1}) which carries translation by 1 to g_1, g_3 (n even, odd) and translation by $\sqrt{2}$ to g_2, g_4 (n even, odd).

By Lemma 5.8, if $\mathbf{a}_1 = (a_0, a_2, a_4, \ldots)$ and $\mathbf{a}_2 = (a_1, a_3, a_5, \ldots)$, there are $b_n \in \Omega_2$ for $n \in \mathbf{Z}$ such that $\{n \geq 0 : a_{2n} p_{\mathbf{a}_1} \neq b_{2n}\} \cup \{n \geq 0 : a_{2n+1} p_{\mathbf{a}_2} \neq b_{2n+1}\}$ is finite. By the techniques already illustrated there are $g_i' \in A(\Omega_2)$ satisfying the corresponding properties with respect to the b_n as g_i do with respect to the a_n. An important point, which may not be entirely clear, is that there are order-isomorphisms of $\mathbf{R}$ to (b_n, b_{n+1}) which carry translations by 1 and $\sqrt{2}$ to g_1', g_3' and g_2', g_4' (depending on n even or odd). The essential reason for this is that $\sqrt{2}$ can be arbitrarily well approximated by rationals, and for any positive integers i and j, $i/j < \sqrt{2} \Leftrightarrow L(\Omega_1).g_1^i < L(\Omega_1).g_2^j$ and similarly for g_3 and g_4. What this argument shows is that if an isomorphism is chosen from $\mathbf{R}$ to (b_{2n}, b_{2n+1}) which carries translation by 1 to g_1', then g_2' arises from translation by r_n where (r_n) is a sequence with limit $\sqrt{2}$. To establish the stronger statement, that $\{n : r_n \neq \sqrt{2}\}$ is *finite* (which is required), we observe that in $A(\Omega_1)/L(\Omega_1)$ there is an element satisfying *pos*1 which commutes with each of $L(\Omega_1).g_1$ and $L(\Omega_1).g_2$, and consequently there must be a corresponding such element in $A(\Omega_2)/L(\Omega_2)$, and this ensures that (r_n) is eventually constant.

Finally letting $\mathbf{c} = (c_n : n \in \omega)$ where $c_n \in (a_n, a_{n+1})$ we can now reach arbitrarily close to any member of (a_0, ∞) by starting from some c_n and applying a member of $L(\Omega_1).\langle g_1, g_2, g_3, g_4\rangle$. Letting $c_n' = c_n p_{\mathbf{c}}$ we define $k : (a_0, \infty) \to (b_0, \infty)$ by

$$\begin{aligned} a_n k &= b_n \\ (c_{2n} g_1^i g_2^j)k &= c_{2n}'(g_1')^i (g_2')^j \\ (c_{2n+1} g_3^i g_4^j)k &= c_{2n+1}'(g_3')^i (g_4')^j. \end{aligned}$$

This defines k on a dense subset of (a_0, ∞), by the remarks made above it is order-preserving, and so extends to a map k as required. Moreover, since the whole procedure is reversible, k is onto.

This tells us how we can choose k. Now we have to show that it has the desired properties. First, for any $\mathbf{e}$, $\{i : e_i p_{\mathbf{e}} \neq e_i k\}$ is finite. For this by use of Lemma 5.8, and without loss of generality, we may restrict to the case where $a_{2n} < e_n < a_{2n+1}$ for each n. By Lemma 5.8 again, $p_{\mathbf{e}}$ may be chosen so that $b_{2n} < e_n' < b_{2n+1}$ for all n. Now e_n is determined by $\{(i,j) : c_{2n} g_1^i g_2^j < e_n\}$, and if this set is the same for all n (or eventually constant) then there is no difficulty. To cope with the (usual) situation where this set is not constant, we have once again to 'store' the e_n as n increases. This is partly why we ensured that $supp(g_1)$, $supp(g_2)$ have unboundedly many gaps; so that there is room to fit the storing mechanism. We choose moieties h_1, h_2 whose supports are $\bigcup_{n\in\omega}(c_{2n}, d_{2n})$ and $\bigcup_{n\in\omega}(e_n, f_n)$, both contained in $\bigcup_{n\in\omega}(a_{2n}, a_{2n+1})$, whose positive orbitals have no a_m as an endpoint, and copy the quadruple (g_1, g_2, h_1, h_2) onto the vacant intervals, so that each contains information on the behaviour of these elements to the left of this point. This is enough now to be able to preserve the truth or falsity of $c_{2n} g_1^i g_2^j < e_n$ under θ. (For instance we may say that there is an element conjugate to h_2 by a negative conjugacy with support disjoint from g_3, whose support is contained in that of $h_1^{g_1^i g_2^j}$ but is disjoint from that of h_2). Hence for all i and j, $c_{2n} g_1^i g_2^j < e_n \Leftrightarrow c_{2n}'(g_1')^i (g_2')^j < e_n'$, and this implies that $\{i : e_i p_{\mathbf{e}} \neq e_i k\}$ is finite.

It remains to show that k induces the isomorphism θ. Since $A(\Omega_1), A(\Omega_2)$ are generated by moieties (at least when we pass to the quotients), it suffices to show that for any moiety $g, (L(\Omega_1).g)^\theta = L(\Omega_2).k^{-1}gk$. Let g have positive orbitals (a_{2n}, a_{2n+1}) for $n \in \omega$, and choose a moiety $g' \in A(\Omega_2)$ having positive orbitals (a'_{2n}, a'_{2n+1}) and such that $(L(\Omega_1).g)^\theta = L(\Omega_2).g'$. Now if $a_{2n+1} \leq x \leq a_{2n+2}, xg = x$, and $a'_{2n+1} \leq xk \leq a'_{2n+2}$, so that $xkg' = xk = xgk$. If there are arbitrarily large x such that $xkg' \neq xgk$ there must therefore be $\mathbf{b}$ such that for each $n, (b_{2n}, b_{2n+1})$ is contained in a distinct positive orbital of g and $b_{2n}kg' \neq b_{2n}gk$. Choose a moiety h having support $\bigcup_{n\in\omega}(b_{2n}, b_{2n+1})$, so that h^g is a moiety having support $\bigcup_{n\in\omega}(b_{2n}g, b_{2n+1}g)$. Let $c_n = b_n g$. Then $\{n : b_{2n}p_{\mathbf{b}} \neq b_{2n}k\} \cup \{n : c_{2n}p_{\mathbf{c}} \neq c_{2n}k\}$ is finite, and for n outside this set, $c'_{2n} = c_{2n}p_{\mathbf{c}} = c_{2n}k = b_{2n}gk \neq b_{2n}kg' = b_{2n}p_{\mathbf{b}}g' = b'_{2n}g'$. But also if $(L(\Omega_1).h)^\theta = L(\Omega_2).h'$, then for large enough n, h' has the (b'_{2n}, b'_{2n+1}) as positive orbitals, and so $(h')^{g'}$ has the $(b'_{2n}g', b'_{2n+1}g')$ as positive orbitals. Hence for large enough $n, c'_{2n} = b'_{2n}g'$.

This contradiction shows that $\{x \in \Omega_1 : xkg' \neq xgk\}$ is bounded above, and hence so is $\{y \in \Omega_2 : yg' \neq yk^{-1}gk\}$. Therefore $(L(\Omega_1).g)^\theta = L(\Omega_2).k^{-1}gk$, as required.

Theorem 5.10 *Let Ω_1, Ω_2 be doubly homogeneous chains having final segments isomorphic to $\mathbf{Q}$. Then for any isomorphism $\theta : A(\Omega_1)/L(\Omega_1) \to A(\Omega_2)/L(\Omega_2)$ there is an isomorphism from a final segment of Ω_1 onto a final segment of Ω_2 which induces θ.*

Proof. This is proved by a very similar method to Theorem 5.9. Since $\mathbf{Q}$ is not order-complete, there are some additional complications. For instance, there are four conjugacy classes of positive elements satisfying *conj*2; two of them consisting of contiguous ones, where the positive orbitals have either all rational or all irrational endpoints; and two of them separated, where the positive orbitals are (a_n, b_n) so that either all the a_n are rational (and all the b_n irrational), or all the a_n are irrational. It is possible to distinguish all four of these classes however by means of a first order formula. The trick again derives from methods given in [7], and so provided we choose one and stick to it, (contiguous with rational endpoints for instance), and make appropriate modifications all through the proof, there is no insurmountable difficulty.

Theorem 5.11 *(i) All automorphisms of $A(\mathbf{R})/L(\mathbf{R})$ or $A(\mathbf{Q})/L(\mathbf{Q})$ are inner.*

(ii) The outer automorphism groups of $A(\mathbf{R})/B(\mathbf{R})$ and $A(\mathbf{Q})/B(\mathbf{Q})$, respectively, have order 2.

The first part is an immediate consequence of Theorems 5.9 and 5.10, and the derivation of (ii) from these is given in [5].

A part of the material I have described in the section is implicit (in more generality) in [12]. Namely, Rubin's techniques enable one to interpret various classes inside $A(\mathbf{R})/B(\mathbf{R})$, such as the family of sets of the form $\bigcup\{(a_{2n,}, a_{2n+1}) : n \in \omega\}$ modulo sets bounded above, where $(a_n)_{n\in\omega}$ is a strictly increasing divergent sequence, and using his results on locally movable groups the action of $A(\mathbf{R})/B(\mathbf{R})$ on the boolean algebra of regular open subsets of $\mathbf{R}$ modulo those bounded above is also interpretable. His techniques do not appear to extend to the part of the argument where all automorphisms of $A(\mathbf{R})/B(\mathbf{R})$ are shown to be inner.

One may enquire to what extent the results indicated here for $\mathbf{R}$ and $\mathbf{Q}$ hold for other chains. Moreover, the form of the results is somewhat unsatisfactory, (though formulated similarly to Theorem 4.13), in that although stating the results in terms of isomorphisms between different $A(\Omega)/L(\Omega)$s makes the results look a little more natural, the restriction on the allowed values of Ω is so stringent as actually to give very little genuinely extra. What would be much more satisfactory would be a result of the form: any isomorphism θ from $A(\Omega_1)/L(\Omega_1)$ to $A(\Omega_2)/L(\Omega_2)$ is induced by an isomorphism between final segments of Ω_1 and Ω_2. This is too ambitious without some further hypotheses on the two chains. With the techniques so far developed the very least that we would expect to have to assume is that they are doubly homogeneous and have countable cofinality. In this case, in view of the 'recognition' results of [7], it seems one can make things at least seem a little more general than I have done here. The case of uncountable cofinality has not be touched on, and so a good test case seems to be to look at the long line $\mathbf{L}$. Does $A(\mathbf{L})/L(\mathbf{L})$ have any outer automorphisms?

References

[1] J. L. Alperin, J. Covington and H. D. Macpherson, *Automorphisms of quotients of symmetric groups*, this volume.

[2] R. N. Ball and M. Droste, *Normal subgroups of doubly transitive automorphism groups of chains*, Trans. Amer. Math. Soc. **290**(1985), 647-664.

[3] N. G. de Bruijn, *Embedding Theorems for Infinite Groups*, Indag. Math. **19**(1975), 560-569.

[4] U. Felgner and F. Haug, *The homomorphic images of infinite symmetric groups*, Forum Mathematicum, **5**(1993), 505-520.

[5] M. Giraudet and J. K. Truss, *On distinguishing quotients of ordered permutation groups*, Quarterly Journal of Mathematics (2), **45**(1994), 181-209.

[6] A. M. W. Glass, *Ordered Permutation Groups*, London Math. Soc. Lecture Notes Series Vol. **55**, Cambridge University Press, 1981.

[7] Y. Gurevich and W. C. Holland, *Recognizing the real line*, Transactions of the American Mathematical Society, **265**(1981), 527-534.

[8] M. Hall, *The Theory of Groups*, Macmillan, 1964.

[9] R. Kaye and D. Macpherson (editors), *Automorphisms of First Order Structures*, Oxford University Press, 1994.

[10] R. McKenzie, *On elementary types of symmetric groups*, Algebra Universalis **1**(1971), 13-20.

[11] M. Rubin, *On the automorphism groups of homogeneous and saturated boolean algebras*, Algebra Universalis **9**(1979), 54-86.

[12] M. Rubin, *The reconstruction of boolean algebras from their automorphism groups*, in *Handbook of Boolean Algebras*, (edited by J. D. Monk, R. Bonnet), Elsevier (1989), 549-606.

[13] M. Rubin, *The reconstruction of topological spaces from their groups of homeomorphisms*, Transactions of the American Mathematical Society **312**(1989), 487-538.

[14] M. Rubin, *The reconstruction of trees from their automorphism groups*, Contemporary Mathematics **151**(1993), American Mathematical Society.

[15] M. Rubin, *Locally moving groups and reconstruction problems*, this volume.

[16] W. R. Scott, *Group Theory*, Prentice Hall Inc, Englewood Cliffs, N.J., 1964.

[17] S. Shelah, *First order theory of permutation groups*, Israel Journal of Mathematics, **14**(1973), 149-162.

[18] S. Shelah, *Errata to: First order theory of permutation groups*, Israel Journal of Mathematics, **15**(1973), 437-441.

[19] S. Shelah and J. K. Truss, *On distinguishing quotients of symmetric groups*, in preparation.

[20] J. K. Truss, *Infinite permutation groups; products of conjugacy classes*, Journal of Algebra, **120**(1989), 454-493.

The Automorphism Groups of Generalized McLain Groups *

Manfred Droste

Institut für Algebra
Technische Universität Dresden
01062 Dresden, Germany
e-mail: droste@math.tu-dresden.de

Rüdiger Göbel

FB 6, Mathematik und Informatik
Universität GHS Essen
45117 Essen, Germany
e-mail: R. Goebel@uni-essen.de

1 Introduction

In 1954, McLain [16] applied the well-known connection between nilpotent algebras (of triangular matrices) and nilpotent groups to establish the existence of particular locally nilpotent groups, now known as McLain groups. McLain groups are important for many questions in group theory, and their construction and properties can be found in most text books of infinite group theory; we refer to [17, pp. 347–349]. The basic ingredients in McLain's construction are the linear ordering $\mathbb{Q}$ of rational numbers and the field F_p of p elements (p a prime), which give rise to McLain groups $G(F_p, \mathbb{Q})$ which consist of upper triangular "$\mathbb{Q} \times \mathbb{Q}$-matrices" over F_p (rows and columns labeled by numbers in $\mathbb{Q}$) with 1's on the diagonal and only finitely many entries from $F_p \backslash \{0\}$ elsewhere. These matrices are clearly invertible and constitute a group $G = G(F_p, \mathbb{Q})$. The field F_p makes G a locally finite p-group and even if F_p is replaced by a field of characteristic 0, matrix multiplication ensures that G is locally nilpotent; moreover $\mathbb{Q}$ is homogeneous, hence G is characteristically simple — in contrast to finite p-groups. McLain's ingenious, but also very transparent construction stimulated further research on McLain groups and their natural extensions. Most notably, J. Roseblade deeply investigated in his Ph. D. thesis under supervision of P. Hall in 1963 the "internal structure" of G and determined its automorphism group; see [18]. Moreover, Wilson [19] studied generalized McLain groups $G(F_p, S)$ replacing $(\mathbb{Q}, \leq)$ by another dense linear ordering S. His extensions of McLain's work are a source for many interesting examples which illustrate the strength and limitation of recent results in group theory.

*This work is supported by GIF project No. G-0294-081.06/93 of the German-Israeli Foundation for Scientific Research & Development.

W. C. Holland (ed.), Ordered Groups and Infinite Permutation Groups, 97–120.

This class of McLain groups can be extended further if the two parameters R and S in the construction of McLain groups $G(R,S)$ range over all rings R and arbitrary partial orderings S; see section 2 for the general definition. Lengthy calculations in [5] show that Roseblade's characterization of the automorphism group $Aut\ G(R,S)$ can be carried over *mutatis mutandis* to arbitrary rings and orderings if and only if R has no zero-divisors $\neq 0$ and $(S,\leq)$ is locally linear and unbounded; see Theorem 2.2. Hence we will assume throughout this paper that the not necessarily commutative ring R has no elements r,s both $\neq 0$ but $rs = 0$. Moreover, we assume for the rest of this paper that $(S,\leq)$ is unbounded and locally linear, i.e., any interval $[x,y] = \{s \in S : x \leq s \leq y\}$ is linear. Domains are well-studied, but also very particular examples of these rings in question and trees ([4],[6]) are obviously locally linear. Just for convenience we will also suppose that the ordering S is connected, i.e., any two points of S can be connected with a finite path with any two consecutive points comparable; see section 2 for details. For such pairs of rings R, R' and orderings S, S' we have shown in [5] that the McLain groups $G(R,S)$ and $G(R',S')$ are isomorphic if and only if the rings R,R' and the orderings S,S' are either isomorphic or anti-isomorphic respectively. This essentially says that the ring structure *and* the structure of the ordering can be reconstructed from the group structure of $G(R,S)$. The answer to this isomorphism problem provides the foundation to the following more general question on automorphism groups $Aut\ G$, which is the topic of this paper.

$$(*)\begin{cases} \textit{Let } G,\ G' \textit{ be two McLain groups and suppose that} \\ Aut\ G \cong Aut\ G'. \textit{ Can we deduce that } G \cong G'? \end{cases}$$

This is a well-known kind of question with many ancestors. It was answered for the classical McLain groups $G(F_p,\mathbb{Q})$ by Roseblade [18]; we will also take the opportunity to fix the flaw in the proof of the crucial Theorem 3 in [18, p. 275] spotted by Falkenberg [11]. $(*)$ also has been answered for other structures than McLain groups in various contexts. The most fundamental relatives of $(*)$ are Dieudonné's results [2] and [3], showing that isomorphisms $GL(F,n) \cong GL(F',n')$ of general linear groups induce isomorphisms of the underlying vector spaces, hence the fields F,F' are isomorphic and the dimensions n, n' (which may be infinite cardinals) are equal. A similar result holds for abelian p-groups G,G'. The case $p \geq 5$ is due to H. Leptin [14] and may be viewed as a type of Ulm's theorem, and the more recent case $p = 3$ is in Liebert [15]. The case $p = 2$ still remains open. If G,G' are torsion-free modules over the p-adic integers, then the positive answer to $(*)$ is the main result of the forth coming paper [1] by Corner and Goldsmith. Many more relatives of $(*)$ and positive answers may be found in the literature, e.g. replacing "*Aut*" by "*End*", the endomorphism ring (monoid) of the underlying structure G. We want to add the positive answer for McLain groups. Our main result (Main Theorem 5.8) will be an isomorphism $G \cong G'$ deduced from $Aut\ G \cong Aut\ G'$ for two McLain groups G,G'. We will have to require that the rings R used for constructing McLain groups have the additional property that their ring-automorphisms are determined by their action on the group of units $U(R)$ of the ring. This is not very surprising, because it is $U(R)$ which plays a central role in decoding R, S from $Aut\ G(R, S)$. Examples of such rings are all (skew) fields, but also many commutative rings including $\mathbb{Z}$. We also note that the isomorphism between G and G' is *not induced* from the isomorphism between $Aut\ G$ and $Aut\ G'$. Finally we want to discuss how to find an isomorphism between G and G'. The strategy is based on a sequence of subgroups of $Aut\ G$, which we explain first:

$$G = Inn\ G \subseteq 1 + \rho T \subseteq \beta(1 + T^*) \subseteq \beta(Aut\ G) \subseteq \sigma(G) \subseteq 1 + T^* = Linn\ G.$$

As G has a trivial centre, we may identify inner automorphisms (= conjugation by group elements) with elements in G, hence $G = Inn\ G \subset Aut\ G$. While elements in G may be written in the form $g = 1+t$ with $t = \sum_{\alpha<\beta} a_{\alpha\beta}e_{\alpha\beta}$, $a_{\alpha\beta} \in R$, $\alpha, \beta \in S$ and $a_{\alpha\beta} = 0$ for almost all $\alpha, \beta \in S$ and multiplication of $S \times S$-matrices $e_{\alpha\beta}$ (having 1 at (α, β) and 0 elsewhere) is given by $e_{\alpha\beta} \cdot e_{\gamma\delta} = \delta_{\beta\gamma} \cdot e_{\alpha\delta}$ ($\delta_{\beta\gamma}$ the Kronecker symbol), some of the automorphisms of G can be expressed similarly: If T denotes the ring of all zero-triangular $S \times S$-matrices over R which are row- and column-finite, then $1+T$ forms a semigroup with respect to multiplication and we consider the maximal group $1 + T^*$ contained in $1 + T$. It is easy to show that $t \in T^*$ if and only if there exists $t^* \in T$ with $t + t^* + tt^* = t + t^* + t^*t = 0$. The arguments given by Roseblade [18, pp. 269,270] show that $1 + T^*$ consists of locally inner automorphisms of G, hence $1 + T^* \subseteq Linn\ G$, where $Linn\ G$ denotes the group of locally inner automorphisms of G. Like in the case of classical McLain groups it can be shown [5] that $1 + T^* = Linn\ G$ and the matrices in question can be viewed group theoretical, cf. Corollary 2.3. We deduce that the Baer radical $\beta(1 + T^*)$ must be contained in the Baer radical $\beta(Aut\ G)$, cf. Corollary 2.4. Recall that $Linn\ G$ is normal in $Aut\ G$ and the Baer radical βH of a group H is the subgroup of H generated by all subnormal nilpotent subgroups (hence by all subnormal cyclic subgroups) of H. Following [18] we also see that $1 + \rho T$ is contained in the Baer radical, cf. Lemma 2.5. Here ρT denotes the Jacobson radical of T comprising all matrices contained in some two-sided nilpotent ideal of the ring T. Obviously $G \subseteq 1 + \rho T$ and trivially $\beta(Aut\ G)$ is contained in the subgroup $\sigma(G)$ of $Aut\ G$ comprising all Baer automorphisms of G. Just note that $\alpha \in Aut\ G$ is a Baer automorphism of G if and only if $\langle\alpha\rangle$ is subnormal in the subgroup $\langle G, \alpha\rangle$ of $Aut\ G$. The above sequence of subgroups of $Aut\ G$ is established after we have shown that $\sigma(G) \subseteq Linn\ G$, which is one of the main tasks of this paper; see Theorem 4.1. For special rings and orderings R, S some of the inclusions of the above sequence are equalities, which however is not needed for our purpose. If φ is now an isomorphism between $Aut\ G$ and $Aut\ G'$ for two McLain groups $G = G(R, S)$ and $G' = G(R', S')$, then $\beta(Aut\ G)$ is mapped onto $\beta(Aut\ G')$ because the Baer radical is a characteristic subgroup of $Aut\ G$. In order to deduce isomorphisms or anti-isomorphisms between the rings R, R' and orderings S, S' respectively we now can use our results from section 3 concerning G-normal closures of subgroups of $Linn\ G$. We conclude with

(3.3), (3.10), and (3.13) that the McLain groups $G(R, S)$ and $G(R', S')$ must be isomorphic as well; see Main Theorem 5.8.

2 Automorphisms of the McLain groups

In this section we recall the definition of generalized McLain groups and the description of their automorphism groups derived in [5].

Let $(S, \leq)$ be a partially ordered set (poset) and R an arbitrary ring with $1 \neq 0$. Let $E = E(R, S)$ be the free R-module with basis $\{e_{\alpha\beta} : \alpha, \beta \in S, \alpha < \beta\}$. Then E becomes an associative R-algebra by defining a multiplication of basis elements via

$$e_{\alpha\beta} \cdot e_{\gamma\delta} = \begin{cases} e_{\alpha\delta} \text{ if } \beta = \gamma \\ 0 \text{ otherwise.} \end{cases}$$

Clearly, all elements of E are nilpotent. Hence the set $G = G(R, S)$ of formal sums $1 + u$ $(u \in E)$ forms a multiplicative group (see, for example [16]). The general element of G is expressible uniquely in the form $g = 1 + a = 1 + \sum\limits_{\alpha<\beta} a_{\alpha\beta} e_{\alpha\beta}$ with $a_{\alpha\beta} \in R$ and $a_{\alpha\beta} = 0$ for almost all $\alpha, \beta \in S$. Subsequently, for short, we will simply write $g = 1 + \sum a_{\alpha\beta} e_{\alpha\beta}$. We write $[g] = \{(\alpha, \beta) : a_{\alpha\beta} \neq 0\}$, the *support* of g, $[g]_1 = \{\alpha \in S : \exists\, \beta \in S\ (\alpha, \beta) \in [g]\}$, the 1-*support* of g, and $[g]_2 = \{\beta \in S : \exists\, \alpha \in S\ (\alpha, \beta) \in [g]\}$, the 2-*support* of g.

We also write $[a] = [g]$ and $[a]_i = [g]_i$ $(i = 1, 2)$, where $g = 1 + a$ as above. For any subset $A \subseteq S$, we let *max* A denote the set of all maximal elements of A. We say that the poset $(S, \leq)$ is *dense*, if for any $\alpha, \beta \in S$ with $\alpha < \beta$ there is $\gamma \in S$ such that $\alpha < \gamma < \beta$.

The following lemma is a straightforward generalization of one due to McLain [16, Lemma 1]. The representation of an arbitrary group element as a product as described below will be crucial throughout.

Lemma 2.1 *G is generated by the set of all elements $1 + ae_{\alpha\beta}$, $(a \in R,\ \alpha, \beta \in S$, with $\alpha < \beta)$.*

Next we turn to the description of the automorphisms of $G(R, S)$. As shown in [5], this requires natural assumptions on R and S, which we introduce first.

A poset $(S, \leq)$ is called *unbounded*, if for each $\sigma \in S$ there are $\alpha, \beta \in S$ with $\alpha < \sigma < \beta$, and *locally linear*, if for any $\alpha, \beta \in S$ with $\alpha < \beta$ the set $\{\sigma \in S : \alpha < \sigma < \beta\}$ is linearly ordered. Two elements $\alpha, \beta \in S$ are called *connected* if there are elements $\alpha_0, \cdots, \alpha_n \in S$ such that $\alpha_0 = \alpha$, $\alpha_n = \beta$, and for each $0 \leq i < n$, either $\alpha_i \leq \alpha_{i+1}$ or $\alpha_{i+1} \leq \alpha_i$. Such a sequence $\alpha_0, \cdots, \alpha_n$ is also called a *path* in S; if here $\alpha_0 = \alpha_n$, we say this path is closed. A *component* of $(S, \leq)$ is a maximal connected subset of S. Clearly, S is the disjoint union of its components. For each component C of S, let $G_C = \{g \in G : [g] \subseteq C \times C\}$. Clearly, G is the direct sum of the McLain groups $G_C \cong G(R, C)$ (C a component of S). Therefore it is natural to assume that the poset S is connected. Furthermore, the description of the automorphisms of $G(R, S)$ in [5] is based on an analysis of the maximal abelian normal subgroups of G. As shown in [5], they can be described order-theoretically if and only if R has no zero-divisors $\neq 0$ and $(S, \leq)$ is unbounded and locally linear. Therefore subsequently we will put the following restriction on R and $(S, \leq)$:

$$(*)\begin{cases} R \textit{ has no zero divisors } \neq 0, \textit{ and} \\ (S,\leq) \textit{ is an unbounded locally linear connected poset.} \\ \textit{We fix } \alpha_0 \in S \textit{ throughout.} \end{cases}$$

Note that this assumption is satisfied trivially for the classical McLain groups where R is the field with p elements (p prime) and $S = \mathbb{Q}$, the rationals, but also for many other rings and posets.

First we describe automorphisms of G induced by the ring and the order. Let φ be an order-automorphism of S and ψ a ring-automorphism of R. For

$$g = 1 + \sum a_{\alpha\beta} e_{\alpha\beta} \in G \text{ let } g\Phi = 1 + \sum (a_{\alpha\beta}\psi) e_{\alpha\varphi, \beta\varphi}.$$

Clearly, Φ is an automorphism of G, called the automorphism *induced* by (φ, ψ).

Next assume that φ is an anti-automorphism of $(S, \leq)$ and ψ an anti-automorphism of the ring R. If

$$g = 1 + \sum a_{\alpha\beta} e_{\alpha\beta} \in G,$$

let

$$g\Phi' = 1 + \sum (a_{\alpha\beta}\psi) e_{\beta\varphi, \alpha\varphi}.$$

If

$$h = 1 + \sum b_{\gamma\delta} e_{\gamma\delta} \in G,$$

we have

$$g \cdot h = 1 + \sum a_{\alpha\beta} e_{\alpha\beta} + \sum b_{\gamma\delta} e_{\gamma\delta} + \sum a_{\alpha\beta} b_{\gamma\delta} e_{\alpha\beta} e_{\gamma\delta},$$

so

$$(g \cdot h)\Phi' = (h\Phi') \cdot (g\Phi').$$

Since inversion is a group-anti-automorphism, it follows naturally that Φ defined by the composition $g\Phi = (g^{-1})\Phi'$ ($g \in G$) is a group-automorphism. We call Φ also the group-automorphism *induced* by (φ, ψ).

Now if $\varphi \in Aut(S, \leq)$, let $\widetilde{\varphi} : G \to G$ be the automorphism induced by (φ, id_R). Similarly, if $\psi \in Aut\ R$, let $\widetilde{\psi} : G \to G$ be the automorphism induced by (id_S, ψ). If S and R are anti-isomorphic to themselves, choose anti-automorphisms φ^* of $(S, \leq)$ and ψ^* of R and let τ be the automorphism of G induced by φ^* and ψ^*. Otherwise, we let $\tau = 1$. Next let $(c_{\alpha\beta})_{\alpha<\beta}$ be a family of units of R satisfying the following identities:

$$(**)\begin{cases} c_{\alpha\beta}\cdot c_{\beta\gamma} = c_{\alpha\gamma} \text{ for all } \alpha<\beta<\gamma \text{ , and} \\ \prod_{i=0}^{n-1} c_{\alpha_i\alpha_{i+1}} \in \zeta(R) \text{ for each closed path } \alpha_0,\cdots,\alpha_n \text{ in } S; \\ \text{here } \zeta(R) \text{ denotes the centre of } R. \end{cases}$$

We put $\psi_{\alpha_0} = id_R$, the identity on R. If $\beta \in S$, choose a path $\alpha_0,\cdots,\alpha_n$ from α_0 to $\alpha_n = \beta$, put $c = \prod_{i=0}^{n-1} c_{\alpha_i\alpha_{i+1}}$, and define ψ_β by $\psi_\beta = c^{-1}\cdot r\cdot c$. Clearly, ψ_β is a well-defined automorphism of R by (**), and $\psi_\alpha^{c_{\alpha\beta}} = \psi_\beta$ for all $\alpha<\beta$ in S, where for $r\in R$, $r(\psi_\alpha^{c_{\alpha\beta}}) = c_{\alpha\beta}^{-1}\cdot(r\psi_\alpha)\cdot c_{\alpha\beta}$.

Now define $\delta : G \to G$ by $(1+\sum a_{\alpha\beta}e_{\alpha\beta})\delta = 1+\sum(a_{\alpha\beta}\psi_\alpha)\cdot c_{\alpha\beta}e_{\alpha\beta}$. It is easy to check that δ is an automorphism of G, called the *dilatation induced by the family* $(c_{\alpha\beta})_{\alpha<\beta}$ *and* α_0.

We note that any one-parameter family $(d_\alpha)_{\alpha\in S}$ of units of R gives rise to a family $(c_{\alpha\beta})_{\alpha<\beta}$ satisfying property (**) by putting $c_{\alpha\beta} = d_\alpha^{-1}\cdot d_\beta$. Conversely, for a large class of posets $(S,\leq)$ (called loop-free in [5]; examples include all trees) any family $(c_{\alpha\beta})_{\alpha<\beta}$ arises in this way from a one-parameter family $(d_\alpha)_{\alpha\in S}$.

We let $\widetilde{S}$ comprise all automorphisms $\widetilde{\varphi}$ of G induced by (φ, id_R) for some $\varphi \in Aut(S,\leq)$; similarly $\widetilde{R} = \{\widetilde{\psi} : \widetilde{\psi}$ induced by (id_S,ψ), $\psi \in Aut\ R\}$ and $D = \{\delta : \delta$ induced by some family $(c_{\alpha\beta})_{\alpha<\beta}$ as above $\}$. Clearly, $\widetilde{S}$ and $\widetilde{R}$ are subgroups of $Aut\ G$, isomorphic to $Aut(S,\leq)$ and $Aut\ R$, respectively. As shown in [5], D is also a subgroup of $Aut\ G$, called the *dilatation group*. Finally, an automorphism of G is called *locally inner*, if it acts by conjugation on all finitely generated subgroups of G. Let $Linn\ G$ comprise all locally inner automorphisms of G; obviously $Linn\ G$ is a normal subgroup of $Aut\ G$.

Now we can state the central result of [5].

Theorem 2.2 [5, Theorem 6.2] *Let R be a ring with no zero-divisors $\neq 0$, and let $(S,\leq)$ be an unbounded locally linear connected poset. Let $G = G(R,S)$. Then we have the following semidirect product decomposition of $Aut\ G$:*

$Aut\ G = (((Linn\ G \rtimes D)\rtimes \widetilde{R})\rtimes \widetilde{S})\cdot\langle\tau\rangle$, *where* $\tau^2 \in \widetilde{R}\times\widetilde{S}$

and

$(Linn\ G \rtimes D)\rtimes \widetilde{R}$ *is normal in* $Aut\ G$.

For any $\alpha<\beta$ in S, let $X_{\alpha\beta}$ comprise all elements $x \in E$ such that any pair $(\gamma,\delta)\in[x]$ satisfies $\gamma\leq\alpha<\beta\leq\delta$ and either $\gamma<\alpha$ or $\beta<\delta$. It is easy to see that if $\psi \in Linn\ G$ and $a \in R$, then $(1+ae_{\alpha\beta})\psi = 1+ae_{\alpha\beta}+x$ for some $x\in X_{\alpha\beta}$.

We will also need another description of locally inner automorphisms. For this, let M be the free R-module with basic elements m_α ($\alpha\in S$). For each $\lambda\in S$, let M_λ be the submodule of M generated by all the elements m_μ with $\lambda\leq\mu$, and let $M_{\lambda+}$ be the submodule generated by $\{m_\mu :\ \lambda<\mu\}$. Let E^+ be the ring of those endomorphisms of M which map M_λ into $M_{\lambda+}$ for each $\lambda \in S$. If $\eta \in E^+$ and $\lambda \in S$, then $m_\lambda \eta$ is a finite linear combination of certain of the m_μ $(\lambda < \mu)$. Therefore E^+ can be thought of as a ring of row-finite zero-triangular matrices, and defining $m_\lambda e_{\lambda\mu} = m_\mu$ and $m_\tau e_{\lambda\mu} = 0$ for $\tau \neq \lambda$, E appears as the subring of E^+ of all zero-triangular matrices with almost all coefficients zero. Let T be the subring of E^+ of all matrices which are also column-finite. Then $1 + T$ forms a semigroup with respect to multiplication. As before, the elements $1 + t$ can be expressed as $1 + \sum_{\alpha<\beta} t_{\alpha\beta} e_{\alpha\beta}$ which, however, now may have infinitely many entries $t_{\alpha\beta} \neq 0$. Let $1+T^*$ be the maximal group contained in $1+T$. Then $1+T^*$ contains all elements of $1 + T$ with a two-sided inverse in $1 + T$, and $t \in T^*$ if and only if there exists $t^* \in T$ with $t+t^*+tt^* = t+t^*+t^*t = 0$. Now if $t \in T^*$, we obtain an automorphism ψ_t of G induced by transforming with $1 + t$. That is, if $1 + t^*$ is the inverse of $1 + t$ in $1+T^*$, we have $(1+re_{\lambda\mu})\psi_t = (1+t^*)(1+re_{\lambda\mu})(1+t)$ for all $r \in R$ and $\lambda < \mu$ in S. We let L_1 comprise all such automorphisms ψ_t $(t \in T^*)$ of G. Subsequently, we identify $L_1 = 1 + T^*$. The argument given by Roseblade [18, pp. 269, 270] shows that $L_1 \subseteq Linn\ G$.

Actually we have:

Corollary 2.3 [5, proof of Theorem 6.2]. *Let R be a ring with no zero-divisors $\neq 0$, and let $(S, \leq)$ be an unbounded locally linear connected poset. Then $L_1 = Linn\ G$.*

As a consequence we obtain:

Corollary 2.4 *Let R be a ring with no zero-divisors $\neq 0$, and let $(S, \leq)$ be an unbounded locally linear connected poset. Then $\beta L_1 \subseteq \beta(Aut\ G(R, S))$.*

Proof. By Corollary 2.3, we have $L_1 = Linn\ G$, a normal subgroup of $Aut\ G$. Hence each subnormal subgroup of L_1 is subnormal in $Aut\ G$, and the result follows by the definition of the Baer radical.

We let ρT denote the Jacobson radical of T comprising all elements contained in some two-sided nilpotent ideal of T. Clearly $E \subseteq \rho T$ and so $G = 1+E \subseteq 1+\rho T$. If $t \in \rho T$, we have $t^n = 0$ for some n, hence $(1 + t)^{-1} = 1 - t + t^2 - \cdots \pm t^{n-1}$ and so $t \in T^*$. This shows that $\rho T \subseteq T^*$ and $1 + \rho T \subseteq 1 + T^* = L_1$. Furthermore, we have:

Lemma 2.5 *Let R have no zero-divisors $\neq 0$ Then $1+\rho T$ is contained in the Baer radical of $1 + T^*$. If $p = char(R) \neq 0$, then $1 + \rho T$ is a locally finite p-group. If $char(R) = 0$, then $1 + \rho T$ is torsionfree.*

Proof. By inspection of [18, p. 269, proof of Lemma 1]. The last two statements are clear.

We will identify elements of $G(R, S)$ with the inner automorphisms induced by conjugation. Hence $G \subseteq L_1$. As for elements of G, for $t = \sum_{\alpha<\beta} t_{\alpha\beta} e_{\alpha\beta} \in T^*$ we write $[\psi_t] = [t] = \{(\alpha,\beta): t_{\alpha\beta} \neq 0\}$, the support of ψ_t, $[\psi_t]_1 = [t]_1 = \{\alpha \in S : \exists\, \beta \in S\ (\alpha,\beta) \in [t]\}$, $[\psi_t]_2 = [t]_2 = \{\beta \in S : \exists\, \alpha \in S\ (\alpha,\beta) \in [t]\}$.

3 Abelian normal subgroups of $Linn(G)$

In this section, we show that the maximal abelian normal subgroups of $Linn(G)$ are of a specific form defined order-theoretically. First we give a characterization of abelian normal closures of cyclic subgroups of $Linn(G)$. For this, the following condition first used in Roseblade [18, p.271] will be crucial.

We will say that an arbitrary $a \in T$ satisfies the condition (ρ) if $a^2 = aba = 0$ for any $b \in E$.

Theorem 3.1 *Let $(S,\leq)$ be unbounded. Let $g = 1 + a$ be an arbitrary element of Linn G and $N = \langle g^G \rangle$ be the G-normal closure of g in Linn G. The following are equivalent:*

1. *N is abelian.*
2. *a satisfies the condition (ρ) in T.*

Proof. Clearly, N is abelian if and only if all generators of N commute, and this is the case precisely if for all $h \in G$, g^h and g commute. With $g = 1 + a$ and $a \in T$, this is equivalent to

$$(*) \qquad h^{-1} \cdot a \cdot h \cdot a = a \cdot h^{-1} \cdot a \cdot h \text{ for all } h \in G.$$

Clearly condition (ρ) implies that $aha = 0$ for all $h = 1 + b \in G$, and hence $(*)$ holds. Conversely assume $(*)$ holds. Suppose $a^2 \neq 0$. Let $h = 1 + e_{\alpha\beta}$ such that $\alpha \in [a^2]_2$ and $\beta \notin [a]_1$. Then $h^{-1} = 1 - e_{\alpha\beta}$, so $h^{-1}aha = a^2$ and $ah^{-1}ah = a^2 + a^2 \cdot e_{\alpha\beta} \neq a^2$, contradicting $(*)$. Hence $a^2 = 0$. Now suppose there is $b \in T$ with $aba \neq 0$. Then there are pairs (α,β), $(\gamma,\delta) \in [a]$ and $(\beta,\gamma) \in [b]$ and coefficients $x \in R$ of $e_{\alpha\beta}$ in a and y of $e_{\beta\gamma}$ in b, and z of $e_{\gamma\delta}$ in a such that $x \cdot y \cdot z \neq 0$. Put $h = 1 + ye_{\beta\gamma} + e_{\delta\eta}$ for some $\eta \notin [a]_1 \cup [a]_2$. Then $h^{-1} = 1 - ye_{\beta\gamma} - e_{\delta\eta}$. Using $a^2 = 0$, we obtain $h^{-1}aha = aha = a \cdot ye_{\beta\gamma} \cdot a$; but $ah^{-1}ah = -a \cdot ye_{\beta\gamma} \cdot a \cdot h$ which has the non-zero coefficient $-x \cdot y \cdot z$ of $e_{\alpha\eta}$, again contradicting $(*)$. The result follows.

Next we derive a commutator formula for elements in $1 + T^*$.

Lemma 3.2 *Let $g = 1 + a = 1 + \sum a_{\alpha\beta}e_{\alpha\beta} \in 1 + T^*$ and $g^{-1} = 1 + a^*$. Let $\gamma < \xi < \zeta < \delta$ and $u = 1 + ze_{\zeta\delta}$, $v = 1 + ye_{\gamma\xi}$. Then $h := [g,u] = 1 - a^*ze_{\zeta\delta} - (1 + a^*)ze_{\zeta\delta}a$ and $[h,v] = 1 - ya_{\xi\zeta}ze_{\gamma\delta} + ye_{\gamma\xi}a^*ze_{\zeta\delta}a$. If g belongs to an abelian subgroup N of $1 + T^*$ such that G is contained in the normalizer of N in $1 + T^*$, we have $[h,v] = 1 - ya_{\xi\zeta}ze_{\gamma\delta}$.*

Proof. Using $a + a^* + a^*a = 0$, we obtain the formula $h = g^{-1}u^{-1}gu = 1 + b$ with $b = -a^*ze_{\zeta\delta} - (1 + a^*)ze_{\zeta\delta}a$. Similarly, $[h, v] = 1 - (1 + b^*)ye_{\gamma\xi}b - b^*ye_{\gamma\xi}$. Under the additional assumption, by condition (ρ) of Theorem 3.1 we have $a^2 = b^2 = 0$, hence $a = -a^*$ and $b = -b^*$, and $a^*ca = b^*cb = 0$ for all $c \in E$. Thus $[h, v] = 1 - ye_{\gamma\xi}b + bye_{\gamma\xi} = 1 - ye_{\gamma\xi}b = 1 - ya_{\xi\zeta}ze_{\gamma\delta}$, as claimed.

We introduce some further notation. Let $X \subseteq S$. We define $\downarrow X = \{s \in S : s \leq x \text{ for some } x \in X\}$ and $\uparrow X$ dually. We call X *downward (upward) closed* if $X = \downarrow X$ ($X = \uparrow X$), respectively. For singletons $x \in S$, let $\downarrow x = \downarrow\{x\}$ and define $\uparrow x$ dually. We say that X is *upper directed* if for any $x, y \in X$ there is $z \in X$ with $x, y \leq z$; *lower directed* is defined dually. X is an *(order-)ideal* of S, if it is non-empty, downward closed and upper directed. Dually, X is an *(order-)filter* of S, if it is non-empty, upward closed and lower directed. Let $A, B \subseteq S$. We write $A < B$ if $a < b$ for all $a \in A$, $b \in B$; and abbreviate $A < \{x\}$ to $A < x$. We call (A, B) a *couple*, if the following three conditions hold:

1. A, B are non-empty and disjoint, and A is downward and B upward closed;
2. for all $a \in A$ there is $b \in B$ with $a \leq b$;
3. for all $b \in B$ there is $a \in A$ with $a \leq b$.

We let $\mathcal{C}$ comprise all couples (A, B) of $(S, \leq)$.

If A is an ideal, B a filter on S and $A < B$, we call (A, B) an *ideal couple*. Let $\mathcal{C}^*$ consist of all ideal couples of $(S, \leq)$. We introduce a partial order on $\mathcal{C}$ by letting $(A, B) \subseteq (A', B')$ if $A \subseteq A'$ and $B \subseteq B'$. We write $J \triangleleft R$ to denote that J is a two-sided ideal of R. We also let $\triangleleft$ denote normal subgroups; the intended meaning will be clear from the context. If (A, B) is a couple and $0 \neq J \triangleleft R$, we call (A, B, J) a *triple*. If, here, (A, B) is an ideal couple, we call (A, B, J) *an ideal triple*. We put $(A, B, J) \subseteq (A', B', J')$ if $(A, B) \subseteq (A', B')$ and $J \subseteq J'$. Each triple (A, B, J) induces a subgroup $N^J_{A,B}$ which consists of all elements of $1 + T^*$ of the form

$$g = 1 + \sum x_{\alpha\beta}e_{\alpha\beta} \text{ with } x_{\alpha\beta} \in J,\ \alpha \in A,\ \beta \in B.$$

Hence $[g]_1 \subseteq A$, $[g]_2 \subseteq B$. We will write $N_{A,B} = N^R_{A,B}$ and call these groups *order-theoretic* normal subgroups of $1 + T^*$.

Lemma 3.3 *Let $(S, \leq)$ be unbounded.*

(a) For any triple (A, B, J), $N^J_{A,B}$ is an abelian normal subgroup of L_1 contained in $1 + \rho T$.

(b) For triples (A, B, J), (A', B', J') we have $(A, B, J) \subseteq (A', B', J')$ if and only if $N^J_{A,B} \subseteq N^{J'}_{A',B'}$.

Proof. (a) Clearly $N^J_{A,B}$ is abelian. Let $1+a \in N^J_{A,B}$ and $(1+t) \in L_1$ with $t \in T^*$ as described at the end of section 2. Then $(1+a)^{(1+t)} = (1+t^*)(1+a)(1+t) = 1+t^*+a+t+t^*a+at+t^*t+t^*at = 1+a+t^*a+at+t^*at \in N^J_{A,B}$, since $[t^*a]$, $[at]$, $[t^*at] \subseteq A \times B$. So $N^J_{A,B} \lhd L_1$. Since $a^2 = 0$, we have $a \in \rho T$ and so $N^J_{A,B} \subseteq 1 + \rho T$.

(b) Let $N^J_{A,B} \subseteq N^{J'}_{A',B'} = N'$, and $\alpha \in A$, $\beta \in B$, $x \in J$ with $\alpha < \beta$ and $x \neq 0$. Then $g = 1 + xe_{\alpha\beta} \in N'$, so it can be expressed as $g = 1 + \sum_{i \in I}^{n} a_i e_{\alpha_i \beta_i}$. The uniqueness of this representation shows $I = \{1\}$, say, and thus $x = a_i \in J'$, $\alpha = \alpha_i \in A'$, $\beta = \beta_i \in B'$ as desired.

Next we show that certain abelian subgroups of $1 + T^*$ are covered by order-theoretic normal subgroups.

Theorem 3.4 *Let $(S, \leq)$ be an unbounded locally linear connected poset and R a ring with no zero-divisors $\neq 0$. Let N be an abelian subgroup of $1 + T^*$ such that the normalizer of N contains G. Then N is contained in an order-theoretic normal subgroup of $1 + T^*$.*

Proof. Given N as above, define $A = \bigcup_{g \in N} [g]_1$ and $B = \bigcup_{g \in N} [g]_2$. We claim that (A, B) is a couple; it will follow that $N \subseteq N^R_{A,B}$. We first show that $A = \downarrow A$. Let $\alpha' \in S$ and $\alpha \in A$ with $\alpha' < \alpha$. There are $g \in N$ and $\beta, \beta' \in S$ with $(\alpha, \beta) \in [g]$, and $\beta < \beta'$. Then $\alpha' \in [g']_1$ for some $g' \in N$ by Lemma 3.2. Similarly $B = \uparrow B$.

Next we observe that there are no elements $g, h \in N$ and $\alpha, \beta, \gamma, \delta \in S$ with $(\alpha, \beta) \in [g]$, $(\gamma, \delta) \in [h]$ and $\alpha < \beta < \gamma < \delta$ in S. For suppose, otherwise. By assumption on $(S, \leq)$, there is $\alpha' \in S$ with $\alpha' < \alpha$ and $(\alpha', \delta) \notin [h]$. By Lemma 3.2, for some $0 \neq a \in R$, the element $g' = 1 + ae_{\alpha'\gamma}$ belongs to N. Since R has no zero-divisors $\neq 0$, we have $(\alpha', \delta) \in [g'h]$; but clearly $(\alpha', \delta) \notin [hg']$, so g' and h do not commute, a contradiction.

Now we show that $A \cap B = \emptyset$. Otherwise, there are $g, h \in N$ and some $\xi \in [g]_2 \cap [h]_1$. Write $g = 1 + a$, $h = 1 + b$ with $a, b \in T^*$. Choose $\alpha, \beta \in S$ with $(\alpha, \xi) \in [g]$, $(\xi, \beta) \in [h]$. By the observation of the preceding paragraph, ξ is a maximal element of $[h]_1$ and a minimal element of $[g]_2$. Now if $\gamma \in S$ with $(\alpha, \gamma) \in [g]$ and $(\gamma, \beta) \in [h]$, then $\gamma \leq \xi$ or $\xi < \gamma$ by local linearity of S, and so $\gamma = \xi$ by the maximality, respectively, minimality property of ξ. Since R has no proper zero-divisors, we have $(\alpha, \beta) \in [ab]$. However, $gh = hg$ implies $ab = ba$. Thus there is $\delta \in S$ with $(\alpha, \delta) \in [b]$ and $(\delta, \beta) \in [a]$. Again, since S is locally linear, we have $\alpha < \delta \leq \xi < \beta$ or $\alpha < \xi < \delta < \beta$. Then $\delta = \xi$ by the observation noted above. Thus (α, ξ), $(\xi, \beta) \in [a]$. By the preceding argument (with $a = b$), we get $(\alpha, \beta) \in [a^2]$. So $a^2 \neq 0$, contradicting Theorem 3.1. Hence (A, B) is a couple.

Now we can characterize group-theoretically the maximal order-theoretic normal subgroups of subgroups U of $1 + T^*$ containing $1 + \rho T$. Their forms turn out to be independent of the choice of U.

Corollary 3.5 *Let U be an arbitrary subgroup of $1+T^*$ with $1+\rho T \subseteq U$, and let $N \subseteq U$. Then the following are equivalent:*

1. *N is maximal with respect to being abelian and containing G in its normalizer.*
2. *N is a maximal abelian normal subgroup of U.*
3. *$N = N_{A,B}$ for some maximal couple (A, B).*

Proof. Subsequently we use that by Lemma 3.3(a) all order-theoretic normal subgroups $N_{A,B}$ are contained in $1+\rho T \subseteq U$. We first show that either of (1) or (2) implies (3). By Theorem 3.4, N is contained in some order-theoretic normal subgroups $N_{A,B}$ of $1+T^*$. Then $N = N_{A,B}$ by maximality of N. By Lemma 3.3, it follows that (A, B) is a maximal couple. Next we show that (3) implies (1) and (2). As noted before, $N_{A,B}$ is an abelian normal subgroup of $1+T^*$ and hence of U. If $N \subseteq M \subseteq 1+T^*$, the normalizer of M contains G, and M is abelian, then $M \subseteq N_{A',B'}$ for some couple (A', B') by Theorem 3.4. Hence $(A, B) \subseteq (A', B')$ by Lemma 3.3. Thus $(A, B) = (A', B')$ and $N = M$.

In order to strengthen the link between the poset of order-theoretic normal subgroups of $G(R, S)$ and the underlying poset $(S, \leq)$, we need to investigate decompositions of couples into ideal couples. We say that two couples (I, F), (I', F') are *disjoint*, if there is no $(\alpha, \beta) \in (I \cap I') \times (F \cap F')$ with $\alpha < \beta$; equivalently, there is no couple (I^*, F^*) contained in both (I, F) and (I', F'). Then, in particular, $I \cap F' = \emptyset = I' \cap F$. Also, two ideal couples (I, F), (I', F') are disjoint iff either $I \cap I' = \emptyset$ or $F \cap F' = \emptyset$.

Now let $\mathcal{A} = \{(I_j, F_j) : j \in J\} \subseteq \mathcal{C}$ be a family of pairwise disjoint couples, and let $A = \bigcup_{j \in J} I_j$, $B = \bigcup_{j \in J} F_j$. Then $(A, B) \in \mathcal{C}$. We will write $(A, B) = \bigoplus_{j \in J} (I_j, F_j)$ if $(I_j, F_j) \subseteq (A, B)$ for each $j \in J$ and for all $\alpha \in A$, $\beta \in B$ with $\alpha < \beta$ there is $j \in J$ with $\alpha \in I_j$, $\beta \in F_j$. Furthermore, if $\emptyset \neq X \lhd R$ is a non-trivial ideal, we let $N^X_{\mathcal{A}}$ comprise all elements $g = 1+T^*$ of the form $g = 1 + \sum x_{\alpha\beta} e_{\alpha\beta}$ such that for each $(\alpha, \beta) \in [g]$ we have $x_{\alpha\beta} \in X$ and $(\alpha, \beta) \in I_j \times F_j$ for some $j \in J$. Then $N^X_{\mathcal{A}} \subseteq N^X_{A,B}$, and we have equality iff $(A, B) = \bigoplus_{j \in J} (I_j, F_j)$. In any case, arguments similar to Lemma 3.3(a) show that $N^X_{\mathcal{A}}$ is an abelian normal subgroup of L_1 contained in $1+\rho T$. We put $N_{\mathcal{A}} = N^R_{\mathcal{A}}$.

Next we show that order-theoretic normal subgroups have a well-behaved decomposition property similar to the famous Azumaya-Krull-Remak-Schmidt result on modules having summands with local endomorphism rings (cf. [12]).

Theorem 3.6 *Let $(A, B) \in \mathcal{C}$ and $\mathcal{A} = \{(I_j, F_j) : j \in J\} \subseteq \mathcal{C}$ such that $(A, B) = \bigoplus_{j \in J} (I_j, F_j)$. Let $0 \neq X \lhd R$ be a non-trivial ideal.*

(a) If $\mathcal{A} = \mathcal{A}_1 \cup \mathcal{A}_2$ (disjoint union), then $N^X_{A,B} = N^X_{\mathcal{A}_1} \oplus N^X_{\mathcal{A}_2}$.

(b) Let $U \subseteq 1+T^*$ be an arbitrary subgroup with $1+\rho T \subseteq U$. Let $\mathcal{A} \subseteq \mathcal{C}^*$ and assume $N^X_{A,B} = N_1 \oplus N_2$ with normal subgroups $N_1, N_2 \lhd U$. Then there exists a decomposition $\mathcal{A} = \mathcal{A}_1 \cup \mathcal{A}_2$ such that $N_i = N^X_{\mathcal{A}_i}$ for $i = 1, 2$.

Proof. (a) If $g = 1 + \sum x_{\alpha\beta} e_{\alpha\beta} \in N^X_{A,B}$ and $(\alpha, \beta) \in [g]$, then $(\alpha, \beta) \in I_j \times F_j$ for some uniquely determined $j \in J$. For $i = 1, 2$, let $g_i = 1 + \sum_{\mathcal{A}_i} x_{\alpha\beta} e_{\alpha\beta}$, meaning that each $(\alpha, \beta) \in [g_i]$ belongs to some $I_j \times F_j$ with $(I_j, F_j) \in \mathcal{A}_i$. Then $g = g_1 \cdot g_2 \in N^X_{\mathcal{A}_1} \cdot N^X_{\mathcal{A}_2}$. Clearly $N^X_{\mathcal{A}_1} \cap N^X_{\mathcal{A}_2} = \{1\}$.

(b) For $i = 1, 2$ let $\mathcal{A}_i$ comprise all couples $(I, F) \in \mathcal{A}$ for which there are $\alpha \in I$, $\beta \in F$ and $g \in N_i$ with $(\alpha, \beta) \in [g]$. Fix some $0 \neq x \in X$. We first show that $\mathcal{A} = \mathcal{A}_1 \cup \mathcal{A}_2$. Let $(I, F) \in \mathcal{A}$ and choose $\alpha \in I$ and $\beta \in F$. Then $1 + xe_{\alpha\beta} = g_1 \cdot g_2$ with $g_i \in N_i$, in particular $[g_i] \subseteq A \times B$ $(i = 1, 2)$. Hence $(\alpha, \beta) \in [g_i]$ for some $i \in \{1, 2\}$, so $(I, F) \in \mathcal{A}_i$.

Next we show that $\mathcal{A}_1$ and $\mathcal{A}_2$ are disjoint. Suppose there was $(I, F) \in \mathcal{A}_1 \cap \mathcal{A}_2$. For $i = 1, 2$, choose $g_i \in N_i$ and $(\alpha_i, \beta_i) \in [g_i]$ with $\alpha_i \in I$, $\beta_i \in F$. Let $r_i \in R$ be the coefficient of $e_{\alpha_i \beta_i}$ in the representation of $g_i = 1 + \sum r_{\alpha\beta} e_{\alpha\beta}$. There is $\alpha \in I$ and $\beta \in F$ with $\{\alpha_1, \alpha_2\} \leq \alpha < \beta \leq \{\beta_1, \beta_2\}$. Then $1 + xe_{\alpha\beta} \in N^X_{A,B}$, so $1 + xe_{\alpha\beta} = h_1 \cdot h_2$ for some $h_i \in N_i$ $(i = 1, 2)$. As above, $(\alpha, \beta) \in [h_1] \cup [h_2]$, and we may assume that $(\alpha, \beta) \in [h_1]$. Let $q \in R$ be the coefficient of $e_{\alpha\beta}$ in the representation of $h_1 = 1 + \sum q_{\sigma\tau} e_{\sigma\tau}$. Now choose $\gamma, \delta \in S$ with $\gamma < \alpha_2 \leq \alpha < \beta \leq \beta_2 < \delta$. Then $q \cdot r_2 \neq 0$ by assumption $(*)$ and by Lemma 3.2 we have $1 \neq 1 + q \cdot r_2 e_{\gamma\delta} \in N_1 \cap N_2$, a contradiction.

By (a) and $\mathcal{A} = \mathcal{A}_1 \cup \mathcal{A}_2$ we have $N^X_{A,B} = N^X_{\mathcal{A}_1} \oplus N^X_{\mathcal{A}_2}$. Clearly, by definition of $\mathcal{A}_i$, each element of N_i belongs to $N^X_{\mathcal{A}_i}$ $(i = 1, 2)$. Hence $N_i = N^X_{\mathcal{A}_i}$ $(i = 1, 2)$.

This result motivates further investigation of $\mathcal{C}^*$ and decomposition of couples into disjoint sums of ideal couples. The following lemma was shown in [5, Lemma 3.3].

Lemma 3.7 *Let $(I, F), (I', F') \in \mathcal{C}^*$ with $I \cap I' \neq \emptyset \neq F \cap F'$. Then $I \subseteq I'$ or $I' \subseteq I$. Also, $F \subseteq F'$ or $F' \subseteq F$. Moreover, if $I \backslash I' \neq \emptyset$, there is $\alpha \in I$ with $I' < \alpha$, and if $I \cap F' = \emptyset$, then $I < F'$.*

The following was shown in [5, Proposition 3.4, Lemma 3.6].

Proposition 3.8
(a) Let $(A, B) \in \mathcal{C}$ and $\mathcal{J} \subseteq \mathcal{C}^*$. The following are equivalent:

(1) $(A, B) = \bigoplus_{(I,F) \in \mathcal{J}} (I, F)$.

(2) $\mathcal{J}$ comprises all ideal couples (I, F) which are maximal with respect to $(I, F) \subseteq (A, B)$.

(b) Let $(A, B) \in \mathcal{C}$ be maximal in $\mathcal{C}$, and let $(I, F) \in \mathcal{C}^*$ be a maximal ideal couple with respect to $(I, F) \subseteq (A, B)$. Then (I, F) is maximal in $\mathcal{C}^*$.

For the rest of this section,

let $U \subseteq 1 + T^*$ be an arbitrary subgroup with $1 + \rho T \subseteq U$.

Let $N \lhd U$. If $N = N_1 \oplus N_2$ with normal subgroups $N_1, N_2 \lhd U$, we call N_i direct summands of N, denoted by $N_i \sqsubseteq N$. If N has no non-trivial proper direct summand, then N will be called *indecomposable* in U.

Corollary 3.9 *Let $(A, B) \in \mathcal{C}$ and $0 \neq X \lhd R$. Then $N^X_{A,B}$ is indecomposable in U if and only if $(A, B) \in \mathcal{C}^*$.*

Proof. Assume $N^X_{A,B}$ is indecomposable. By Proposition 3.8(a), $(A, B) = \bigoplus_{(I,F)\in\mathcal{J}}(I, F)$ for some subset $\mathcal{J} \subseteq \mathcal{C}^*$. Then by Theorem 3.6(a) we get $|\mathcal{J}| = 1$, so $(A, B) \in \mathcal{C}^*$. The converse is immediate by Theorem 3.6(b) with $\mathcal{A} = \{(A, B)\}$.

Recall that Corollary 3.5 gave a group-theoretic characterization of the order-theoretic normal subgroups $N_{A,B}$ where (A, B) is maximal couple in $\mathcal{C}$. Now we characterize the normal subgroups $N_{I,F}$ for which (I, F) is a maximal ideal couple in $\mathcal{C}^*$. This provides a strong link between the order-structure of $(S, \leq)$, represented by $(\mathcal{C}^*, \leq)$, and the poset of abelian normal subgroups of U. We will apply this result to the Baer radical of *Aut* G, which will be shown later (see Corollary 4.2) to be contained in $1 + T^*$.

Theorem 3.10 *Let $U \subseteq 1 + T^*$ be an arbitrary subgroup with $1 + \rho T \subseteq U$, and let $N \lhd U$. The following are equivalent:*

(1) $N = N_{I,F}$ for some maximal ideal couple (I, F) in $\mathcal{C}^*$.

(2) N is abelian, indecomposable and a direct summand of some maximal abelian normal subgroup $M \lhd U$.

(3) N is abelian, indecomposable and a direct summand of any abelian normal subgroup of U containing it.

Proof. (1) $\to$ (2). By Lemma 3.3 N is abelian, and by Corollary 3.9 it is indecomposable. By Zorn's lemma, there is a maximal couple (A, B) in $\mathcal{C}$ containing (I, F). Then $M = N_{A,B}$ is a maximal abelian normal subgroup of U by Corollary 3.5. Trivially, (I, F) is maximal in $\mathcal{C}^*$ with respect to $(I, F) \subseteq (A, B)$. By Proposition 3.8(a), we have $(A, B) = \bigoplus_{(I',F')\in\mathcal{J}}(I', F')$ for some $\mathcal{J} \subseteq \mathcal{C}^*$ with $(I, F) \in \mathcal{J}$. Let $\mathcal{J}_1 = \{(I, F)\}$ and $\mathcal{J}_2 = \mathcal{J}\backslash\mathcal{J}_1$. Then Theorem 3.6 implies $N = N_{\mathcal{J}_1} \sqsubseteq N_{\mathcal{J}_1} \oplus N_{\mathcal{J}_2} = M$.

(2) $\to$ (1). By assumption, we have $M = N \oplus K$ for some $K \lhd U$. By Corollary 3.5 and Proposition 3.8, we obtain $M = N_{A,B} = \bigoplus_{(I,F)\in\mathcal{J}} N_{I,F}$ for some maximal couple (A, B) in $\mathcal{C}$ and some subset $\mathcal{J} \subseteq \mathcal{C}^*$ containing all ideal couples (I, F) which are maximal in (A, B). The exchange property, Theorem 3.6 implies that $N = N_{\mathcal{J}'}$ for some $\mathcal{J}' \subseteq \mathcal{J}$. Then $|\mathcal{J}'| = 1$, since N is indecomposable. So, $N = N_{(I,F)}$ where $\mathcal{J}' = \{(I, F)\}$, and (I, F) is maximal in $\mathcal{C}^*$ by Proposition 3.8.

(2) $\to$ (3). Clear by using the modular law.

(3) $\to$ (2). Apply Zorn's lemma.

Next we characterize "principal" couples of the form $(\downarrow\alpha, \uparrow\beta)$, $(\alpha, \beta \in S,\ \alpha < \beta)$ by intersection properties of maximal ideal couples.

Lemma 3.11 *Let $(I, F) \in \mathcal{C}^*$ be an ideal couple.*

(a) There exist maximal ideal couples $(I_j, F_j) \in \mathcal{C}^*$ $(j = 1, 2)$ such that $I = I_1 \cap I_2$ and $F = F_1 \cap F_2$; they coincide iff (I, F) is a maximal ideal couple.

(b) The following are equivalent:

(1) $(I, F) = (\downarrow\alpha, \uparrow\beta)$ for some $\alpha, \beta \in S$ with $\alpha < \beta$.

(2) There are at least two distinct ideal couples $(I_j, F_j) \in \mathcal{C}^*$ $(j = 1, 2)$ maximal with respect to being properly contained in (I, F) such that $I_1 \cap I_2 \neq \emptyset$ and $F_1 \cap F_2 \neq \emptyset$.

Proof. (a) Put $I_1 = I$ and, by Zorn's lemma, let F_1 be a filter in S maximal with $F \subseteq F_1$ and $I < F_1$. Assume that (I', F_1) is an ideal couple with $I_1 \subset I'$. By Lemma 3.7, there is $\alpha \in I'$ with $I_1 < \alpha$, so $(I_1, \uparrow\alpha)$ is an ideal couple with $F_1 \subset \uparrow\alpha$, contradicting the maximality of F_1. Hence (I_1, F_1) is a maximal ideal couple. Dually, let I_2 be an ideal in S maximal with $I \subseteq I_2$ and $I_2 < F$, and put $F_2 = F$. The result follows.

(b) (1) $\rightarrow$ (2): By Zorn's lemma, let I_1 be an ideal in S maximal with $I_1 < \alpha$, and put $F_1 = \uparrow\beta$. Dually, let $I_2 = \downarrow\alpha$ and F_2 a filter in S maximal with $\beta < F_2$.

(2) $\rightarrow$ (1): Let $(I_j, F_j) \in \mathcal{C}^*$ $(j = 1, 2)$ be two distinct ideal couples maximal with respect to being properly contained in (I, F) such that $I_1 \cap I_2 \neq \emptyset \neq F_1 \cap F_2$. We may assume that $I_1 \subset I$. By Lemma 3.7, there is $\alpha \in I$ with $I_1 < \alpha$. So, $(\downarrow\alpha, F_1)$ is an ideal couple with $(I_1, F_1) \subsetneq (\downarrow\alpha, F_1) \subseteq (I, F)$. Hence $(\downarrow\alpha, F_1) = (I, F)$. Dually, if $F_2 \subset F$, we obtain that $F = \uparrow\beta$ for some $\beta \in S$ and thus the result. Therefore suppose now that $F_2 = F = F_1$. Since $I_1 \cap I_2 \neq \emptyset$, by Lemma 3.7 we have either $I_1 \subset I_2$ or $I_2 \subset I_1$, which implies a contradiction to the maximality of (I_1, F_1) or (I_2, F_2).

The following shows that non-trivial intersections of ideal couples are ideal couples.

Lemma 3.12 *Let (I, F), $(I', F') \in \mathcal{C}^*$ be ideal couples with $I \cap I' \neq \emptyset$ and $F \cap F' \neq \emptyset$. Then $(I \cap I',\ F \cap F')$ is an ideal couple.*

Proof. We only have to show that $I \cap I'$ is upper directed and $F \cap F'$ is lower directed. Choose any $\alpha, \beta \in I \cap I'$. There are $\gamma \in I$ and $\delta \in I'$ with $\{\alpha, \beta\} \leq \{\gamma, \delta\}$. Let $\varepsilon \in F \cap F'$. Then $\{\gamma, \delta\} \leq \varepsilon$. Since $(S, \leq)$ is locally linear, we have $\gamma \leq \delta$ or $\delta \leq \gamma$, hence $\gamma \in I \cap I'$ or $\delta \in I \cap I'$. Dually, $F \cap F'$ is lower directed.

Let us call a normal subgroup $N \lhd U$ a *(maximal) ideal normal subgroup*, if $N = N_{I,F}$ for some (maximal) ideal couple $(I, F) \in \mathcal{C}^*$, respectively. Observe that

Theorem 3.10 gave a group-theoretic characterization of the maximal ideal normal subgroups of U. Now we use this result and Lemma 3.11 to derive a group-theoretic characterization of the ideal normal subgroups of U, and of the normal subgroups $N_{\downarrow\alpha,\uparrow\beta}$ with $\alpha, \beta \in S$.

Corollary 3.13 *Let $N \lhd U$ be a normal subgroup.*

(a) N is an ideal normal subgroup if and only if $N = N_1 \cap N_2$ for some maximal ideal normal subgroups N_1, N_2 of U.

(b) The following are equivalent:

(1) $N = N_{\downarrow\alpha,\uparrow\beta}$ for some $\alpha, \beta \in S$ with $\alpha < \beta$.

(2) N is an ideal normal subgroup, and the collection of all ideal normal subgroups properly contained in N contains at least two maximal elements N_1, N_2 such that $N_1 \cap N_2 \neq 1$.

Proof. If (I_1, F_1), $(I_2, F_2) \in C^*$ with $I_1 \cap I_2 \neq \emptyset$ and $F_1 \cap F_2 \neq \emptyset$, we have $(I_1 \cap I_2,\ F_1 \cap F_2) \in C^*$ by Lemma 3.12 and $N_{I_1,F_1} \cap N_{I_2,F_2} = N_{I_1 \cap I_2, F_1 \cap F_2}$. Now part (a) easily follows from Lemma 3.11(a), and part (b) from Lemma 3.11(b), using also Lemma 3.3(b).

This result will be crucial in the following sections for the transfer of group-isomorphisms to order-theoretic isomorphisms.

4 Baer-automorphisms of the McLain group

In this section we show that all Baer automorphisms of the McLain group $G(R, S)$ are locally inner, under suitable assumptions on R and S. This also corrects a flaw in the proof of the corresponding result for $G(F_p, \mathbb{Q})$ in [18, Theorem 3].

Theorem 4.1 *Let R be a ring with no zero-divisors $\neq 0$ such that each automorphism of R is determined by its action on $U(R)$, and let $(S, \leq)$ be an unbounded locally linear connected poset. Then $\sigma(G) \subseteq Linn\ G$.*

Proof. Let $\vartheta \in \sigma(G)$. By Theorem 2.2, we may write ϑ in the form $\vartheta = \sigma^* \cdot \eta^* \cdot \delta \cdot \psi \cdot \tau$ with either $\tau = 1$ or τ induced by an order-anti-isomorphism $\overline{\tau}$ of $(S, \leq)$ and a ring-anti-automorphism of R, and elements $\sigma \in Aut\ S$, $\eta \in Aut\ R$, $\delta \in D$ and $\psi \in Linn\ G$. By assumption, $\langle\vartheta\rangle$ is subnormal in $G_\vartheta = G \rtimes \langle\vartheta\rangle$ and hence

$$(*) \qquad [G, \underbrace{\langle\vartheta\rangle, \ldots, \langle\vartheta\rangle}_{n \text{ times}}] = 1 \text{ for some } n \in \mathbb{N}.$$

Suppose $\tau \neq 1$, hence $\overline{\tau} \neq 1$. In order to find a contradiction, in a first step we show that $N := [G, \langle\vartheta\rangle]$ contains elements $1 + e_{\lambda\mu}$ for suitable $\lambda < \mu$ in S. Because of the commutator formula $[x,y]^z = [x \cdot z, y] \cdot [z,y]^{-1}$, clearly N is normal in G. We abbreviate $\alpha' = \alpha\sigma$ ($\alpha \in S$) and $\rho := \overline{\tau}^2 \in Aut\ S$. Let $\nu \in S$. Since $(S, \leq)$ is unbounded, we can find elements $\lambda, \mu \in S$ such that $\lambda < \nu < \mu$, $\lambda\overline{\tau} \neq \nu\rho\sigma$, and $\nu\overline{\tau} \neq \mu\rho\sigma$. Then $\alpha := \nu\overline{\tau}\sigma^{-1}$ and $\beta := \lambda\overline{\tau}\sigma^{-1}$ satisfy $\alpha < \beta$ by $\tau \neq 1$ and $\alpha \neq \mu\rho$, $\beta \neq \nu\rho$. Now choose $a \in R$. We compute

$$\begin{aligned}
[1 + ae_{\alpha\beta}, \vartheta] &= (1 - ae_{\alpha\beta}) \cdot (1 + ae_{\alpha'\beta'})^{\eta^* \delta\psi_\tau} \\
&= (1 - ae_{\alpha\beta}) \cdot (1 + a'e_{\alpha'\beta'})^{\psi_\tau} \\
&= (1 - ae_{\alpha\beta}) \cdot (1 + a'e_{\alpha'\beta'})^\tau \cdot (1 + x_{\alpha'\beta'})^\tau
\end{aligned}$$

where $a' \in R$ and $x_{\alpha'\beta'} \in X_{\alpha'\beta'}$ is obtained as in the discussion of the locally inner automorphisms of G following Theorem 2.2. Recall that $X_{\alpha',\beta'}$ contains only the elements $x_{\alpha',\beta'} \in E$ where each pair $[\delta, \varepsilon] \in [x_{\alpha',\beta'}]$ satisfies $\delta \leq \alpha' < \beta' \leq \varepsilon$ and either $\delta < \alpha'$ or $\beta' < \varepsilon$. Also observe that the correspondence $a \to a'$ is a composition of two ring-automorphisms and multiplication with a unit of R, hence it is a bijection of R which leaves $U(R)$ invariant. Also, since $\beta' = \lambda\overline{\tau} > \mu\overline{\tau}$, for any $b \in R$ we have

$$\begin{aligned}
[(1 + x_{\alpha'\beta'})^\tau, 1 + be_{\nu\rho,\mu\rho}] &= [1 + x_{\alpha'\beta'}, 1 - be_{\mu\overline{\tau},\nu\overline{\tau}}]^\tau \\
&= ((1 - x_{\alpha'\beta'})(1 + be_{\mu\overline{\tau},\nu\overline{\tau}})(1 + x_{\alpha'\beta'})(1 - be_{\mu\overline{\tau},\nu\overline{\tau}}))^\tau \\
&= (1 + be_{\mu\overline{\tau},\nu\overline{\tau}} \cdot x_{\nu\overline{\tau},\lambda\overline{\tau}})^\tau \\
&= 1 + x'_{\lambda\rho,\nu\rho} \cdot b'e_{\nu\rho,\mu\rho}
\end{aligned}$$

where $x'_{\lambda\rho,\nu\rho}$ and b', respectively, have entries obtained by transforming the ones of $x_{\nu\overline{\tau},\lambda\overline{\tau}}$, respectively, b with the ring-anti-automorphism inducing τ, and $x'_{\lambda\rho,\nu\rho} \cdot e_{\nu\rho,\mu\rho} \in X_{\lambda\rho,\mu\rho}$. Note that in contrast to the claim of [18, p. 275], this may be non-trivial even in the case $S = \mathbb{Q}$, $R = F_p$, as pointed out in Falkenberg [11, p. 77].

We have $[1 - ae_{\alpha\beta}, 1 + be_{\nu\rho,\mu\rho}] = 1$ since $\beta \neq \nu\rho$ and $\alpha \neq \mu\rho$. Using the identities $[x \cdot y, z] = [y, z]$ if $[x, z] = 1$ and $[x \cdot y, z] = [x, z]^y \cdot [y, z]$, we obtain

$$\begin{aligned}
[[1 + ae_{\alpha\beta}, \vartheta], 1 + be_{\nu\rho,\mu\rho}] &= [(1 - ae_{\alpha\beta}) \cdot (1 + a'e_{\alpha'\beta'})^\tau \cdot (1 + x_{\alpha'\beta'})^\tau, 1 + be_{\nu\rho,\mu\rho}] \\
&= [(1 + a'e_{\alpha'\beta'})^\tau \cdot (1 + x_{\alpha'\beta'})^\tau, 1 + be_{\nu\rho,\mu\rho}] \\
&= [(1 + a'e_{\alpha'\beta'})^\tau, 1 + be_{\nu\rho,\mu\rho}]^{(1+x_{\alpha'\beta'})^\tau} \cdot [(1 + x_{\alpha'\beta'})^\tau, 1 + be_{\nu\rho,\mu\rho}] \\
&= [1 - a''e_{\lambda\rho,\nu\rho}, 1 + be_{\nu\rho,\mu\rho}]^{(1+x^*_{\lambda\rho,\mu\rho})} \cdot (1 + x'_{\lambda\rho,\nu\rho} \cdot be_{\nu\rho,\mu\rho}) \\
&= (1 - a''be_{\lambda\rho,\mu\rho})^{(1+x^*_{\lambda\rho,\mu\rho})} \cdot (1 + x'_{\lambda\rho,\nu\rho} \cdot b'e_{\nu\rho,\mu\rho}) \\
&= (1 - a''be_{\lambda\rho,\mu\rho}) \cdot (1 + x'_{\lambda\rho,\nu\rho} \cdot b'e_{\nu\rho,\mu\rho}) \\
&= 1 - a''be_{\lambda\rho,\mu\rho} + x'_{\lambda\rho,\nu\rho} \cdot b'e_{\nu\rho,\mu\rho}.
\end{aligned}$$

The correspondence $a \to a''$ is again a bijection of R preserving units. In particular, if a is a unit and $b = -(a'')^{-1}$, we have

$$(**) \qquad [[1 + ae_{\alpha\beta}, \vartheta], 1 + be_{\nu\rho,\mu\rho}] = 1 + e_{\lambda\rho,\mu\rho} + x''_{\lambda\rho,\nu\rho} \cdot e_{\nu\rho,\mu\rho}.$$

Observe that

$$(1+e_{\lambda\rho,\mu\rho})^{(1+ye_{\gamma,\lambda\rho})} = 1+e_{\lambda\rho,\mu\rho} - y\cdot e_{\gamma,\mu\rho}.$$

Moreover, conjugating this further with $1+y'e_{\gamma',\lambda\rho}$ gives

$$1+e_{\lambda\rho,\mu\rho} - y\cdot e_{\gamma,\mu\rho} - y'\cdot e_{\gamma',\mu\rho}, \text{ if } \gamma,\ \gamma' < \lambda\rho.$$

Together with (**), this shows that a suitable conjugate of $[[1+ae_{\alpha\beta},\vartheta],1+be_{\nu\rho,\mu\rho}]$ equals $1+e_{\lambda\rho,\mu\rho}$. Hence $1+e_{\lambda\rho,\mu\rho}\in N$.

Now let $N' := [N,\langle\vartheta\rangle]$ and choose $\lambda_1<\nu_1<\mu_1$ and $\lambda_2=\lambda_1\rho\sigma\overline{\tau}^{-1}<\nu_2<\mu_2$ in S such that

(I) $\lambda_i\overline{\tau}\neq\nu_i\rho\sigma,\ \nu_i\overline{\tau}\neq\mu_i\rho\sigma$ for $i=1,2$;

(II) $\mu_1\neq\lambda_2$ and $\lambda_1\neq\mu_2$.

By (I) and the argument above, we have $1+e_{\lambda_1\rho,\mu_1\rho}\in N$ and $1+e_{\lambda_2\rho,\mu_2\rho}\in N$. Now define $\alpha=\lambda_1\rho,\ \beta=\mu_1\rho,\ \lambda_3=\beta\sigma\overline{\tau}^{-1}=\mu_1\rho\sigma\overline{\tau}^{-1},\ \nu_3=\alpha\sigma\overline{\tau}^{-1}=\lambda_1\rho\sigma\overline{\tau}^{-1}=\lambda_2,\ \mu_3=\mu_2$. Then $\lambda_3<\nu_3<\mu_3$ and $\lambda_3\overline{\tau}=\mu_1\rho\sigma\neq\lambda_2\rho\sigma=\nu_3\rho\sigma,\ \nu_3\overline{\tau}=\lambda_2\overline{\tau}=\lambda_1\rho\sigma\neq\mu_2\rho\sigma=\mu_3\rho\sigma$ by (II). Also $\alpha=\lambda_1\rho=\nu_3\overline{\tau}\sigma^{-1}$ and $\beta=\mu_1\rho=\lambda_3\overline{\tau}\sigma^{-1}$. Hence, by the argument above, N' contains the element $[[1+e_{\alpha\beta},\vartheta],1+e_{\nu_3\rho,\mu_3\rho}]$ and so also $1+e_{\lambda_3\rho,\mu_3\rho}$. Applying the same argument $(n-2)$ times again, we obtain that $[G,\langle\vartheta\rangle,\ldots,\langle\vartheta\rangle]\neq 1$, a contradiction to (*). Thus $\tau=1$.

Next we show that $\sigma=1$. Suppose $\sigma\neq 1$. There is $\lambda\in S$ with $\lambda\neq\lambda\sigma$, and we may assume that $\lambda<\lambda\sigma$. For $n\in\mathbb{N}$ let $\lambda_n=\lambda\sigma^n$. Then $\lambda=\lambda_0<\lambda_1<\lambda_2<\ldots$. Let $t=1+e_{\lambda\lambda_1}$ and $t_n=t^{\vartheta^n}$. Then we can write t_n in the form $t_n=1+b_ne_{\lambda_n\lambda_{n+1}}+x_n$ with $b_n\in R$ and $x_n\in X_{\lambda_n,\lambda_{n+1}}$. Also

$$t_{n+1}=t_n^{\vartheta}=(1+b_ne_{\lambda_{n+1},\lambda_{n+2}}+y_n)^{\eta^*\delta\psi}$$

with $y_n\in X_{\lambda_{n+1},\lambda_{n+2}}$. Comparing this with the coefficient of $e_{\lambda_{n+1},\lambda_{n+2}}$ in t_{n+1} shows that b_{n+1} is obtained by applying to b_n two ring-automorphisms and multiplication with a unit of R. Inductively, we obtain $b_n\neq 0$ for each $n\in\mathbb{N}$. Now consider the element $[t,t_1,\ldots,t_n]$. By the special form of t_n, Lemma 3.2, and the induction, we obtain that $[t,t_1,\ldots,t_n]\in N_{\lambda,\lambda_n}$ and the coefficient of $e_{\lambda,\lambda_{n+1}}$ is $b_1\cdot b_2\cdot\ldots\cdot b_n\neq 0$. Hence $[t,t_1,\ldots,t_n]\neq 1$ for each $n\in\mathbb{N}$. Thus the subgroup $\langle t,\vartheta\rangle$ of G_ϑ is not nilpotent and consequently G_ϑ is not locally nilpotent. However, since G is a Baer-group (which coincides with its Baer radical) and ϑ a Baer-automorphism, G_ϑ is a Baer-group and hence locally nilpotent, a contradiction. Thus $\sigma=1$.

Finally we show that $\eta^*\delta=1$. Since $\vartheta=\eta^*\cdot\delta\cdot\psi$, for any $a\in R$, $\lambda<\mu$ in S and $x\in X_{\lambda\mu}$ with $e=e_{\lambda\mu}$ we have $(1+ae+x)^{\vartheta}=1+a'e+x'$, and $x'\in X_{\lambda\mu}$, where the correspondence $a\to a'=a\rho\cdot c$ is the composition of a ring-automorphism $\rho=\eta\cdot\psi_\lambda$ and multiplication with a unit $c=c_{\lambda\mu}$ of R. In particular, $a\neq 0$ implies $a'\neq 0$. Thus $[1+ae+x,\vartheta]=(1-ae-x)\cdot(1+a'e+x')=1+(a'-a)e-x+x'$.

Suppose ϑ is not locally inner. Then there is $a_0 \in R$ with $a_0' \neq a_0$. Let $G_n = [G, \underbrace{\langle\vartheta\rangle, ..., \langle\vartheta\rangle}_{n \text{ times}}]$. Since $\langle\vartheta\rangle$ is subnormal in G_ϑ, we have $G_n = 1$ for some $n \in \mathbb{N}$. We now construct a sequence of elements $y_i \in G$ as follows. Put $y_0 = 1 + (a_0' - a_0)e$. Suppose we have constructed y_i of the form $y_i = 1 + (a' - a)e + x$ for some $a \in R$ and $x \in X_{\lambda\mu}$ such that $w := a' - a \neq 0$. Our first choice for y_{i+1} is $y_{i+1} := [y_i, \vartheta] = 1 + (w' - w)e + x'$, provided $w \neq w'$. If $w = w'$, then we consider $y(d) := 1 + w \cdot de + x_d$ where $d \in U(R)$, $x_d \in X_{\lambda\mu}$. Observe that $y(d) = y_i^\alpha$ for a suitable choice of $\alpha \in Aut\ G$ and $x_d \in X_{\lambda\mu}$. Then consider $[y(d), \vartheta] = 1 + ((w \cdot d)' - w \cdot d)e + x_d'$. If $(w \cdot d)' \neq w \cdot d$ for some $d \in U(R)$, then we choose such an element d and call $y_{i+1} := [y(d), \vartheta]$. In any case $y_{i+1} \in G_{i+1}^\beta$ for a suitable $\beta \in Aut\ G$ (composed of some of the α's). Since $G_n = 1$, it follows that the above construction process stops, say at step $i + 1$. So, for some $a \in R$ with $w = a' - a \neq 0$ and all $d \in U(R)$ we have

(1) $w = w'$ and

(2) $w \cdot d = (w \cdot d)'$.

Next we determine the solutions of these equations. We have

$(w \cdot d)' = w \cdot d = w' \cdot d$, and

$w' = w^\rho c$, $(w \cdot d)' = w^\rho \cdot d^\rho \cdot c$.

Hence

$w^\rho \cdot d^\rho \cdot c = w^\rho \cdot c \cdot d$.

Since $w \neq 0$ and R has no zero-divisors, it follows that $d^\rho = cdc^{-1}$, for all $d \in U(R)$. Since, by assumption on R, ring-automorphisms are determined uniquely by their action on $U(R)$, ρ is conjugation with c^{-1}. Hence, for any $r \in R$ we have $r' = r^\rho c = c \cdot r$. Now equation (1) implies $a'' - 2a' + a = 0$, so $(c^2 - 2c + 1)a = 0$. Since $a \neq 0$ and R has no zero-divisors, this implies $c = 1$. Hence also $\rho = 1$ and in particular $a_0' = a_0$, a contradiction.

As a consequence, we obtain the following chain of inclusions for the group $1+\rho T$ induced by the Jacobson radical, the Baer-radicals $\beta(1 + T^*)$ and $\beta(Aut\ G)$, the group $\sigma(G)$ of Baer-automorphisms and the locally inner automorphisms $Linn\ G = 1 + T^*$ of $Aut\ G$.

Corollary 4.2 *Let R be a ring with no zero-divisors $\neq 0$ such that each automorphism of R is determined by its action on $U(R)$. Let $(S, \leq)$ be an unbounded locally linear connected poset. Then*

$$G \subseteq 1 + \rho T \subseteq \beta(1 + T^*) \subseteq \beta(Aut\ G) \subseteq \sigma(G) \subseteq 1 + T^* \subseteq Aut\ G.$$

Proof. We noted before that clearly $G = 1+E \subseteq 1+\rho T$. By Lemma 2.5, we have $1 + \rho T \subseteq \beta(1 + T^*)$. The next inclusion is immediate by Corollary 2.4. The Baer radical of $Aut\ G$ is generated by cyclic subnormal subgroups $\langle z\rangle$ of $Aut\ G$, which are clearly subnormal in $\langle z, G\rangle$ and hence contained in $\sigma(G)$. Finally, $\sigma(G) \subseteq 1+T^*$ by Theorem 4.1 and Corollary 2.3.

This result will be crucial in the proof of our main theorem.

5 Proof of the main result

In this section we will show that if two McLain groups $G(R,S)$ and $G(R',S')$ have isomorphic automorphism groups, then the McLain groups themselves are isomorphic. In fact, we show that the underlying posets and rings are either isomorphic or anti-isomorphic. The argument will be based on Corollary 4.2 and the results of section 3, in particular the purely group-theoretic characterization of ordered pairs $(\alpha,\beta) \in S \times S$ given in Theorem 3.10 and Corollary 3.13.

Therefore we make the following assumption throughout this section:

(5.0) R, R' are rings with no zero-divisors $\neq 0$ such that each automorphism of R (resp. R') is determined by its action on $U(R)$ (resp. $U(R')$), and $(S,\leq)$, $(S',\leq)$ are unbounded locally linear connected posets.

Furthermore, let $G = G(R,S)$, $G' = G(R',S')$, and $A = Aut\ G$, $A' = Aut\ G'$ and assume $\varphi : A \to A'$ is a fixed isomorphism. We will abbreviate $N_{\alpha\beta} = N_{\downarrow\alpha,\uparrow\beta}$ for $\alpha,\beta \in S$, $\alpha < \beta$. Note that by Lemma 3.3(b), $N_{\alpha\beta} = N_{\gamma\delta}$ iff $\alpha = \gamma$ and $\beta = \delta$.

Proposition 5.1 *Let $\alpha,\beta \in S$ with $\alpha < \beta$. Then $N_{\alpha\beta}\varphi = N_{\alpha'\beta'}$ for some $\alpha',\beta' \in S$ with $\alpha' < \beta'$.*

Proof. Clearly $(\beta A)\varphi = \beta A'$, i.e. φ maps the Baer-radical of $Aut\ G$ onto that of $Aut\ G'$. By Corollary 4.2, we have $1 + \rho T \subseteq \beta A \subseteq 1 + T^* \subseteq A$, and by Theorem 3.10 and Corollary 3.13 the groups $N_{\alpha\beta}$ can be characterized group-theoretically within βA. This implies the result.

Now put $\mathcal{J} = \mathcal{J}(S) = \{[\alpha,\beta] : \alpha,\beta \in S,\ \alpha < \beta\}$ where $[\alpha,\beta] = \{\sigma \in S : \alpha \leq \sigma \leq \beta\}$, an interval in S, and similarly $\mathcal{J}' = \mathcal{J}(S')$. We define a mapping $\overline{\varphi} : \mathcal{J} \to \mathcal{J}'$ by

$$[\alpha,\beta]\overline{\varphi} = [\alpha',\beta'] \quad \text{if} \quad N_{\alpha\beta}\varphi = N_{\alpha'\beta'}.$$

By Proposition 5.1 and Lemma 3.3(b), $\overline{\varphi}$ is well-defined and bijective.

We will show that this map is induced by an isomorphism or anti-isomorphism between S and S'. For this, we need the following commutator formulas:

Lemma 5.2 *Let $\alpha,\beta,\gamma,\delta \in S$ with $\alpha < \beta$ and $\gamma < \delta$.*

(a) Let $a = 1 + \sum_i a_i e_{\alpha_i\beta_i} \in N_{\alpha\beta}$ and $b = 1 + \sum_j b_j e_{\gamma_j\delta_j} \in N_{\gamma\delta}$. Then

$$[a,b] = 1 + \sum_{i,j} a_i b_j \cdot e_{\alpha_i\beta_i} \cdot e_{\gamma_j\delta_j} - \sum_{i,j} b_j \cdot a_i \cdot e_{\gamma_j\delta_j} e_{\alpha_i\beta_i};$$

(b) $[N_{\alpha\beta}, N_{\gamma\delta}] \neq 1$ *if and only if* $\beta \leq \gamma$ *or* $\delta \leq \alpha$.

Proof. (a) Easy, using $a^{-1} = 1 - \sum_i a_i e_{\alpha_i\beta_i}$, similarly b^{-1}.

(b) If $\beta \leq \gamma$, we have $1 \neq 1 + e_{\alpha\delta} = [1 + e_{\alpha\beta}, 1 + e_{\beta\delta}] \in [N_{\alpha\beta}, N_{\gamma\delta}]$ by (a). The converse is also clear by (a).

Lemma 5.3 *(a) Let* $I, J \in \mathcal{J}$. *If* $I \leq J$ *then either* $I\overline{\varphi} \leq J\overline{\varphi}$ *or* $J\overline{\varphi} \leq I\overline{\varphi}$. *Moreover* $I \subseteq J$ *iff* $I\overline{\varphi} \subseteq J\overline{\varphi}$.

(b) Let $\alpha < \beta < \gamma$ *in* S *and* $[\alpha', \beta'] = [\alpha, \beta]\overline{\varphi}$. *Then there is* $\gamma' \in S'$ *such that either* $[\beta, \gamma]\overline{\varphi} = [\beta', \gamma']$ *and* $[\alpha, \gamma]\overline{\varphi} = [\alpha', \gamma']$ *or* $[\beta, \gamma]\overline{\varphi} = [\gamma', \alpha']$ *and* $[\alpha, \gamma]\overline{\varphi} = [\gamma', \beta']$.

(c) Let $\{\alpha, \delta\} < \beta$ *in* S *and let* $[\alpha', \beta'] = [\alpha, \beta]\overline{\varphi}$ *and* $[\delta', \beta''] = [\delta, \beta]\overline{\varphi}$. *Then either* $\beta' = \beta''$ *or* $\delta' = \alpha'$.

Proof. (a) Consider $I = [\alpha, \beta]$ and $J = [\gamma, \delta]$, first with $\beta \leq \gamma$. Let $[\alpha', \beta'] = I\overline{\varphi}$ and $[\gamma', \delta'] = J\overline{\varphi}$. By Lemma 5.2, we have $[N_{\alpha\beta}, N_{\gamma\delta}] \neq 1$. Applying φ, we obtain $[N_{\alpha'\beta'}, N_{\gamma'\delta'}] \neq 1$. Again by Lemma 5.2, this shows that either $\beta' \leq \gamma'$ or $\delta' \leq \alpha'$. Hence $I\overline{\varphi} \leq J\overline{\varphi}$ or $J\overline{\varphi} \leq I\overline{\varphi}$. Now let $I \subseteq J$. Then $\gamma \leq \alpha < \beta \leq \delta$, so $N_{\gamma\delta} \subseteq N_{\alpha\beta}$. Hence $N_{\gamma'\delta'} \subseteq N_{\alpha'\beta'}$, showing $\gamma' \leq \alpha' < \beta' \leq \delta'$ and thus $I\overline{\varphi} \subseteq J\overline{\varphi}$. The converse follows dually.

(b) Let $[\alpha^*, \gamma'] = [\alpha, \gamma]\overline{\varphi}$. By (a), we have $[\alpha', \beta'] \subseteq [\alpha^*, \gamma']$, so $\alpha^* \leq \alpha' < \beta' \leq \gamma'$. Suppose we had $\alpha^* < \alpha'$ and $\beta' < \gamma'$. Then by (a), $J' = [\beta, \gamma]\overline{\varphi}$ satisfies $J' \subseteq [\alpha^*, \gamma']$ and either $J' \leq [\alpha', \beta']$ or $[\alpha', \beta'] \leq J'$. Thus either $J' \subseteq [\alpha^*, \alpha']$ or $J' \subseteq [\beta', \gamma']$. In any case there is an interval contained in $[\alpha^*, \gamma']$ and disjoint (except, possibly, for endpoints) to J' and $[\alpha', \beta']$. But there is no interval contained in $[\alpha, \gamma]$ disjoint (except for endpoints) to $[\beta, \gamma]$ and $[\alpha, \beta]$, a contradiction to the properties of $\overline{\varphi}$.

Hence either $\alpha^* = \alpha'$ or $\beta' = \gamma'$, but not both. If $\alpha^* = \alpha'$, we obtain $[\beta, \gamma]\overline{\varphi} \subseteq [\alpha', \gamma']$ and $[\alpha', \beta'] \leq [\beta, \gamma]\overline{\varphi}$, so $[\beta, \gamma]\overline{\varphi} \subseteq [\beta', \gamma']$. By a similar argument as before, it follows that this inclusion cannot be proper, proving our claim. If $\alpha^* < \alpha'$ and $\beta' = \gamma'$, similarly we get $[\beta, \gamma]\overline{\varphi} = [\alpha^*, \alpha']$, and our claim.

(c) Choose $\gamma \in S$ with $\beta < \gamma$. We may assume, by (b), that $[\beta, \gamma]\overline{\varphi} = [\beta', \gamma']$ and $[\alpha, \gamma]\overline{\varphi} = [\alpha', \gamma']$ where $\alpha' < \beta' < \gamma'$ in S' (in the other case argue dually). Again by (b), we have either $\beta'' = \beta'$ or $\delta' = \gamma'$. The latter case, however, is impossible since then $\alpha' < \beta' < \gamma' < \beta''$, so $[\alpha, \beta] \leq [\delta, \beta]$ or $[\delta, \beta] \leq [\alpha, \beta]$ by (a), a contradiction.

As a consequence we obtain:

Corollary 5.4 *Either for all* $I, J \in \mathcal{J}$, $I \subseteq J$ *implies* $I\overline{\varphi} \subseteq J\overline{\varphi}$, *or for all* $I, J \in \mathcal{J}$, $I \subseteq J$ *implies* $J\overline{\varphi} \subseteq I\overline{\varphi}$.

Proof. Let $\alpha < \beta < \gamma < \delta$ in S and put $I = [\alpha, \beta]$, $J = [\beta, \gamma]$, $K = [\gamma, \delta]$. Assume that $I\overline{\varphi} \leq J\overline{\varphi}$. Lemma 5.3 shows that then $J\overline{\varphi} \leq K\overline{\varphi}$. Since S and S'

are connected and unbounded, all intervals can be connected and we obtain that $I' \subseteq J'$ implies $I'\overline{\varphi} \subseteq J'\overline{\varphi}$ for all $I', J' \in \mathcal{J}$. If $J\overline{\varphi} \leq I\overline{\varphi}$, we argue similarly.

We now define a mapping $\varphi' : S \to S'$ *induced* by φ as follows. Assume first that $I\overline{\varphi} \leq J\overline{\varphi}$ for all $I, J \in \mathcal{J}$ with $I \leq J$. Let $\alpha \in S$. Choose any $\beta \in S$ with $\alpha < \beta$ and let $[\alpha', \beta'] = [\alpha, \beta]\overline{\varphi}$. We put $\alpha\varphi' = \alpha'$. By Lemma 5.3(c) φ' is well-defined. If $\gamma \in S$ with $\beta < \gamma$, then $[\beta, \gamma]\overline{\varphi} = [\beta', \gamma']$ by (5.3)(b), so $\beta\varphi' = \beta' > \alpha' = \alpha\varphi'$ and φ' is order-preserving. Since φ is an isomorphism, φ' is an order-isomorphism.

Now assume, by Corollary 5.4, the alternative case that $J\overline{\varphi} \leq I\overline{\varphi}$ for all $I, J \in \mathcal{J}$ with $I \leq J$. Let $\alpha \in S$. Choose $\beta \in S$ with $\alpha < \beta$ and let $[\alpha', \beta'] = [\alpha, \beta]\overline{\varphi}$. We put $\alpha\varphi' = \beta'$. A similar argument shows that $\varphi' : S \to S'$ is an order-anti-isomorphism. Hence we have:

Proposition 5.5 *Under the assumption (5.0), the induced mapping $\varphi' : S \to S'$ is either an order-isomorphism or an order-anti-isomorphism satisfying, correspondingly, $N_{\alpha\beta}\varphi = N_{\alpha\varphi',\beta\varphi'}$ for all $\alpha < \beta$ in S, or $N_{\alpha\beta}\varphi = N_{\beta\varphi',\alpha\varphi'}$ for all $\alpha < \beta$ in S.*

Subsequently, we will write $\alpha' = \alpha\varphi'$, $\beta' = \beta\varphi'$ for $\alpha < \beta$ in S. We let $Y_{\alpha\beta}$ comprise all elements $y \in T^*$ such that for any $(\gamma, \delta) \in [y]$ we have $\gamma \leq \alpha < \beta \leq \delta$ and either $\gamma < \alpha$ or $\beta < \delta$. (Thus $Y_{\alpha\beta} \cap E = X_{\alpha\beta}$ defined before.) Next we turn to the proof that R and R' are isomorphic or anti-isomorphic. Let $\alpha < \beta$ in S. We define a mapping $\varphi_{\alpha\beta} : R \to R'$ as follows. For $a \in R$, we have $g = 1 + ae_{\alpha\beta} \in N_{\alpha\beta}$. First let φ' be order-preserving. Then $g\varphi \in N_{\alpha'\beta'}$. So, $g\varphi = 1 + a'e_{\alpha'\beta'} + y$ where $y \in Y_{\alpha'\beta'}$. Then put $a\varphi_{\alpha\beta} = a'$. Secondly, if φ' is order-reversing, we have $g\varphi \in N_{\beta',\alpha'}$. Hence $g\varphi = 1 + a'e_{\beta'\alpha'} + y$ where $y \in Y_{\beta'\alpha'}$. Again put $a\varphi_{\alpha\beta} = a'$.

Proposition 5.6 *(a) For any $\alpha < \beta$ in S, $\varphi_{\alpha\beta} : R^+ \to R'^+$ is an isomorphism.*
(b) For any $\alpha < \beta < \gamma$ in S and $a, b \in R$, the following equation holds:

$$\begin{aligned}(a\varphi_{\alpha\beta}) \cdot (b\varphi_{\beta\gamma}) &= (ab)\varphi_{\alpha\gamma}, \text{ if } \varphi' \text{ is order-preserving, and} \\ (b\varphi_{\beta\gamma}) \cdot (a\varphi_{\alpha\beta}) &= -(ab)\varphi_{\alpha\gamma}, \text{ if } \varphi' \text{ is order-reversing.}\end{aligned}$$

Proof. (a) For any $a, b \in R$ we have $(1 + ae_{\alpha\beta}) \cdot (1 + be_{\alpha\beta}) = 1 + (a + b)e_{\alpha\beta}$. Applying φ and computing the product in G', we get $(a + b)\varphi_{\alpha\beta} = a\varphi_{\alpha\beta} + b\varphi_{\alpha\beta}$. We now show that $\varphi_{\alpha\beta}(\varphi^{-1})_{\alpha'\beta'} = id_R$. Let $a \in R$ and $g = 1 + ae_{\alpha\beta}$. Then, as noted above, $g\varphi = 1 + (a\varphi_{\alpha\beta})e_{\alpha'\beta'} + y'$ where $y' \in Y'_{\alpha'\beta'}$. Since $e_{\alpha'\beta'} \cdot y = 0$, we have $g\varphi = (1 + (a\varphi_{\alpha\beta})e_{\alpha'\beta'}) \cdot (1 + y')$. Applying φ^{-1} to this equation, we obtain $g = (1 + a\varphi_{\alpha\beta}(\varphi^{-1})_{\alpha'\beta'}e_{\alpha\beta} + y) \cdot (1 + y')\varphi^{-1}$ where $y \in Y_{\alpha\beta}$ and $(1 + y')\varphi^{-1} \in 1 + Y_{\alpha\beta}$. So $a\varphi_{\alpha\beta}(\varphi^{-1}_{\alpha'\beta'}) = a$. It follows that $\varphi_{\alpha\beta}$ is an isomorphism.

(b) We have $[1 + ae_{\alpha\beta}, 1 + be_{\beta\gamma}] = 1 + abe_{\alpha\gamma}$ by Lemma 5.2(a). First let φ' be order-preserving. Applying φ, we get $(1 + ae_{\alpha\beta})\varphi = 1 + (a\varphi_{\alpha\beta})e_{\alpha'\beta'} + y'$ and $(1 + be_{\beta\gamma})\varphi = 1 + (b\varphi_{\beta\gamma})e_{\beta'\gamma'} + z'$. So, computing the commutator in G' with Lemma 5.2(a), we have $(1 + abe_{\alpha\gamma})\varphi = [(1 + ae_{\alpha\beta})\varphi, (1 + be_{\beta\gamma})\varphi] = 1 + (a\varphi_{\alpha\beta})(b\varphi_{\beta\gamma})e_{\alpha'\gamma'} +$

x' where $x' = (a\varphi_{\alpha\beta}e_{\alpha'\beta'}) \cdot z' + y' \cdot (b\varphi_{\beta\gamma})e_{\beta'\gamma'} + y' \cdot z'$ and $x' \in Y'_{\alpha'\beta'}$. Hence $(a\varphi_{\alpha\beta})(b\varphi_{\beta\gamma}) = (ab)\varphi_{\alpha\gamma}$. If φ' is order-reversing, perform a similar calculation.

For any $\alpha < \beta$ in S, we put $c_{\alpha\beta} = 1\varphi_{\alpha\beta} \in R'$.

Proposition 5.7 *For any $\alpha < \beta$ in S, $c_{\alpha\beta}$ is a unit of R'. For $\alpha < \beta$ in S, define $\psi_{\alpha\beta} : R \to R'$ by $a\psi_{\alpha\beta} = (a\varphi_{\alpha\beta}) \cdot c_{\alpha\beta}^{-1}$ for all $a \in R$. If φ' is order-preserving, then $\psi_{\alpha\beta}$ is a ring-isomorphism. If φ' is order-reversing, then $\psi_{\alpha\beta}$ is a ring-anti-isomorphism.*

Proof. Let $\delta < \alpha < \beta < \gamma$ and assume φ' is order-preserving. By Proposition 5.6(a), $\varphi_{\alpha\gamma}$ and $\varphi_{\delta\beta}$ are bijective, so there are $y', y \in R$ with $y\varphi_{\alpha\gamma} = 1 = y'\varphi_{\delta\beta}$. Then by (5.6)(b) we obtain $1 = (1 \cdot y)\varphi_{\alpha\gamma} = c_{\alpha\beta} \cdot (y\varphi_{\beta\gamma})$ and $1 = (y' \cdot 1)\varphi_{\delta\beta} = (y'\varphi_{\delta\alpha}) \cdot c_{\alpha\beta}$. Clearly, $y'\varphi_{\delta\alpha} = (y'\varphi_{\delta\alpha}) \cdot c_{\alpha\beta} \cdot (y\varphi_{\beta\gamma}) = y\varphi_{\beta\gamma}$. So, $c_{\alpha\beta}$ is a unit in R'. By (a) and Proposition 5.6(a), $\psi_{\alpha\gamma} : R^+ \to R^+$ is an isomorphism. Now let $a, b \in R$. Then using the functional equation (5.6)(b),

$$\begin{aligned}(a\psi_{\alpha\gamma}) \cdot (b\psi_{\alpha\gamma}) &= (a \cdot 1)\varphi_{\alpha\gamma} \cdot c_{\alpha\gamma}^{-1} \cdot (1 \cdot b)\varphi_{\alpha\gamma} \cdot c_{\alpha\gamma}^{-1} \\ &= (a\varphi_{\alpha\beta}) \cdot c_{\beta\gamma} \cdot c_{\alpha\gamma}^{-1} \cdot c_{\alpha\beta} \cdot (b\varphi_{\beta\gamma}) \cdot c_{\alpha\gamma}^{-1} \\ &= (ab)\varphi_{\alpha\gamma} \cdot c_{\alpha\gamma}^{-1} \\ &= (ab)\psi_{\alpha\gamma}.\end{aligned}$$

So, $\psi_{\alpha\gamma}$ is a ring-isomorphism. If φ' is order-reversing, a similar computation applies.

Now we can summarize our results:

Main Theorem 5.8 *Let R, R' be rings with no zero-divisors $\neq 0$ such that each automorphism of R (resp. R') is determined by its action on $U(R)$ (resp. $U(R')$), and let $(S, \leq)$, $(S', \leq)$ be unbounded locally linear connected posets. Let Aut $G(R, S)$ and Aut $G(R', S')$ be isomorphic. Then $G(R, S)$ and $G(R', S')$ are isomorphic.*

Proof. By Propositions 5.5 and 5.7, we obtain either a pair of isomorphisms $\varphi' : S \to S'$, $\psi' : R \to R'$, or a pair of such anti-isomorphisms. Similarly as described in section 2, such pairs induce an isomorphism between $G(R, S)$ and $G(R', S')$.

References

[1] A. L. S. Corner, B. Goldsmith, *Isomorphic automorphism groups of torsion-free p-adic modules*, submitted.

[2] J. Dieudonné, *La géométrie des groupes classiques*, Vol. 5, Ergebnisse der Mathematik, Springer, Berlin, Heidelberg, New York, 1991.

[3] J. Dieudonné, *On the automorphisms of classical groups*, Memoirs Amer. Math. Soc. **2**(1950).

[4] M. Droste, *Structure of partially ordered sets with transitive automorphism groups*, Memoirs Amer. Math. Soc. **334**(1985).

[5] M. Droste and R. Göbel, *McLain groups over arbitrary rings and orderings*, Math. Proc. Camb. Phil. Soc. **117**(1995), 439–467.

[6] M. Droste, C. Holland, and D.Macpherson, *Automorphism groups of infinite semilinear orders (I), (II)*, Proc. London Math. Soc. (3) **58**(1989), 454–478 and 479–494.

[7] M. Dugas and R. Göbel, *Automorphism groups of torsion-free nilpotent groups of class two*, Trans. American Math. Soc. **332**(1992), 633–646.

[8] M. Dugas and R. Göbel, *All infinite groups are Galois groups over any fields*, Trans. American Math. Soc. **304**(1987), 355–384.

[9] M. Dugas and R. Göbel, *On locally finite p-groups and a problem of Philip Hall's*, J. Algebra **159**(1993), 115–138.

[10] M. Dugas and R. Göbel, *Automorphism groups of fields*, Manuscripta Mathematicae **85**(1994), 227–242.

[11] H. Falkenberg, *Die McLain-Gruppen über allgemeinen Ketten und ihre Automorphismengruppen*, Diplomarbeit, Universität Essen, 1993.

[12] R. Göbel, *Wie weit sind Moduln vom Satz von Krull-Remark-Schmidt entfernt?*, Jahresber. DMV **88**(1986), 11–49.

[13] P. Hall, *Wreath powers and characteristically simple groups*, Math. Proc. Camb. Phil. Soc. **58**(1962), 170–184.

[14] H. Leptin, *Abelsche p-Gruppen und ihre Automorphismengruppen*, Math. Zeitschr. **73**(1960), 235–253.

[15] W. Liebert, *Isomorphic automorphism groups of primary abelian p-groups*, in *Abelian Group Theory*, (R. Göbel, E. A. Walker, ed.), Gordon and Breach, London, 1987, 9–13.

[16] D. H. McLain, *A characteristically-simple group*, Math. Proc. Camb. Phil. Soc. **50**(1954), 641–642.

[17] D. J. S. Robinson, *A Course in the Theory of Groups*, Vol. **80**, Springer Graduate Texts, 1980.

[18] J. E. Roseblade, *The automorphism group of McLain's characteristically simple group*, Math. Zeitschr. **82**(1963), 267–282.

[19] J.S. Wilson, *Groups with many characteristically simple subgroups*, Math. Proc. Camb. Phil. Soc. **86**(1979), 193–197.

Locally Moving Groups and Reconstruction Problems

Matatyahu Rubin*

Department of Mathematics
Ben Gurion University of the Negev
Beer Sheva
Israel

Abstract

Let B be a complete atomless Boolean algebra. For $g \in Aut(B)$ let $var(g) \stackrel{def}{=} \sum\{a \in B \mid g(a) \cdot a = 0\}$. Let $G \leq Aut(B)$. We say that G is a locally moving subgroup of $Aut(B)$ if $\{var(g) \mid g \in G\}$ is dense in B. We show (Corollary 1.4) that for every complete atomless Boolean algebra B and a locally moving subgroup G of $Aut(B)$, the group G determines B up to isomorphism. We then use this theorem to prove that certain mathematical structures are determined by their automorphism group. In Section 3 we show such a reconstruction theorem for a certain class of locally compact topological spaces. Then we prove a reconstruction theorem for subspaces of the foliated torus. In Section 4 we present a reconstruction theorem for 2-homogeneous linear orders and circular orders.

1 Introduction

The goal of this paper is to prove a new theorem about locally moving groups, and demonstrate the role of such groups in certain reconstruction problems.

A reconstruction theorem is a result of the following form. A class K of mathematical structures is given, and it is proved that for every $M, N \in K$ and isomorphism φ between $Aut(M)$ and $Aut(N)$ there is an isomorphism τ between M and N such that for every $f \in Aut(M)$, $\varphi(f) = \tau \circ f \circ \tau^{-1}$.

We next present the notations needed for the definition of a locally moving group. A more detailed treatment of this notion and of some prerequisites will be given in Section 2.

*This work was prepared in 1994–95 while the author was a visitor in Bowling Green State University, Bowling Green, Ohio and in the University of Colorado, Boulder, Colorado. The author is most grateful to both of these institutions.

W. C. Holland (ed.), Ordered Groups and Infinite Permutation Groups, 121–157.

Notation and Definitions 1.1 Let B be a Boolean algebra (BA).

(a) The operations, units and partial ordering of B are denoted respectively by $+^B$, $\cdot^B$, $-^B$, 0^B, 1^B and $\leq^B$. If A is a subset of B, then $\sum^B A$ and $\prod^B A$ denote respectively the supremum and infimum of A in B whenever they exist.

The superscript ^{B} is omitted whenever no confusion may arise.

(b) B is *atomless*, if $\langle B - \{0\}, \leq \rangle$ has no minimal elements.

(c) B is *complete*, if $\langle B, \leq \rangle$ is a complete lattice. That is, every subset of B has an infimum and a supremum.

(d) Let $a \in B$, then $B|a \overset{def}{=} \{b \in B \mid b \leq a\}$.

(e) Let $Aut(B)$ denote the automorphism group of B. Let $g \in Aut(B)$ and $a \in B$. Then $g|a \overset{def}{=} g|(B|a)$.

(f) Suppose that B is a complete BA. Let $g \in Aut(B)$, then $fix(g) \overset{def}{=} \sum\{a \in B \mid g|a = Id\}$ and $var(g) \overset{def}{=} -fix(g)$.

(g) A subset D of B is *dense* in B, if for every $b \in B - \{0\}$ there is $d \in D - \{0\}$ such that $d \leq b$.

Definition 1.2 Let B be a complete atomless BA and $G \leq Aut(B)$. G is a *locally moving subgroup* of $Aut(B)$ if $\{var(g) \mid g \in G\}$ is dense in B. The pair $\langle B, G \rangle$ is then called a *local movement system.*

A group H is a *locally moving group* if for some complete atomless BA B, H is isomorphic to a locally moving subgroup of $Aut(B)$.

Theorem 1.3 (The Expressibility Theorem)
There are first order formulas $\varphi_{Eq}(x, y)$, $\varphi_{\leq}(x, y)$ and $\varphi_{Ap}(x, y, z)$ in the language of groups such that for every local movement system $\langle B, G \rangle$, the following holds. For every $f, g, h \in G$: $G \models \varphi_{Eq}[f, g]$ iff $var(f) = var(g)$, $G \models \varphi_{\leq}[f, g]$ iff $var(f) \leq var(g)$ and $G \models \varphi_{Ap}[f, g, h]$ iff $f(var(g)) = var(h)$.

The Expressibility Theorem can be used in answering two closely related types of questions. The first one concerns reconstruction questions, and the second one is about elementary equivalence of automorphism groups.

The following corollary is a reconstruction theorem that follows quite easily from the Expressibility Theorem. This corollary is later used as a first step in the proof of other reconstruction theorems.

Corollary 1.4 (The Reconstruction Theorem for Local Movement Systems)
For $i = 1, 2$ let $\langle B_i, G_i \rangle$ be a local movement system. Suppose that φ is an isomorphism between G_1 and G_2. Then there is an isomorphism τ between B_1 and B_2 such that for every $f \in G_1$, $\varphi(f) = \tau \circ f \circ \tau^{-1}$.

We shall discuss the elementary equivalence questions later. That discussion requires some more definitions and notation. Next we mention an isolated result (it has so far no continuations), that seems to be of interest. It also follows relatively easily from the Expressibility Theorem.

Corollary 1.5 *There is a first order sentence φ_{LM} in the language of groups such that for every group H: $H \models \varphi_{LM}$ iff H is locally moving.*

Weaker variants of the expressibilty theorem appear in [12, 13, 14, 15]. Those variants include more assumptions on G on top of its "local movingness". These weaker theorems sufficed for the various applications that were proved there. However, the present expressibility theorem has new applications that could not be proved using the weaker variants.

In the remaining part of the Introduction we describe some settings in which locally moving groups occur. In these settings the Expressibility Theorem can be used in the proof of reconstruction and elementary equivalence results. In the last part of the introduction we discuss the topic of elementary equivalence of automorphism groups.

The article is organized in such a way that the reader can skip the definitions and theorems concerning elementary equivalence, and the remaining part of the article which concerns only with reconstruction problems remains self-contained.

The proof of the Expressibility Theorem and its immediate corollaries will be presented in Section 2.

Section 3 describes two applications in topology. The first one concerning locally compact spaces is old. The second one on foliations of the torus is new.

Section 4 describes an application for linear and circular orders. All the results in that section are old. However, we believe that the unified treatment of several different structures is new, and adds something to the understanding of the phenomenon of reconstructibility of such structures. Section 4 does not use the results of Section 3. Once the part of the Introduction about homeomorphism groups has been read, Section 4 can be read independently of Section 3.

We now describe a number of settings in which locally moving groups occur. In these settings locally moving groups were used in order to prove various reconstruction theorems.

In what follows, it will be convenient to use the notion of a "faithful class".

Definition 1.6 (a) Let K be a class of pairs of the form $\langle M, G\rangle$, where M is structure and $G \leq Aut(M)$. The class K is *faithful*, if for every $\langle M_1, G_1\rangle, \langle M_2, G_2\rangle \in K$ and an isomorphism $\varphi : G_1 \cong G_2$ there is an isomorphism $\tau : M_1 \cong M_2$ such that τ induces φ.

(b) A class K of structures is said to be *faithful* if $\{\langle M, Aut(M)\rangle \mid M \in K\}$ is faithful.

Homeomorphism groups of topological spaces

Let $H(X)$ denote the group of self-homeomorphisms of a topological space X. We wish to show that under appropriate assumptions on X, X can be recovered from $H(X)$. Indeed, we shall be able to recover X from various subgroups of $H(X)$.

We associate with every topological space a complete Boolean algebra

Definition 1.7 Let X be a topological space. (a) For $U \subseteq X$, $int(U)$ and $cl(U)$ denote respectively the interior and closure of U.

A subset U of X is *regular open* in X, if $U = int(cl(U))$. Let $Ro(X)$ denote the set of regular open subsets of X.

(b) The poset $\langle Ro(X), \subseteq \rangle$ is the partial ordering of a complete BA. The operations of this BA are $U \cdot V = U \cap V$, $U + V = int(cl(U \cup V))$ and $\sim U = int(X - U)$. We thus regard $Ro(X)$ as a Boolean algebra.

(c) For $g \in H(X)$ let $g^{Ro} : Ro(X) \to Ro(X)$ be defined as follows: $g^{Ro}(U) = \{g(x) \mid x \in U\}$. For $G \subseteq H(X)$ let $G^{Ro} = \{g^{Ro} \mid g \in G\}$.

We next describe the connection between locally moving groups and the action of $H(X)$ on X.

Proposition 1.8 *Let X be a Hausdorff space.*

(a) The function $g \mapsto g^{Ro}$ is an embedding of $H(X)$ into $Aut(Ro(X))$.

We shall later regard $H(X)$ as a subgroup of $Aut(Ro(X))$.

(b) For $g \in H(X)$, $fix(g^{Ro}) = int(\{x \in X \mid x = g(x)\})$ and $var(g^{Ro}) = int(cl(\{x \in X \mid x \neq g(x)\}))$.

(c) Let G be a subgroup of $H(X)$; then G^{Ro} is a locally moving subgroup of $Aut(Ro(X))$ iff

(∗) *for every nonempty open subset U of X there is $g \in G - \{Id\}$ such that $g \restriction (X - U) = Id$.*

Note that if X is an open subset of $\mathbb{R}^n$, then $H(X)$ and many of its subgroups have property (∗) from the above proposition. This is also true for many other topological spaces.

In Theorem 3.1 we shall use this fact in order to conclude that the class of open subsets of $\mathbb{R}^n$ regarded as topological spaces is faithful. In fact, Theorem 3.1 is considerably stronger than this. The faithfulness of Euclidean manifolds is due to Whittaker [20].

Reconstruction theorems in topology have been proved by quite a number of people. The works of Whittaker [20], Ling [8] and the author [12] can be mentioned. This should not be regarded however as a description of the history of this topic.

As an example of such a result we mention here a theorem from [16].

Theorem 1.9 *Let $LIP(X)$ denote the group of bi-Lipschitz self-homeomorphisms of a metric space X. Let $K = \{\langle X, G \rangle \mid X$ is an open subset of a Banach space, and $LIP(X) \leq G \leq H(X)\}$. Then K is faithful.*

Automorphism groups of linearly ordered sets

Definition 1.10 A linear ordering $\langle L, < \rangle$ is *2-homogeneous* if $|L| > 2$ and for every $x_1 < x_2$ and $y_1 < y_2$ in L there is $g \in Aut(\langle L, < \rangle)$ such that for $i = 1, 2$, $g(x_i) = y_i$.

We regard linear orderings as topological spaces equipped with their order topology. Then $Aut(\langle L, <\rangle)$ is a subgroup of $H(L)$. It is easy to see that if $\langle L, <\rangle$ is 2-homogeneous, then $Aut(\langle L, <\rangle)$ is a locally moving subgroup of $Aut(Ro(L))$.

It will follow easily from this fact that up to the reversal of its ordering, the Dedekind completion of a 2-homogeneous linear ordering is determined by its automorphism group.

The Reconstruction problem of 2-homogeneous linear orders has a long history. Holland [6] proved the reconstructibility of the completion of a chain from its automorphism group regarded as lattice ordered group. Then Rabinovich [11] 1975 and McCleary [10] 1978 independently proved the reconstructibility of the completion of a chain from the automorphism group regarded only as a group. McCleary also showed there that the chain is determined by rich enough subgroups of its full automorphism group.

The question of distinguishing between automorphism groups of chains by first order sentences was dealt with by Gurevich and Holland in [4]. Among other things they proved that there are first order sentences in the language of group theory that distinguish $Aut(\mathbb{R})$ and $Aut(Q)$ from the automorphism groups of all other 2-homogeneous chains (the irrationals excluded).

In Section 4 we also deal with circular orders. These are structures which are obtained by turning linear orders into circles. The exact definition appears in 4.1. I do not know whether the reconstruction result for 2-homogeneous circular orders has ever been explicitly published. In any case, it is a special case of [12] Theorem 3.5. It may have been known before.

Recently McCleary and the author strengthened the known reconstruction results on orders and circular orders. We proved that the Dedekind completion of such structures is determined up to order or orientation reversal by any subgroup of the automorphism group of the structure which is approximately 2-transitive and locally moving. The reconstruction uses only first order formulas. These results will appear elsewhere. (A group of automorphisms of a chain or a circular order L is *approximately 2-transitive* if for every distinct $x_1, x_2 \in L$ and open intervals $I_1, I_2 \subseteq L$: if there are $y_1 \in I_1$ and there $y_2 \in I_2$ such that the map $x_i \mapsto y_i$, $i = 1, 2$ is a partial isomorphism, then there is $g \in G$ such that for every $i = 1, 2$, $g(x_i) \in I_i$.) A variant of the notion of approximate transitivity was defined by Bieri and Strebel in [2]. There, they reconstruct the real line from subgroups that have stronger approximate transitivity properties.

Automorphism groups of trees

There are different mathematical objects that are called trees. Here the notion of a tree stands for a certain type of a partially ordered sets. However, graphs which are trees and structures with a betweenness relation which are trees also have locally moving automorphism groups if they are sufficiently homogeneous.

Definition 1.11 (a) Let $M = \langle T, < \rangle$ be a partially ordered set. M is a *tree* if for every $s \in T$ the set $\{t \in T \mid t < s\}$ is linearly ordered by $<$.

(b) Let $M = \langle T, < \rangle$ be a tree. M is an *ever-splitting tree*, if for every $s \in T$ there are $t_1, t_2 > s$ such that t_1 and t_2 are incomparable.

In order to keep the presentation simple, we shall deal only with ever-splitting trees. We shall associate with every such tree a complete atomless BA.

Definition 1.12 Let $M = \langle T, < \rangle$ be an ever-splitting tree. For $s \in T$ let $M^{\geq s} = \{t \in T \mid t \geq s\}$. Let $B_1(M)$ be the subalgebra of the power set of T generated by $\{M^{\geq s} \mid s \in T\}$. Let $I_1(M)$ be the ideal of $B_1(M)$ consisting of all finite sets belonging to $B_1(M)$, and let $B_2(M) = B_1(M)/I_1(M)$. Let $B(M)$ be the completion of $B_2(M)$.

It is obvious that if M is an ever-splitting tree, then $Aut(M)$ can be regarded as a subgroup of $Aut(B(M))$. We shall next find a condition which assures that $Aut(M)$ is a locally moving subgroup of $Aut(B(M))$.

Proposition 1.13 *Let $M = \langle T, < \rangle$ be an ever-splitting tree.*

(a) $B(M)$ is atomless.

(b) $Aut(M)$ is a locally moving subgroup of $Aut(B(M))$ iff

(**) *for every $s \in T$ there is $g \in Aut(M)$ and $t > s$ such that $g(s) = s$ and $g(t) \neq t$.*

Note that the binary tree, i.e., the tree of all finite $\{0,1\}$-sequences has property (**) from the above.

The reconstruction problem for trees is dealt with extensively in [15]. There, a nearly complete answer is given to the question of when two trees have isomorphic automorphism groups. The analysis of the linear case is included there as a special case.

Automorphism groups of incomplete Boolean algebras

If B is a BA, then every automorphism of B can be extended in a unique way to an automorphism of the completion of B. Let $\bar{B}$ denote the completion of B. We may thus regard $Aut(B)$ as a subgroup of $Aut(\bar{B})$.

Proposition 1.14 *Let B be an atomless BA. Then $Aut(B)$ is a locally moving subgroup of $Aut(\bar{B})$ iff for every $a \in B - \{0\}$ there are distinct $b_1, b_2 \leq a$ such that $B|b_1 \cong B|b_2$.*

The reconstruction question for incomplete Boolean algebras is dealt with in [13, 14]. The results in [14] precede those of [13], but they already include strong reconstruction theorems for homogeneous Boolean algebras. The results in Theorem

4.5 of [13] are much stronger. They do not assume the homogeneity of B, but rather a much weaker assumption. Let us state here a special case of Theorem 4.5 of [13]. This will indicate the type of assumptions that suffice for a reconstruction theorem in this context.

Definition 1.15 (a) Let B be a BA and $G \leq Aut(B)$. G is *restriction-closed* (R-closed) if (i) for every $g \in G$ and $a \in B$, if $g(a) = a$, then the automorphism of B extending $g\lceil a \cup Id\lceil - a$ belongs to B; and (ii) for every $g \in G$ and $a \in B$, if $g(a) \cdot a = 0$, then the automorphism of B extending $g\lceil a \cup g^{-1}\lceil g(a) \cup Id\lceil - (a + g(a))$ belongs to B.

(b) For a BA B let $Ult(B)$ denote the set of ultrafilters of B. Note that $Aut(B)$ acts on $Ult(B)$.

Theorem 1.16 *Let K be the class of all pairs $\langle B, G\rangle$ such that: (i) B is an atomless BA, and $G \leq Aut(B)$; (ii) G is a locally moving subgroup of $Aut(\bar{B})$; and (iii) for every $p \in Ult(B)$, $3 \leq |\{g(p) \mid g \in G\}| \neq 6$. Then K is faithful.*

Groups of measure preserving automorphisms

Definition 1.17 (a) Let B be an atomless σ-complete BA, (that is, a Boolean algebra in which every countable subset has a supremum and an infimum). A function μ from B to the non-negative extended reals is a *strictly positive measure* on B, if the following holds.

(1) For every $a \in B$, $\mu(a) = 0$ iff $a = 0$.

(2) For every countable subset A of B of pairwise disjoint elements, $\mu(\sum A) = \sum\{\mu(a) \mid a \in A\}$.

The pair $\langle B, \mu\rangle$ is then called a measure algebra.

(b) Let $\langle B, \mu\rangle$ be a measure algebra.

$Aut(\langle B, \mu\rangle) \stackrel{def}{=} \{g \in Aut(B) \mid \text{for every } a \in B,\ \mu(g(a)) = \mu(a)\}$.

Measure algebras can be obtained by dividing the σ-algebra of a measure space by its ideal of measure zero sets.

It follows from a theorem of Maharam [9] that for every measure algebra $\langle B, \mu\rangle$, $Aut(\langle B, \mu\rangle)$ is a locally moving subgroup of $Aut(\bar{B})\}$.

Section 6 of [13] describes many results, and several open questions in this topic.

Quotient groups acting on quotient algebras

Recently a new type of reconstruction theorem was considered. We demonstrate this type of question by an example.

Question 1.1 Let $BH(\mathbb{R}^n)$ denote the group of self-homeomorphisms of $\mathbb{R}^n$ whose support is bounded, that is, $BH(\mathbb{R}^n) \stackrel{def}{=} \{g \in H(\mathbb{R}^n) \mid \{x \mid x \neq g(x)\}$ is bounded$\}$. Let $QBH(\mathbb{R}^n) = H(\mathbb{R}^n)/BH(\mathbb{R}^n)$.

Is it true that for every automorphism φ of $QBH(\mathbb{R}^n)$ there is $g \in H(\mathbb{R}^n)$ such that $g \cdot BH(\mathbb{R}^n)$ induces φ?

In this setting the group $QBH(\mathbb{R}^n)$ acts in a locally moving fashion on the quotient of $Ro(\mathbb{R}^n)$ over the ideal of bounded regular open sets. So the Reconstruction Theorem for Local Movement Systems can be invoked.

This situation occurs also in other settings, e.g. the setting of linear orderings. However, only a few final results are presently known.

The first results in this direction are theorems on the recovery of the quotients of the power set of κ, $P(\kappa)$, from the corresponding quotients of the symmetric group $Sym(\kappa)$. Two such theorems concerning reconstruction and elementary equivalence appear in [14] Theorems 4.2 and 4.3. The full answer to the elementary equivalence question of quotients of the symmetric group was recently found by Shelah. It proves the cases that were not covered in [14] Theorem 4.3. It will appear in [17].

A complete reconstruction result for the quotient of $P(\omega)$ is due to Alperin, Covington, and Macpherson [1] in this volume.

Another reconstruction result from quotients is due to Giraudet and Truss [5]. It concerns the quotient groups of $Aut(\langle \mathbb{R}, < \rangle)$ and $Aut(\langle \mathbb{Q}, < \rangle)$. The exact statement of this theorem appears in the article of Truss [19] in this volume.

The author has recently generalized Giraudet and Truss' result to the class of 2-homogeneous linear orderings whose cofinality is $\aleph_0$. However, the general question for 2-homogeneous linear orderings is still open.

Elementary equivalence of automorphism groups

The Expressibility Theorem can be used in proving that certain automorphism groups are not elementarily equivalent. We start with some definitions.

Definition 1.18 We use the terminology of model theory. We let M, N denote structures. $|M|$ denotes the universe of M. If R is a relation or a function symbol in the language of M, then its interpretation in M is denoted by R^M. If $\varphi(x_1, \ldots, x_n)$ is a formula in the language of M whose free variables are among $x_1, \ldots, x_n$, and $a_1, \ldots, a_n \in |M|$, then $M \models \varphi[a_1, \ldots, a_n]$ means that φ holds in M in the assignment $x_1 \mapsto a_1, \ldots, x_n \mapsto a_n$.

(a) Let M and N be structures in the same language L. We say that M and N are *elementarily equivalent* ($M \equiv N$) if for every first order sentence φ in L: $M \models \varphi$ iff $N \models \varphi$.

(b) Let M be a structure in a language L and G be a group of permutations of $|M|$. (G will usually be a group of automorphisms of M.) The structure N defined below is called *the action structure* of M and G. $|N| = |M| \cup G$. The language of N is $L \cup \{P, \circ, Ap\}$, where P is a unary predicate, and $\circ$ and Ap are binary function symbols. For every relation symbol $R \in L$, $R^N = R^M$. If $F \in L$ is a k-place function symbols, then $F^N \lceil |M|^k = F^M$; we extend this to $|N|^k$ in some trivial way; say, for every $\vec{a} = \langle a_1, \ldots, a_k \rangle \in |N|^k - |M|^k$, $F^N(\vec{a}) = a_1$. The interpretation of the new symbols is as follows. $P^N = |M|$. $\circ^N$ is the composition operation on G; and $\circ^N$ is defined in some trivial way on $|N|^2 - G^2$. Ap^N is the application function from $G \times |M|$ to $|M|$. That is, if $g \in G$ and $a \in |M|$, then $Ap^N(g, a) = g(a)$. Ap^N is defined in some trivial way on $|N|^2 - G \times |M|$. We denote N by $Act(M, G)$.

(c) For a local movement system $\langle B, G \rangle$, let $Var(B, G) \stackrel{def}{=} \langle \{var(g) | g \in G\}, \leq^B \rangle$. We use the notation $Var(B, G)$ to also denote just the set $\{var(g) \mid g \in G\}$.

Part (a) of the following corollary demonstrates the way that the Expressibility Theorem can be used in showing that certain automorphism groups are not elementarily equivalent. However, in applications, it is instead Part (b) which is being used. It is also used in the proof of Part (a).

Corollary 1.19 *(a) For $l = 1, 2$ let $\langle B_l, G_l \rangle$ be local movement systems. Then if $G_1 \equiv G_2$, then $Act(Var(B_1, G_1), G_1) \equiv Act(Var(B_2, G_2), G_2)$.*

(b) For every first order formula $\varphi(x_1, \ldots, x_k, y_1, \ldots, y_m)$ in the language of $Act(Var(B, G), G)$ there is a first order formula $\varphi^*(x_1, \ldots, x_k, y_1, \ldots, y_m)$ in the language of groups such that for every local system $\langle B, G \rangle$, $\vec{h} \in G^k$ and $\vec{g} \in G^m$, $Act(Var(B, G), G) \models \varphi[var(h_1), \ldots, var(h_k), g_1, \ldots, g_m]$ iff $G \models \varphi^*[\vec{h}, \vec{g}]$.

The fact that (a) follows from (b) is trivial. It is quite easy to see how Part (b) follows from the Expressibility Theorem. So we shall not return to the proof of this corollary.

Theorem 4.6(b) is an example of the use of the above corollary in answering questions concerning the elementary equivalence of automorphism groups of linear orders and related structures.

2 Locally moving groups

In order to make this section self-contained, we repeat some of the notions that were defined telegraphically in the introduction. Locally moving groups are defined in 2.4, and the Expressibility Theorem is restated in 2.5.

Notation 2.1 Let B be a BA.

(a) Let $a \in B$. $B \lceil a \stackrel{def}{=} \{b \in B \mid b \leq a\}$.

(b) Let $g \in Aut(B)$ and $a \in B$; then $g \lceil a \stackrel{def}{=} g \lceil (B \lceil a)$.

(c) Suppose that B is complete. Let $g \in Aut(B)$; then $fix(g) \stackrel{def}{=} \sum \{a \in B \mid g \lceil a = Id\}$.

(d) $var(g) \stackrel{def}{=} -fix(g)$.
(e) For a group G and $g, h \in G$ let $g^h = hgh^{-1}$ and $[g,h] = ghg^{-1}h^{-1}$.

Proposition 2.2 *Let B be a BA, $A \subseteq B$ and $b \in B$. Then the following are equivalent:*
(1) For every $b_1 \in B|b - \{0\}$ there is $a \in A$ such that $b_1 \cdot a \neq 0$.
(2) For every upper bound a of A, $b \leq a$.

Proof: (1) $\Rightarrow$ (2). Let a be an upper bound of A. Let $c = b - a$. So c is disjoint from a. So it is disjoint from every $a' \leq a$. So for every $a' \in A$, $c \cdot a' = 0$. Since $c \leq b$, $c = 0$. So $b \leq a$.

(2) $\Rightarrow$ (1). Let $b' \in B|b - \{0\}$, and suppose by contradiction that for every $a \in A$, $b' \cdot a = 0$. Hence $-b'$ is an upper bound of A. So $b \leq -b'$, a contradiction. □

Proposition 2.3 *(a) Let B be a BA, $g \in Aut(B)$, and $a \in B$ be such that $g(a) \neq a$. Then there is $0 < b \leq a$ such that $g(b) \cdot b = 0$.*

(b) Let B be a complete BA and $g \in Aut(B)$. Then $var(g) = \sum\{a \in B \mid a \cdot g(a) = 0\}$.

Proof: (a) If $a - g(a) \neq 0$, then $a - g(a)$ is the required b. Otherwise $g(a) > a$. Let $c = g(a) - a$ and $b = g^{-1}(c)$. Then b is as required.

(b) Let B and g be as in part (b). Let $A = \{a \in B \mid g(a) \cdot a = 0\}$ and $a_0 = \sum A$. For every $a \in A$, $a \cdot fix(g) = 0$. So $a_0 \cdot fix(g) = 0$. So $a_0 \leq -fix(g) = var(g)$. We show that $var(g) \leq a_0$. Let $0 \neq b \leq var(g)$. So $g|b \neq Id$. Let $c \in B|b$ be such that $g(c) \neq c$. Hence there is $0 \neq a \leq c$ such that $g(a) \cdot a = 0$. So $a \in A$. By Proposition 2.2, $var(g) \leq a_0$. □

Definition 2.4 Let B be an atomless complete BA and $G \leq Aut(B)$. G is a *locally moving subgroup* of $Aut(B)$, if for every $a \in B - \{0\}$ there is $g \in G - \{Id\}$ such that $var(g) \leq a$. The pair $\langle B, G\rangle$ is called a *local movement system*.

We say that a group G is a *locally moving group*, if there is a complete atomless BA B such that G is isomorphic to a locally moving subgroup of $Aut(B)$.

Theorem 2.5 (The Expressibility Theorem)
There are first order formulas $\varphi_{Eq}(x,y)$, $\varphi_{\leq}(x,y)$ and $\varphi_{Ap}(x,y,z)$ in the language of groups such that for every complete atomless BA B and locally moving subgroup $G \leq Aut(B)$ the following holds. For every $f, g \in G$: $G \models \varphi_{Eq}[f,g]$ iff $var(f) = var(g)$; $G \models \varphi_{\leq}[f,g]$ iff $var(f) \leq var(g)$; and $G \models \varphi_{Ap}[f,g,h]$ iff $f(var(g)) = var(h)$.

The proof of Theorem 2.5 will be completed at the end of this section.

Definition 2.6 Let B be a BA and $D \subseteq B$. D is *dense* in B if for every $b \in B - \{0\}$ there is $d \in D - \{0\}$ such that $d \leq b$.

Remark 2.7 The definition of a locally moving subgroup of $Aut(B)$ can be rephrased as follows. $G \leq Aut(B)$ is locally moving if $\{var(g) \mid g \in G\}$ is dense in B.

Proposition 2.8 *(a) If D is dense in B, then for every $a \in B$, $a = \sum\{d \in D \mid d \leq a\}$.*

(b) For $i = 1,2$ let B_i be a complete BA and D_i be a dense subset of B_i. Let $\alpha : \langle D_1, \leq\rangle \cong \langle D_2, \leq\rangle$. Then α can be extended to an isomorphism between B_1 and B_2, and this extension is unique.

Proof: The proposition is well-known, and its proof is easy. □

Notation 2.9 Let B be a Boolean algebra and $G \leq Aut(B)$.
$M(B,G) \stackrel{def}{=} \langle B, G; \leq, \circ, Ap\rangle$, where $Ap : G \times B \to B$ and $Ap(g,a) = g(a)$.

We next rephrase Corollary 1.4, and show how it follows from the Expressibility Theorem.

Corollary 2.10 *(The Reconstruction Theorem for Local Movement Systems) For $i = 1,2$ let $\langle B_i, G_i\rangle$ be a local movement system. Suppose that $\alpha : G_1 \cong G_2$. Then there is $\beta \supseteq \alpha$ such that $\beta : M(B_1, G_1) \cong M(B_2, G_2)$.*

Proof: Let $P_i = \{var(g) | g \in G_i\}$. For $a \in P_1$ let $g \in G_1$ be such that $var(g) = a$, and define $\gamma_1(a) = var(\alpha(g))$. If $h \in G_1$ and $var(h) = a$, then $G_1 \models \varphi_{Eq}[g,h]$. So $G_2 \models \varphi_{Eq}[\alpha(g), \alpha(h)]$. Hence $var(\alpha(h)) = var(\alpha(g))$. This shows that the definition of γ_1 does not depend on the choice of g. A similar argument shows that γ_1 is injective.

If $b \in P_2$, then for some $g \in G_2$, $b = var(g)$. Hence $b = \gamma_1(var(\alpha^{-1}(g)))$. So $Rng(\gamma_1) = P_2$.

Let $a, b \in P_1$ and $a \leq b$. Let $g, h \in G_1$ be such that $var(g) = a$ and $var(h) = b$. Hence $G_1 \models \varphi_{\leq}[g,h]$. So $G_2 \models \varphi_{\leq}[\alpha(g), \alpha(h)]$, and thus $var(\alpha(g)) \leq var(\alpha(h))$. This shows that γ_1 is order preserving.

Let $a, b \in P_2$ and $a \leq b$. Let $g, h \in G_2$ be such that $var(g) = a$ and $var(h) = b$. Hence $G_2 \models \varphi_{\leq}[g,h]$. So $G_1 \models \varphi_{\leq}[\alpha^{-1}(g), \alpha^{-1}(h)]$, and thus $var(\alpha^{-1}(g)) \leq var(\alpha^{-1}(h))$. So $\gamma_1^{-1}(a) \leq \gamma_1^{-1}(b)$. We have shown that $\gamma_1 : P_1 \cong P_2$.

By the fact that G_1 and G_2 are locally moving, P_1 and P_2 are dense in B_1 and B_2 respectively. So γ_1 extends to an isomorphism γ between B_1 and B_2. Let $\beta = \alpha \cup \gamma$.

We show that $\beta : M(B_1, G_1) \cong M(B_2, G_2)$. It remains to show that β preserves Ap.

Let $f \in G_1$, $a \in P_1$ and $b = f(a)$. Let $g, h \in G_1$ be such that $var(g) = a$ and $var(h) = b$. So $G_1 \models \varphi_{Ap}[f, g, h]$, and hence $G_2 \models \varphi_{Ap}[\alpha(f), \alpha(g), \alpha(h)]$. So $\alpha(f)(var(\alpha(g))) = var(\alpha(h))$. So $\gamma(f)(\gamma(a)) = \gamma(b)$.

Suppose now that $a \in B_1$ and $g \in G_1$.

So $Ap(g,a) = g(a) = \sum\{g(b) \mid b \in P_1 \cap B_1 \restriction a\}$. Hence

$$\begin{aligned}\beta(Ap(g,a)) &= \beta(\textstyle\sum\{g(b) \mid b \in P_1 \cap B_1 \restriction a\}) \\ &= \textstyle\sum\{\beta(g(b)) \mid b \in P_1 \cap B_1 \restriction a\} \\ &= \textstyle\sum\{\beta(g)(\beta(b)) \mid b \in P_1 \cap B_1 \restriction a\} \\ &= \textstyle\sum\{\beta(g)(c) \mid c \in P_2 \cap B_2 \restriction \beta(a)\} \\ &= \beta(g)(\beta(a)) = Ap(\beta(g), \beta(a)).\end{aligned}$$

□

Another seemingly interesting corollary of the Expressibility Theorem is the fact that the class of locally moving groups is elementary. That is, there is a first order sentence in the language of groups which holds exactly in the locally moving groups. We state this as Corollary 2.12 below and show how it follows from Theorem 2.5. The following lemma due to Büchi is the main additional claim needed in the proof of 2.12.

Lemma 2.11 [3] *Let $\langle P, <\rangle$ be a poset without a minimum. We write $a \parallel b$ if there is $c \in P$ such that $c \le a, b$. We write $a \perp b$ to mean $a \not\parallel b$. We say that $\langle P, <\rangle$ is* separative *if for every $a, b \in P$: if $a \not\le b$, then there is $c \in P$ such that $c \le a$ and $c \perp b$. Then the following are equivalent.*
(1) P is embeddable as a dense subset of a complete BA.
(2) P is separative.

Proof: A complete proof can be found in [7] Chapter 2 Proposition 4.16. Let us indicate another proof.

The fact that (1) ⇒ (2) is trivial.

(2) ⇒ (1). Let P be a poset without a minimum which is separative. Let $A \subseteq P$ and $b \in P$. We say that b is covered by A, if for every $b' \le b$ there is $a \in A$ such that $b' \parallel a$. We say that a subset $A \subseteq P$ is complete, if for every $b \in P$, if b is covered by A, then $b \in A$. Let $\mathcal{B}$ be the set of complete sets ordered by inclusion.
Claim 1: $\mathcal{B}$ is a complete BA.
Claim 2: For $a \in P$ let $P^{\le a} \stackrel{def}{=} \{b \in P \mid b \le a\}$. Then for every $a \in P$, $P^{\le a} \in \mathcal{B}$, and the function $a \mapsto P^{\le a}$ is an embedding of P onto a dense subset of $\mathcal{B}$.

The proofs of both claims are easy. The first proof though, is a bit long. □

Corollary 2.12 *There is a first order sentence φ_{LM} in the language of groups such that for every group G: $G \models \varphi_{LM}$ iff G is a locally moving group.*

Proof: Note that the fact that the poset P is separative is expressible by a first order sentence. Let G be any group. For $g \in G$ let $V(g) \stackrel{def}{=} \{h \in G \mid G \models \varphi_{\le}[h, g]\}$. Let $\mathcal{V}(G) \stackrel{def}{=} \{V(g) \mid g \in G - \{1_G\}\}$.

Let φ_{LM} be the conjunction of the following first order statements.
(1) $\langle \mathcal{V}(G), \subseteq\rangle$ has no minimal elements.
(2) $\langle \mathcal{V}(G), \subseteq\rangle$ is separative.

(3) For every $g \in G$, if for every $h \in G$, $V(h^g) = V(h)$, then $g = 1_G$.
(4) For every $h \in G - \{1_G\}$ there is $g \in G - \{1_G\}$ such that for every $k \in G - \{1_G\}$, if in $\langle \mathcal{V}(G), \subseteq \rangle$, $V(k) \perp V(h)$, then $V(k^g) = V(k)$,

Note that the conjunction of (1) and (2) says that $\langle \mathcal{V}(G), \subseteq \rangle$ is embeddable as a dense subset of a complete atomless Boolean algebra. (3) says that the action of G on $\mathcal{V}(G)$ by conjugation is faithful. (4) says that the above action is a locally moving action.

It follows from the Expressibility Theorem that if G is a locally moving group, then $G \models \varphi_{LM}$.

Clearly, φ_{LM} describes a faithful action of G on a dense subset of some atomless complete BA B. By 2.8(b), such an action can be extended to an action on B, and by (4), it is a locally moving action. So if $G \models \varphi_{LM}$, then G is locally moving. $\square$

The rest of this section is devoted to the proof of the Expressibility Theorem. In what follows, B denotes an atomless complete BA and G denotes a locally moving subgroup of $Aut(B)$.

Notation 2.13 *For $a \in B$ let $Sp_G(a) \stackrel{def}{=} \{g \in G \mid var(g) \leq a\}$.*

Lemma 2.14 *(a) Let $0 \neq a \leq \prod_{i=1}^{n} var(g_i)$; then there is $0 \neq b \leq a$ such that $b \cdot \sum_{i=1}^{n} g_i(b) = 0$.*

(b) Let $a \neq 0$ and $n \in \mathbb{N}^+$. Then there is $h \in Sp_G(a)$ such that $h^n \neq Id$.

(c) Let $0 \neq a \leq var(f) \cdot var(g)$. Then there is $h \in Sp_G(a)$ such that $[f^h, g] \neq Id$.

(d) Let $g \in G$ and $0 \neq a \leq var(g)$. Then there is $k \in Sp_G(a)$ such that $[k, g] \neq Id$.

Proof: (a) is proved by a trivial induction on n.

(b) We prove by induction on n that there are $h \in Sp_G(a)$ and $0 \neq b \leq a$ such that $b, h(b), \ldots, h^n(b)$ are pairwise disjoint. For $n = 1$, let $h \in Sp_G(a) - \{Id\}$. So there is $0 \neq b \leq var(h)$ such that $b \cdot h(b) = 0$. Suppose the claim is true for n. Let $h \in Sp_G(a)$ and $0 \neq b \leq a$ be as assured by the induction hypothesis. If $h^{n+1}|b \neq Id$, let $0 \neq c \leq b$ be such that $c \cdot h^{n+1}(c) = 0$. So h and c are as required. Suppose that $h^{n+1}|b = Id$. Let $k \in Sp_G(b) - \{Id\}$ and $0 \neq c \leq b$ be such that $c \cdot k(c) = 0$. Let $g = kh$. So for $i \leq n$, $g^i(c) = h^i(c) \leq h^i(b)$. Hence $c, \ldots, g^n(c)$ are pairwise disjoint. Since $g^{n+1}(c) = (kh)^{n+1}(c) = kh(kh)^n(c) = kh(h^n(c)) = k(c)$, then $g^{n+1}(c) \cdot c = 0$. Since $g^{n+1}(c) \leq b$, for every $1 \leq i \leq n$, $g^{n+1}(c) \cdot g^i(c) = 0$.

(c) Let $0 \neq a \leq var(f) \cdot var(g)$. If $[f, g] \neq Id$, then $h = Id$ is as required. So we assume that $[f, g] = Id$. Let $0 \neq b_1 \leq a$ be such that $b_1 \cdot (f(b_1) + g(b_1)) = 0$. If $b_1 \cdot var(fg) \neq 0$, then let $0 \neq b \leq b_1$ be such that $b \cdot fg(b) = 0$. Otherwise, let $b = b_1$. Let $h \in Sp_G(b)$ be such that $h^2 \neq Id$. Let $c \leq b$ be such that $c, h(c), h^2(c)$ are distinct.

Then,

(1) $$gf^h(h(c)) = ghfh^{-1}h(c) = ghf(c) = gf(c).$$

(2) $$f^h g(h(c)) = hfh^{-1}gh(c) = hfgh(c).$$

Case 1: $fg\restriction b \neq Id$. Then by the choice of b, $fg(b) \cdot b = 0$. Hence in (2) we obtain that $hfgh(c) = fg(h(c)) = gf(h(c))$. Since $gf(h(c)) \neq gf(c)$, f^h and g do not commute.

Case 2: $fg\restriction b = Id$. Then in (1) $gf(c) = c$ and in (2) $hfgh(c) = h^2(c)$. So g and f^h do not commute.

(d) Let $0 \neq a \leq var(g)$. Let $f \in Sp_G(a) - \{Id\}$. By part (c), there is $h \in Sp_G(var(f))$ such that $[f^h, g] \neq Id$. It follows that f^h is as required. □

Notation 2.15 Let $C_G(f) = \{g \in G \mid [g, f] = Id\}$. We abbreviate $C_G(f)$ by $C(f)$.

The next lemma says that for $f \in Aut(B)$ and $a \in B$, no sum of a pairwise disjoint collection of $n+1$ images $f^{k_i}(a)$ can be covered by any sum of n images $h_i(a)$, when the h_i's commute with f.

Lemma 2.16 *Let $k_0, \ldots, k_n \in \mathbb{Z}$, $f \in Aut(B)$ and $a \in B$. Suppose that $f^{k_0}(a), \ldots, f^{k_n}(a)$ are pairwise disjoint. Then for every $h_1, \ldots, h_n \in C(f)$ and $0 \neq b \leq a$,*

$$\sum_{i=0}^{n} f^{k_i}(b) \not\leq \sum_{i=1}^{n} h_i(a).$$

Proof: By induction on n. Let $n = 1$. If $f^{k_0}(b) + f^{k_1}(b) \leq h_1(a)$, then $0 \neq b \leq f^{-k_0}(h_1(a)) \cdot f^{-k_1}(h_1(a))$. Hence since $h_1 \in C(f)$, $0 \neq f^{-k_0}(a) \cdot f^{-k_1}(a)$. Apply $f^{k_0+k_1}$ to both sides. So $f^{k_1}(a) \cdot f^{k_0}(a) \neq 0$, a contradiction.

Suppose the claim is true for n. Let $k_0, \ldots, k_{n+1}$, $h_1, \ldots, h_{n+1}$, a and b be as in the lemma. If $h_{n+1}(a) \cdot \sum_{i=0}^{n+1} f^{k_i}(b) = 0$, then by the induction hypothesis

$$\sum_{i=1}^{n+1} h_i(a) \not\geq \sum_{i=0}^{n+1} f^{k_i}(b).$$

Suppose otherwise, and let j be such that $h_{n+1}(a) \cdot f^{k_j}(b) \stackrel{def}{=} c \neq 0$. We may assume that $j = n+1$. We will define by induction for $0 \leq i \leq n$, $0 \neq b_i \leq b$ such that (i) $h_{n+1}(a) \cdot f^{k_i}(b_i) = 0$, and such that (ii) $b_{i+1} \leq b_i$. Let $b' = f^{-k_{n+1}}(c)$. Since $0 \neq b' \leq a$, and by the case $n = 1$, $h_{n+1}(a) \not\geq f^{k_0}(b') + f^{k_{n+1}}(b') = f^{k_0}(b') + c$. Since $h_{n+1}(a) \geq c$, $h_{n+1}(a) \not\geq f^{k_0}(b')$.

Let $b_0 = f^{-k_0}(f^{k_0}(b') - h_{n+1}(a))$. It follows that $0 \neq b_0 \leq b$. Also, (i) holds. Suppose b_i has been defined. So $b_i \leq b'$. Hence $h_{n+1}(a) \not\geq f^{k_{i+1}}(b_i) + f^{k_{n+1}}(b_i)$.

But $h_{n+1}(a) \geq f^{k_{n+1}}(b_i)$. Let $b_{i+1} = f^{-k_{i+1}}(f^{k_{i+1}}(b_i) - h_{n+1}(a))$. So (i) and (ii) hold.

We obtain that $h_{n+1}(a) \cdot \sum_{i=0}^{n} f^{k_i}(b_n) = 0$. By the external induction hypothesis,

$$\sum_{i=1}^{n} h_i(a) \not\geq \sum_{i=0}^{n} f^{k_i}(b_n)$$

and so

$$\sum_{i=1}^{n+1} h_i(a) \not\geq \sum_{i=0}^{n} f^{k_i}(b_n).$$

Hence

$$\sum_{i=1}^{n+1} h_i(a) \not\geq \sum_{i=0}^{n+1} f^{k_i}(b).$$

□

Lemma 2.17 *Let*

$$\varphi_1(f, f') \equiv \forall g(([g, f] \neq Id) \rightarrow$$
$$(\exists f_1, f_2 \in C(f'))(([g, f_1, f_2] \neq Id) \wedge ([[g, f_1, f_2], f'] = Id))).$$

(a) For every local movement system $\langle B, G\rangle$ and $f, f' \in G$: if $var(f) \cdot var(f') = 0$, then $G \models \varphi_1[f, f']$.

(b) For every local movement system $\langle B, G\rangle$ and $f, f' \in G$: if $var(f) \cdot var((f')^{12}) \neq 0$, then $G \not\models \varphi_1[f, f']$.

Proof: (a) Let $\langle B, G\rangle$ be a local movement system. Let $f, f' \in G$ and suppose that $var(f) \cdot var(f') = 0$. Let $g \in G$ be such that $[g, f] \neq Id$.

So $var(g) \cdot var(f) \neq 0$. By proposition 2.3(b), there is $a \in B|(var(f) - \{0\})$ such that $g(a) \cdot a = 0$. Let $f_1 \in Sp_G(a) - \{Id\}$. Hence $f_1 \in C(f')$. Let $g_1 = [g, f_1]$. So $g_1 = g f_1 g^{-1} f_1^{-1} = (f_1)^g \cdot f_1^{-1}$. Recall that $g(var(f_1)) \cdot var(f_1) = 0$. So $var(g_1) = var(f_1) + g(var(f_1))$ and $g_1(var(f_1)) = var(f_1)$.

Let $b \in B|(var(f_1) - \{0\})$ be such that $g_1(b) \cdot b = 0$. Let $f_2 \in Sp_G(b) - \{Id\}$. Since $b \cdot var(f') = 0$, $f_2 \in C(f')$. Let $g_2 = [g_1, f_2]$. So $g_2 = f_2^{g_1} \cdot f_2^{-1}$. $g_1(var(f_2)) \cdot var(f_2) = 0$. So $var(g_2) = var(f_2) + g_1(var(f_2)) \neq 0$. Also $var(f_2) \leq var(f_1)$ and $g_1(var(f_1)) = var(f_1)$. So $var(g_2) \leq var(f_1)$. Hence $var(g_2) \cdot var(f') = 0$. So $g_2 \in C(f')$. We have shown that $G \models \varphi_1[f, f']$.

(b) Suppose that $a \stackrel{def}{=} var(f) \cdot var((f')^{12}) \neq 0$. So for every $i = 1, \ldots, 4$, $a \leq var((f')^i)$. Hence by Lemma 2.14(a), there is $b \in B|a - \{0\}$ such that

$$b \cdot \sum_{i=1}^{4} (f')^i(b) = 0.$$

Hence $\{(f')^i(b) \mid 0 \le i \le 4\}$ are pairwise disjoint. By Lemma 2.14(d), there is $g \in Sp_G(b)$ such that $[g, f] \ne Id$.

Let $f_1, f_2 \in C(f')$ be such that $g_2 \stackrel{def}{=} [[g, f_1], f_2] \ne Id$. We show that $[g_2, f'] \ne Id$. Now,

$$\begin{aligned} g_2 &= [g \cdot (g^{-1})^{f_1}, f_2] = g(g^{-1})^{f_1} \cdot ((g \cdot (g^{-1})^{f_1})^{-1})^{f_2} \\ &= g(g^{-1})^{f_1} \cdot (g^{f_1} g^{-1})^{f_2} = g(g^{-1})^{f_1} g^{f_2 f_1} (g^{-1})^{f_2}. \end{aligned}$$

So $0 \ne var(g_2) \le b + f_1(b) + f_2 f_1(b) + f_2(b)$.

Let $h \in \{Id, f_1, f_2 f_1, f_2\}$ be such that $c \stackrel{def}{=} var(g_2) \cdot h(b) \ne 0$. Since $\{(f')^i(b) \mid 0 \le i \le 4\}$ is pariwise disjoint, and $h \in C(f')$, $\{(f')^i(h(b)) \mid 0 \le i \le 4\}$ is pairwise disjoint.

Suppose by contradiction that $[g_2, f'] = Id$. Hence for every $i = 0, \ldots, 4$, $(g_2)^{(f')^i} = g_2$. Recall also that $var(g_2) \ge c$. So

$$(1) \qquad var(g_2) = \sum_{i=0}^{4} var((g_2)^{(f')^i}) = \sum_{i=0}^{4} (f')^i(var(g_2)) \ge \sum_{i=0}^{4} (f')^i(c).$$

But $var(g_2) \le b + f_1(b) + f_2(b) + f_2 f_1(b)$. By lemma 2.16, and since h^{-1}, $f_1 h^{-1}$, $f_2 h^{-1}$ and $f_2 f_1 h^{-1}$ commute with f',

$$\sum_{i=0}^{4} (f')^i(c) \not\le h^{-1}(h(b)) + f_1 h^{-1}(h(b)) + f_2 h^{-1}(h(b)) + f_2 f_1 h^{-1}(h(b)) =$$

$$b + f_1(b) + f_2(b) + f_2 f_1(b) \ge var(g_2).$$

Hence $\sum_{i=0}^{4} (f')^i(c) \not\le var(g_2)$. This contradicts (1), and so $[g_2, f'] \ne Id$. We have shown that $G \not\models \varphi_1[f, f']$. □

Definition 2.18 Let $D_1(f) \stackrel{def}{=} \{f' \mid G \models \varphi_1[f, f']\}$ and $V(f) \stackrel{def}{=} C(\{(f')^{12} \mid f' \in D_1(f)\})$.

Proposition 2.19 *For every $f \in G$, $V(f) = Sp_G(var(f))$.*

Proof: Let $g \in Sp_G(var(f))$. Let $f' \in D_1(f)$. So $var((f')^{12}) \cdot var(f) = 0$, and hence $var(g) \cdot var((f')^{12}) = 0$. Hence $[g, (f')^{12}] = Id$. We have shown that $Sp_G(var(f)) \subseteq V(f)$.

Next let $g \in G - Sp_G(var(f))$. Hence $a \stackrel{def}{=} var(g) - var(f) \ne 0$. By Lemma 2.14(b), there is $f' \in Sp_G(a)$ such that $(f')^{12} \ne Id$. By Lemma 2.14(c), there is $h \in Sp_G(var((f')^{12}))$ such that $[((f')^{12})^h, g] \ne Id$. Now $var(f') \cdot var(f) = 0$ and $var(h) \le var((f')^{12}) \le var(f')$, so $var((f')^h) = var(f')$, and hence $var((f')^h) \cdot var(f) = 0$. It follows that $(f')^h \in D_1(f)$. So $g \notin V(f)$. We have shown that $V(f) \subseteq Sp_G(var(f))$. □

Proof of Theorem 2.5:
Let $\varphi_{\le}(f,g) \equiv V(f) \subseteq V(g)$. Clearly, $\varphi_{\le}(f,g)$ can be written as a first order formula. If $var(f) \le var(g)$, then by the above proposition, $G \models \varphi_{\le}[f,g]$. If on the other hand, $var(f) \not\le var(g)$, then since G is locally moving, there is $h \in Sp_G(var(f)) - Sp_G(var(g))$. By the above proposition, $V(f) \not\subseteq V(g)$. So $\varphi_{\le}(f,g)$ is as required.

Let $\varphi_{Eq}(f,g) \equiv \varphi_{\le}(f,g) \wedge \varphi_{\le}(g,f)$ and $\varphi_{Ap}(f,g,h) \equiv \varphi_{Eq}(g^f,h)$. It is trivial that φ_{Eq} and φ_{Ap} are as required. □

3 Reconstruction of topological spaces

In this section we describe two applications of the Reconstruction Theorem (Corollary 2.10) in topology. The first one, Theorem 3.1 appeared in [12]. The second one, Theorem 3.8 is new.

We shall use the the notions defined in Definition 1.7. We shall also use Proposition 1.8. Its proof is very easy, and is left to the reader. By Proposition 1.8(a), $H(X)$ can be regarded as a subgroup of $Aut(Ro(X))$.

Let K be a class of pairs of the form $\langle X, G\rangle$, where X is a topological space and G is a subgroup of the group $H(X)$ of all self-homeomorphisms of X. We say that K is *faithful* if for every $\langle X_l, G_l\rangle \in K$, $l = 1,2$ and every isomorphism φ between G_1 and G_2 there is a homeomorphism $\tau : X_1 \cong X_2$ which induces φ, that is, for every $g \in G_1$, $\varphi(g) = \tau \circ g \circ \tau^{-1}$.

For $G \subseteq H(X)$ and $x \in X$, let $G(x) \stackrel{def}{=} \{g(x) \mid g \in G\}$. A set $D \subseteq X$ is *somewhere dense* if there is a nonempty open set U such that $D \cap U$ is dense in U. For an open set $V \subseteq X$, let $Sp_G(V) \stackrel{def}{=} \{g \in G \mid var(g^{Ro}) \subseteq V\}$. We leave it to the reader to check that $var(g^{Ro}) \subseteq V$ iff $g\lceil(X - V) = Id$.

Let $K_{LCG} \stackrel{def}{=} \{\langle X, G\rangle \mid X$ is a locally compact Hausdorff space, and for every open $V \subseteq X$ and $x \in V$, $Sp_G(V)(x)$ is somewhere dense$\}$.

The following theorem appears in [12]. It is the conjunction of parts (a), (b) and (c) of theorem 3.5 there.

Theorem 3.1 *K_{LCG} is faithful.*

In the proof of the second result in this section we shall need a theorem somewhat stronger than 3.1. We next define the notions needed in the statement of this theorem.

Definition 3.2 Let X be a topological space.

(a) For $x \in X$ let $Nbr^X(x)$ denote the set of regular open neighborhoods of x in X. The superscript X is usually omitted.

(b) Let $G \le H(X)$. The set of *densely conjugated points of* $\langle X, G\rangle$ is defined as $Dc(X,G) = \{x \in X \mid G(x)$ is somewhere dense$\}$.

The set of *good points of* $\langle X,G\rangle$ is defined as
$Gd(X,G)=\{x\in X \mid \text{ for every } U\in Nbr(x),\ Sp_G(U)(x) \text{ is somewhere dense}\}$.

(c) X is *regionally compact* if $\bigcup\{T\subseteq X \mid T \text{ is open and } cl(T) \text{ is compact}\}$ is dense in X.

(d) Let
$K_{RCM} \stackrel{def}{=} \{\langle X,G\rangle \mid X$ is Hausdorff and regionally compact, $G \leq H(X)$ is a locally moving subgroup of $Aut(Ro(X))$ and $Dc(X,G)=Gd(X,G)\}$.

(e) For $\langle X,G\rangle\in K_{RCM}$ let
$Mdc(X,G)\stackrel{def}{=}\langle Ro(X),Dc(X,G),G;\leq,\varepsilon,\circ,Ap\rangle$, where
$\varepsilon=\{\langle x,U\rangle \mid x\in Dc(X),\ U\in Ro(X) \text{ and } x\in U\}$.

Note that by definition, $Gd(X,G)\subseteq Dc(X,G)$. Note also the following trivial observations.

Proposition 3.3 *(a) Let X be Hausdorff and regionally compact. Then for every nonempty open $U\subseteq X$, there is nonempty $V\in Ro(X)$ such that $cl(V)\subseteq U$.*

(b) $K_{LCG}=\{\langle X,G\rangle\in K_{RCM} \mid Gd(X,G)=X\}$.

Proof: The proof of part (a) is similar to the proof that a locally compact Hausdorff space is regular. The proof of part (b) is easy. □

We can now state the strengthening of Theorem 3.1.

Theorem 3.4 *For $l=1,2$ let $\langle X_l,G_l\rangle\in K_{RCM}$, and φ be an isomorphism between G_1 and G_2. Then there is $\psi: Mdc(X_1,G_1)\cong Mdc(X_2,G_2)$ such that ψ extends φ.*

Let us first see how Theorem 3.1 follows from Theorem 3.4.

Proof of 3.1: We assume Theorem 3.4. For $l=1,2$ let $\langle X_l,G_l\rangle\in K_{LCG}$ and φ be an isomorphism between G_1 and G_2. By Theorem 3.4, there is

$$\psi: Mdc(X_1,G_1)\cong Mdc(X_2,G_2).$$

Then $Dc(X_l,G_\ell)=X_l$. So $\psi|X_1$ is a bijection between X_1 and X_2. Since ψ preserves ε, for every $U\in Ro(X_1)$, $\{\psi(x)\mid x\in U\}=\psi(U)$. So $\psi|X_1$ takes the basis $Ro(X_1)$ of X_1 to the basis $Ro(X_2)$ of X_2. So $\psi|X_1: X_1\cong X_2$. Since ψ preserves Ap and ψ extends φ, for every $g\in G_1$ and $U\in Ro(X_2)$, $\varphi(g)(U)=g^{(\psi\lceil Ro(X_1))}(U)$, and since for every $x\in X_2$, $\{x\}=\bigcap Nbr(x)$, $\varphi(g)(x)=g^{(\psi\lceil X_1)}(x)$. So $\psi|X_1$ induces φ. □

The following notions will be needed in the proof of Theorem 3.4.

Definition 3.5 (a) A subset p of a BA B is called an *ultrafilter* if: (i) $0\notin p$; (ii) if $a_1,\ldots,a_n\in p$, then $\prod_{i=1}^n a_i\in p$; (iii) if $a\in p$ and $b\geq a$, then $b\in p$; and (iv) p is not properly contained in a subset of B satisfying (i)–(iii).

The set of ultrafilters of B is denoted by $Ult(B)$.

A subset q of B has the *finite intersection property* if for every finite $\sigma \subseteq q$, $\bigcap \sigma \neq 0$. Note that by Zorn's lemma, every subset of B with the finite intersection property is contained in an ultrafilter.

Let $\langle X, G\rangle \in K_{RCM}$.

(b) We say that an ultrafilter p in $Ro(X)$ is *densely conjugated* if there is $W \in Ro(X) - \{\emptyset\}$ such that for every $V \in Ro(X)|W - \{\emptyset\}$ there is $g \in G$ such that $g(V) \in p$. Let $Dcu(X, G)$ denote the set of densely conjugated ultrafilters.

(c) For $p \in Ult(Ro(X))$, let $A_p \stackrel{def}{=} \bigcap\{cl(V) \mid V \in p\}$. If A_p is a singleton, let x_p be such that $A_p = \{x_p\}$.

Theorem 3.4 follows from a certain "Expressibility Theorem" in very much the same way that the Reconstruction Theorem for local movement systems followed from the Expressibility Theorem for locally moving groups. That is, we shall find certain objects in $M(Ro(X), G)$ which will represent the densely conjugated points of X. In fact, each densely conjugated point $x \in X$ will be represented by any densely conjugated ultrafilter p such that $x_p = x$. We shall then show that the fact that two densely conjugated ultrafilters p and q represent the same point x, is equivalent to a property of $\langle p, q\rangle$ expressible in $M(Ro(X), G)$. Similarly the fact that $x_p \in U$ will be shown to be equivalent to a property of $\langle p, U\rangle$ expressible in $M(Ro(X), G)$.

The above explanation may be a little vague, because ultrafilters are not members of $M(Ro(X), G)$ but rather subsets of $M(Ro(X), G)$. We shall later formalize the situation by referring to formulas in second order logic.

Proposition 3.6 *Let $\langle X, G\rangle \in K_{RCM}$ and $p \in Ult(Ro(X))$.*

(a) $|A_p| \leq 1$.

(b) If A_p is a singleton, then $p \supseteq Nbr(x_p)$.

(c) If $x \in X$ and $p \supseteq Nbr(x)$, then A_p is a singlton, and $x_p = x$.

(d) If p is densely conjugated, then A_p is a singleton, and $x_p \in Dc(X, G)$.

(e) If $x \in Dc(X, G)$ and $p \supseteq Nbr(x)$, then $p \in Dcu(X, G)$.

Proof: Part (a) follows from the fact that X is a Hausdorff space.

(b) Suppose that $A_p = \{x\}$. Then $p \cup Nbr(x)$ has the finite intersection property. By the maximality of p, $p \supseteq Nbr(x_p)$.

(c) Let $U \in p$. So $X - cl(U) \notin p$. So $X - cl(U) \notin Nbr(x)$. So $x \in cl(U)$.

(d) Let p be densely conjugated. Let $W \in Ro(X) - \{\emptyset\}$ be such that for every $V \in Ro(X)|W - \{\emptyset\}$ there is $g \in G$ such that $g(V) \in p$. Let $V \in Ro(X)|W - \{\emptyset\}$ be such that $cl(V)$ is compact, and let $g \in G$ be such that $V' \stackrel{def}{=} g(V) \in p$. So $cl(V')$ is compact. Hence $A_p = \bigcap\{cl(U) \cap cl(V') \mid U \in p\} \neq \emptyset$. So by part (a), A_p is a singlton.

We show that $G(x_p) \cap W$ is dense in W. Let $U \in Ro(X)|W - \{\emptyset\}$. Let $V \in Ro(X) - \{\emptyset\}$ be such that $cl(V) \subseteq U$. For some $g \in G$, $g(V) \in p$. So $x_p \in cl(g(V))$. Hence $g^{-1}(x_p) \in cl(V) \subseteq U$. So $G(x_p) \cap W$ is dense in W.

(e) Let $x \in Dc(X,G)$ and $p \supseteq Nbr(x)$. Let $W \in Ro(X) - \{\emptyset\}$ be such that $G(x) \cap W$ is dense in W. Let $V \in Ro(X)|W - \{\emptyset\}$. Then there is $g \in G$ such that $g(x) \in V$. So $g^{-1}(V) \in Nbr(x) \subseteq p$. Hence p is densely conjugated. □

We shall next present three formulas in second order logic with variables ranging over subsets of $M(Ro(X), G)$. Note first that the fact that p is an ultrafilter can be expressed by such a formula. That is, p is an ultrafilter iff it satisfies in $M(Ro(X), G)$ the formula $\psi_{Ult}(p)$ defined as follows:

$$\begin{aligned}\psi_{Ult}(p) \overset{def}{\equiv}\ & (p \subseteq Ro(X)) \wedge (0 \notin p) \wedge \\ & (\forall U \in Ro(X))(\forall V \in Ro(X))((U, V \in p) \rightarrow (U \cdot V \in p)) \wedge \\ & (\forall U \in Ro(X))(\forall V \in Ro(X))((U \in p \wedge U \leq V) \rightarrow (V \in p)) \wedge \\ & (\forall U \in Ro(X))((U \in p) \vee (-U \in p));\end{aligned}$$

$$\begin{aligned}\psi_{Dc}(p) \overset{def}{\equiv}\ & \psi_{Ult}(p) \wedge \\ & (\exists W \in Ro(X) - \{\emptyset\})(\forall V \in Ro(X)|W - \{\emptyset\})(\exists g \in G) \\ & (g(V) \in p);\end{aligned}$$

$$\begin{aligned}\psi_{Eq}(p,q) \overset{def}{\equiv}\ & \neg(\exists U, U_0 \in p - q)(\forall V \in Ro(X)|U_0 - \{0\})(\exists g \in Sp_G(U)) \\ & (V \in f(p));\end{aligned}$$

and $\psi_{\varepsilon}(p, U) \overset{def}{\equiv} (\forall q)((\psi_{Dc}(q) \wedge \psi_{Eq}(p,q)) \rightarrow (U \in q))$.

The important property of these formulas is that they involve only the relations and functions of the structure $M(Ro(X), G)$. So if ψ is any of the above formulas, $\tau : M(Ro(X_1), G_1) \cong M(Ro(X_2), G_2))$ and p is a subset of $M(Ro(X_1), G_1)$, then $M(Ro(X_1), G_1) \models \psi[p]$ iff $M(Ro(X_2), G_2) \models \psi[\tau(p)]$.

Proposition 3.7 *Let $\langle X, G\rangle \in K_{RCM}$, $p, q \in Ult(Ro(X))$ and $U \in Ro(X)$. Then*
(a) $M(Ro(X), G) \models \psi_{Dc}[p]$ iff $p \in Dcu(X, G)$.

(b) If $p, q \in Dcu(X, G)$, then x_p and x_q exist, and $M(Ro(X), G) \models \psi_{Eq}[p, q]$ iff $x_p = x_q$.

(c) If $p \in Dcu(X, G)$, then $M(Ro(X), G) \models \psi_{\varepsilon}[p, U]$ iff $x_p \in U$.

Proof: Part (a) is completely trivial because $\psi_{Dc}(p)$ just states the definition of being a densely conjugated ultrafilter.

(b) Let $p, q \in Dc(X, G)$. It is proved in Proposition 3.6(d) that x_p and x_q exist.

Suppose first that $x_p = x_q$. Suppose by contradiction that $M(Ro(X), G) \not\models \psi_{Eq}[p, q]$. Let U and U_0 be as assured by $\neg\psi_{Eq}[p, q]$. So $U, U_0 \notin q$, and hence $U, U_0 \notin Nbr(x_p)$. Let $V \in Ro(X) - \{\emptyset\}$ be such that $cl(V) \subseteq U_0$. Let $g \in Sp_G(U)$. Since $x_p \notin U$, $g(x_p) = x_p \notin U_0$, and hence $g(x_p) \notin cl(V)$. But $x_{g(p)} = g(x_p)$, so $V \notin g(p)$. This means that U and U_0 do not satisfy the requirement of $\neg\psi_{Eq}[p, q]$, a contradiction.

Suppose next that $x_p \neq x_q$. By Proposition 3.6(d), $x_p \in Dc(X, G)$. Since $\langle X, G\rangle \in K_{RCM}$, $x_p \in Gd(X, G)$. Let $U \in Nbr(x_p)$ be such that $x_q \notin cl(U)$. Let $U_0 \subseteq U$ be such that $Sp_G(U)(x_p) \cap U_0$ is dense dense in U_0. Let $V \in Ro(X)|U_0 - \{\emptyset\}$.

Then there is $g \in Sp_G(U)$ such that $g(x_p) \in V$. So $V \in Nbr(g(x_p)) = Nbr(x_{g(p)}) \subseteq g(p)$. Hence U and U_0 are as required in $\neg\psi_{Eq}[p,q]$.

(c) Let $p \in Dcu(X,G)$ and $U \in Ro(X)$. Suppose first that $x_p \in U$. Let q be such that $M(Ro(X),G) \models \psi_{Dc}[q] \wedge \psi_{Eq}[p,q]$. By parts (a) and (b), $q \in Dcu(X,G)$ and $x_p = x_q$. So $U \in Nbr(x_q) \subseteq q$. We have shown that $M(Ro(X),G) \models \psi_\varepsilon[p,U]$.

Suppose next that $x_p \notin U$. So $Nbr(x_p) \cup \{-U\}$ has the finite intersection property. Let q be an ultrafilter containing $Nbr(x_p) \cup \{-U\}$. By Proposition 3.6(c), $x_q = x_p$. Also, $x_p \in Dc(X,G)$, hence by by Proposition 3.6(d), $q \in Dcu(X,G)$, and so $M(Ro(X),G) \models \psi_{Dc}[q]$. By part (b), $M(Ro(X),G) \models \psi_{Eq}[p,q]$. But $U \notin q$. So $M(Ro(X),G) \not\models \psi_\varepsilon[p,U]$. □

Proof of Theorem 3.4: For $l = 1,2$ let $\langle X_l, G_l\rangle \in K_{RCM}$, and φ be an isomorphism between G_1 and G_2. Since $\langle Ro(X_l), G_l\rangle$ are local movement systems, by 2.10, there is $\alpha \supseteq \varphi$ such that $\alpha : M(Ro(X_1),G_1) \cong M(Ro(X_2),G_2)$. We define $\beta : Dc(X_1,G_1) \to Dc(X_2,G_2)$. Let $x \in Dc(X_1,G_1)$. Let p be such that $Nbr(x) \subseteq p \in Ult(Ro(X_1))$. Then by 3.6(d), 3.7(a) and 3.6(c), $x_{\alpha(p)}$ exists and it belongs to $Dc(X_2,G_2)$. Let $\beta(x) = x_{\alpha(p)}$.

The proof that β is well-defined and that $\alpha \cup \beta$ is an isomorphism between $Mdc(X_1,G_1)$ and $Mdc(X_2,G_2)$ is similar to the proof of the parallel parts in the Reconstruction Theorem for Local Movement Systems (Corollary 2.10). □

Foliated Spaces

Previous versions of the theorem on locally moving groups were not general enough to handle the homeomorphism groups associated with certain foliated spaces. The simplest such example was the torus with an irrational foliation.

Let us describe the setting. We represent the torus $\boldsymbol{T}$ as $[0,1) \times [0,1)$. Let $frc(a)$ denote the fractional part of the real number a. Let $cvr(a,b) \stackrel{def}{=} (frc(a), frc(b))$. So cvr is the covering map from $\mathbb{R}^2$ to $\boldsymbol{T}$. A set $\mathcal{L}$ of straight lines in $\mathbb{R}^2$ is called a *parallel family* if every two distinct members of $\mathcal{L}$ are parallel. For a parallel family $\mathcal{L}$ let $\mathcal{F}(\mathcal{L}) \stackrel{def}{=} \{cvr(L) \mid L \in \mathcal{L}\}$ and $\boldsymbol{T}_\mathcal{L} \stackrel{def}{=} \bigcup \mathcal{F}(\mathcal{L})$. Note that if $\mathcal{L}$ consists of a single line with an irrational slope, then $\boldsymbol{T}_\mathcal{L}$ is dense in $\boldsymbol{T}$. We consider two possible groups. Let

$H_0(\boldsymbol{T}_\mathcal{L}) \stackrel{def}{=} \{h \in H(\boldsymbol{T}_\mathcal{L}) \mid \text{for every } L \in \mathcal{F}(\mathcal{L}),\ h(L) = L\}$; and

$H_1(\boldsymbol{T}_\mathcal{L}) \stackrel{def}{=} \{h \in H(\boldsymbol{T}_\mathcal{L}) \mid \text{for every } L \in \mathcal{F}(\mathcal{L}),\ h(L) \in \mathcal{F}(\mathcal{L})\}$.

Note that if the members of $\mathcal{L}$ have an irrational slope, and $\boldsymbol{T}_\mathcal{L} \neq \boldsymbol{T}$, then $H_1(\boldsymbol{T}_\mathcal{L}) = H(\boldsymbol{T}_\mathcal{L})$.

As a corollary of a more general statement, we shall obtain the following result.

Theorem 3.8 *For $i = 1,2$ let $\mathcal{L}_i$ be parallel families and $m_i \in \{0,1\}$. Suppose that $\varphi : H_{m_1}(\boldsymbol{T}_{\mathcal{L}_1}) \cong H_{m_2}(\boldsymbol{T}_{\mathcal{L}_2})$. Then there is $\tau : \boldsymbol{T}_{\mathcal{L}_1} \cong \boldsymbol{T}_{\mathcal{L}_2}$ such that τ induces φ, and either $m_1 = m_2$ or $H_{m_1}(\boldsymbol{T}_{\mathcal{L}_1}) = H_{m_2}(\boldsymbol{T}_{\mathcal{L}_1})$.*

We shall next define a notion of a foliated space. The reconstruction problem for this class of spaces is still open. We can only prove a special case. Another important special case was proved by W. Ling in [8]. He proved the reconstruction theorem for Euclidean manifolds with a foliation. His method is entirely different from the method presented here. Ling's proof requires that the space in question be locally compact and connected.

Definition 3.9 (a) Let E and F denote normed vector spaces over $\mathbb{R}$. B^E, $\bar{B}^E$ and S^E denote respectively the unit open ball, the unit closed ball and the unit sphere of E. $B^E(x,r)$ denotes the open ball with center at x and radius r. We use the same notation for $\bar{B}$. Let $B^{E,F}$ and $\bar{B}^{E,F}$ denote respectively $B^E \times B^E$ and $\bar{B}^E \times \bar{B}^E$. Let $\sim^{(E,F)} \overset{def}{=} \{\langle (y,z_1),(y,z_2)\rangle \mid y \in \bar{B}^E, z_1, z_2 \in \bar{B}^F\}$.

(b) Let $\alpha(y,z) = (\alpha_1(y), \alpha_2(y,z))$ be a map with the following properties: (1) $Dom(\alpha), Rng(\alpha) \subseteq \bar{B}^{E,F}$; (2) $\alpha : Dom(\alpha) \cong Rng(\alpha)$; and (3) $\alpha_1 : Dom(\alpha_1) \cong Rng(\alpha_1)$. Then α is called an *(E,F)-selfmap*.

(c) Let X be a topological space, and $\gamma : \bar{B}^{E,F} \to X$ be a map with the following properties: (1) $\gamma : \bar{B}^{E,F} \cong Rng(\gamma)$; (2) $\gamma(\bar{B}^{E,F})$ is closed in X; and (3) $\gamma(B^{E,F})$ is open in X. Then γ is called a *normal (E,F)-map*.

(d) Let $\mathcal{M} = \langle X, \Gamma\rangle$ be as follows: (1) X is a topological space; (2) Γ is a set of normal (E,F)-maps, and for every $\gamma \in \Gamma$, $Rng(\gamma) \subseteq X$; (3) For every $x \in X$ there is $\gamma \in \Gamma$ such that $x \in \gamma(B^{E,F})$; (4) For every $\beta, \gamma \in \Gamma$, $\beta^{-1} \circ \gamma$ is an (E,F)-selfmap; and (5) There is $\Gamma_0 \subseteq \Gamma$ such that $|\Gamma_0| \leq \aleph_0$, and for every $\gamma \in \Gamma$ there is $\gamma_0 \in \Gamma_0$ such that $Rng(\gamma) \subseteq Rng(\gamma_0)$. Then $\mathcal{M}$ is called a *foliated manifold (Fl-manifold) based on* (E,F).

We call Γ_0 a *dominating set* for Γ, and in fact, the Fl-manifold obtained from Γ_0 is identical to the one obtained from Γ.

(e) Let $\mathcal{M} = \langle X, \Gamma\rangle$ be an Fl-manifold based on (E,F). For $\gamma \in \Gamma$ let $\sim_\gamma = \{\langle \gamma(u), \gamma(v)\rangle \mid u \sim^{(E,F)} v\}$. Let $\sim^{\mathcal{M}}$ be the transitive closure of $\bigcup\{\sim_\gamma \mid \gamma \in \Gamma\}$.

(f) Let $\mathcal{M} = \langle X_0, \Gamma\rangle$ be an Fl-manifold, and $\mathcal{F} = \langle X, \sim\rangle$ be such that: (1) $X \subseteq X_0$; (2) for every $x \in X$ and $y \sim^{\mathcal{M}} x$, $y \in X$; (3) $\sim \, = \, \sim^{\mathcal{M}} \lceil X$; and (4) for every $\gamma \in \Gamma$, $\gamma^{-1}(X)$ is dense in $\bar{B}^{E,F}$.
Then $\mathcal{F}$ is said to be an *Fl-subspace* of $\mathcal{M}$. We say that $\mathcal{F}$ is an *Fl-space* if it is an Fl-subspace of some Fl-manifold $\mathcal{M}$.

(g) For a topological space Z and $U \subseteq Z$, let $bd^Z(U)$ denote the boundary of U in Z. We say that a topological space Z is *thin* if for every $z \in Z$ there are $U \in Nbr(z)$ and $V \in Ro(Z)|U$ such that $bd^U(V) = \{z\}$.

(h) Let $p_{E,F}$ be the projection of $\bar{B}^{E,F}$ on $\bar{B}^E$. Let $\mathcal{F}$ and $\mathcal{M}$ be as in (f). For $\gamma \in \Gamma$ let $Y_\gamma^{\mathcal{F}} = p_{E,F}(\gamma^{-1}(X))$. We say that $\mathcal{F}$ is thin, if for every $\gamma \in \Gamma$, $Y_\gamma^{\mathcal{F}}$ is thin.

(i) For a foliated space $\mathcal{F} = \langle X, \sim\rangle$, let

$$H_1(\mathcal{F}) \overset{def}{=} \{h \in H(X) \mid h \text{ preserves } \sim\},$$

and

$$H_0(\mathcal{F}) \stackrel{def}{=} \{h \in H(\mathcal{F}) \mid (\forall x \in X)(h(x) \sim x)\}.$$

Theorem 3.10 *For $l = 1, 2$ let $\mathcal{F}_l$ be thin Fl-spaces based on E_l, F_l, and $\varphi : H_{m_1}(\mathcal{F}_1) \cong H_{m_2}(\mathcal{F}_2)$. Then there is $\tau : \mathcal{F}_1 \cong \mathcal{F}_2$ such that τ induces φ.*

We shall prove here the special case when the F_l's are finite dimensional. In order to prove the full theorem, it is necessary to first prove the reconstruction theorem for infinite dimensional normed spaces. This reconstruction theorem is proved in [12]. See Part II of Section 3 there, and in particular Corollaries 3.20 and 3.21.

From here on $\mathcal{F} = \langle X, \sim \rangle$ denotes a foliated space, and H is either $H_1(\mathcal{F})$ or $H_0(\mathcal{F})$. Let $\mathcal{M} = \langle X_0, \Gamma \rangle$ denote an Fl-manifold such that $\mathcal{F}$ is an Fl-subspace of $\mathcal{M}$. We assume that $\mathcal{M}$ is based on E, F.

Proposition 3.11 *$\langle Ro(X), H \rangle$ is a local movement system.*

Proof: The easy proof is left to the reader. □

We shall represent the points of X by some objects of $M(Ro(X), H)$. This is quite similar to the way that good ultrafilters represented points when we dealt with K_{LCG}. In the present case however, the representing objects are more complex, and the representation requires more preparation.

Proposition 3.12 *(a) Let $\gamma, \delta \in \Gamma$. Then $\sim_\gamma \restriction (Rng(\gamma) \cap Rng(\delta)) = \sim_\delta \restriction (Rng(\gamma) \cap Rng(\delta))$.*

(b) Let $\gamma \in \Gamma$ and $L \in X_0/\sim^{\mathcal{M}}$. Then $L \cap Rng(\gamma)$ is the union of $\leq \aleph_0$ equivalence classes of $\sim_\gamma$.

Proof: (a) follows trivially from the definition of an Fl-manifold.

(b) follows easily from Part (a) and the fact that Γ is dominated by a countable set. □

Remark: We do not know how to get rid of the artificial assumption that Γ has a countable dominating set.

It will be convenient to assume that Γ is closed in the following sense. Suppose that $\bar{B}^E(y_0, r) \times \bar{B}^F(z_0, s) \subseteq \bar{B}^{E,F}$. Let $\alpha : \bar{B}^{E,F} \to \bar{B}^E(y, r) \times \bar{B}^F(z, s)$ be defined by $\alpha((y, z)) = (y_0 + \frac{1}{r}y, z_0 + \frac{1}{s}z)$. We assume that if $\gamma \in \Gamma$, then $\gamma \circ \alpha \in \Gamma$.

Definition 3.13 Let $U \in Ro(X)$.

(a) Let $INV(U) = \{V \in Ro(X) | U \mid (\forall g \in Sp_H(U))(g(V) = V)\}$.

(b) For $V \in INV(U)$ and $g \in Sp_H(U)$ let $L(V; U) = bd^U(V)$, $g_{V,U} = g \restriction L(V; U)$, $G(V; U) = \{g_{V,U} \mid g \in Sp_H(U)\}$ and $G_{Id}(V; U) = \{g \in Sp_H(U) \mid g_{V,U} = Id\}$.

(c) Let $H(E, F) = \{\alpha \in H(\bar{B}^{E,F}) \mid \alpha$ is an (E, F)-selfmap and $\alpha \restriction (\bar{B}^{E,F} - B^{E,F}) = Id\}$.

(d) For $\gamma \in \Gamma$ let $\bar{B}_\gamma^{\mathcal{F}} = Rng(\gamma) \cap X$ and $B_\gamma^{\mathcal{F}} = \gamma(B^{E,F}) \cap X$. We shall usually omit the superscript $\mathcal{F}$.

Proposition 3.14 *(a) For $\gamma \in \Gamma$ and $h \in H(E,F)$ let $h_\gamma = \gamma \circ h \circ \gamma^{-1} \lceil B_\gamma \cup Id \lceil (X - B_\gamma)$ and $H_\gamma = \{h_\gamma \mid h \in H(E,F)\}$. Then $H_\gamma \subseteq Sp_H(B_\gamma)$,*

(b) Let $\gamma \in \Gamma$. Let $L' \subseteq \bar{B}^E$ and $L \stackrel{def}{=} L' \times \bar{B}^F$. Let $h \in H$ be such that $h(\gamma(L)) = \gamma(L)$ and $h \lceil \gamma(L' \times S^F) = Id$. Then for every $x \in \gamma(L)$, $h(x) \sim_\gamma x$. In particular, for every $h \in Sp_H(B_\gamma)$ and $x \in B_\gamma$, $h(x) \sim_\gamma x$.

(c) Let $U \in Ro(X)$ and $V \in INV(U) - \{\emptyset\}$. Then $\langle L(V;U), G(V;U)\rangle$ is a local movement system.

Proof: (a) It is trivial that $H_\gamma \subseteq Sp_H(B_\gamma)$.

(b) For notational simplicity we assume that $\gamma = Id$. Let L' and h be as in (b). Suppose by contradiction that $x \in L$ is such that $h(x) \not\sim_\gamma x$. Let $x = (y,z)$ and $h(x) = (y',z')$. Hence $y \neq y'$. Let $z_1 \in S^F$ and $x_1 = (y, z_1)$. Then $h(x_1) = x_1$. Let J be the line segment connecting x and x_1. $A \stackrel{def}{=} p_{E,F}(h(J)) \ni y, y'$. So $|A| = 2^{\aleph_0}$. So $h(J)$ intersects $2^{\aleph_0}$ equivalence classes of $\sim_\gamma$. For every $u \in h(J)$, $u \sim x_1$. So $x_1/\!\sim$ intersects $2^{\aleph_0}$ equivalence classes of $\sim_\gamma$. This contradicts Proposition 3.12.

(c) Let $L = L(V;U)$ and $G = G(V;U)$. Since $V \in INV(U)$, L is invariant under $Sp_H(U)$. So G is a group. Let $x \in L$ and $T \in Nbr^L(x)$. We show that there is $g \in Sp_G(T)$ such that $g(x) \neq x$. Let $\gamma \in \Gamma$ be such that $x \in B_\gamma$. For notational simplicity we assume that $\gamma = Id$. We may also assume that $\bar{B}_\gamma \subseteq U$, and that $\bar{B}_\gamma \cap L \subseteq T$. Since L is invariant under $Sp_H(U)$, $L \cap \bar{B}_\gamma$ has the form $A \times \bar{B}^F$, where $A \subseteq \bar{B}^E$. Obviously, there is $h \in H(E,F)$ such that $h(x) \neq x$. It follows that $g \stackrel{def}{\equiv} h_\gamma \lceil L \in G$, and that $g \lceil (L - T) = Id$. So g is as required. □

Proposition 3.15 *(a) If $x \sim y$, then there is an arc $J \subseteq X$ connecting x and y such that for every $z \in J$, $x \sim z$.*

(b) Let $h \in H$ and $W \in Ro(X)$ be such that $g \stackrel{def}{=} h \lceil W \cup Id \lceil (X - W) \in H(X)$. Then $g \in H$.

Let $U \in Ro(X) - \{\emptyset\}$ and $V \in INV(U)$.

(c) The function $g/G_{Id}(V;U) \mapsto g \lceil L(V;U)$ is an isomorphism between $Sp_H(U)/G_{Id}(V;U)$ and $G(V;U)$.

(d) $G_{Id}(V;U) = \{gh \mid g \in Sp_H(V),\ h \in Sp_H(U -^{Ro} V)\}$.
($-^{Ro}$ denotes the difference operation in $Ro(X)$).

Remark: Part (d) has a real shortcoming. It does not transfer to the case when H is a group of diffeomorphisms. We did not find a way to fix this at this time. However, the method seems to work for the case that H is a group of bi-Lipschitz homeomorphisms.

Proof: The easy proofs of parts (a) and (c) are left to the reader.

(b) Let h, W and g be as in part (b). We show that $g \in H$. If $H = H_0(\mathcal{F})$, then clearly for every $x \in X$, $g(x) \sim x$. So $g \in H$.

Suppose that $H = H_1(\mathcal{F})$. Suppose by contradiction that $g \notin H$. Then by replacing g by g^{-1} if needed, we may assume that there are $x, y \in X$ such that $x \sim y$ and $g(x) \not\sim g(y)$. We may assume that $x \in W$ and $y \in X - W$. So $x \not\sim h(x)$. Let J be an arc connecting x and y such that for every $z \in J$, $z \sim x$. Then $h(J) \cap J = \emptyset$. Also $g(J) \subseteq J \cup h(J)$, $g(x) \in g(J) \cap h(J)$ and $y \in g(J) \cap J$. So g(J) is not connected, a contradiction.

(d) Let $f \in G_{Id}(U; V)$. We define $g = (f\restriction V) \cup (Id\restriction(X - V))$ and $h = fg^{-1}$. Clearly $g \in H(X)$. So by part (b), $g \in H$. Hence $h \in H$. Obviously, $g \in Sp_H(V)$ and $h \in Sp_H(U -^{Ro} V)$. So $G_{Id}(V; U) \subseteq \{gh \mid g \in Sp_H(V),\ h \in Sp_H(U -^{Ro} V)\}$. The inclusion in the other direction is trivial. □

By combining parts (c) and (d) of Proposition 3.15 together with part (b) of Proposition 3.14, we obtain an expressibility result. Loosely speaking, it says that there is a first order formula in the language of $M(Ro(X), H)$ that says about $U \in M(Ro(X), H)$, $V \in INV(U) - \{\emptyset\}$ and $g, h \in Sp_H(U)$ that $var(g_{V,U}) \leq var(h_{V,U})$. Similarly, there are formulas expressing the facts that $var(g_{V,U}) = var(h_{V,U})$ and that $f_{V,U}(var(g_{V,U})) = var(h_{V,U})$. As expected, these expressibilty facts translate into certain intermediate reconstruction results.

There is a formal framework in which these expressibility and reconstruction facts can be stated neatly and precisely. However, we decided to spare the reader from these formalities. A detailed exposition of this framework is presented e.g. in [15] Chapter 1. We shall proceed somewhat informally, relying on the reader's intuition to translate expressibility facts into reconstruction results.

Definition 3.16 Let $U \in M(Ro(X), H)$, $V \subset INV(U) - \{\emptyset\}$ and $L = L(V; U)$. We say that $\langle U, V\rangle$ is a *usable pair* if for every $T \in Ro(L) - \{\emptyset\}$ there are $\gamma \in \Gamma$, $y \in B^E$, $z \in B^F$ and $r > 0$ such that
$T \cap \gamma(B^E(y, r) \times B^F(z, r)) = \gamma(\{y\} \times B^F(z, r))$.

We shall prove three main facts about usable pairs. (1) The property of being a usable pair is expressible in $M(Ro(X), H)$. (2) If F is finite dimensional, and $\langle U, V\rangle$ is a usable pair, then $\langle L(V; U), G(V; U)\rangle \in K_{RCM}$, and $Dc(L(V; U), G(V; U))$ is dense in $L(V; U)$. (3) For every $x \in X$ there is a usable pair $\langle U, V\rangle$ such that $x \in Dc(L(V; U))$.

Definition 3.17 (a) Let B be a complete atomless BA and $G \leq Aut(B)$. Let $a, b \in B - \{0\}$. Then $a \prec^G b$, if for every $c \in B\restriction b - \{0\}$ there is $g \in Sp_G(b)$ such that $g(a) \leq c$.

(b) let ψ_{sml} be the following sentence in the language of $M(B, G)$. $\psi_{sml} \equiv (\forall a \in B - \{0\})(\exists b, c \in B\restriction a - \{0\})(b \prec c)$.

Lemma 3.18 *(a) Let $U \in Ro(X)$ and $V \in INV(U) - \{\emptyset\}$. Then the following conditions are equivalent.*

(1) $\langle U, V\rangle$ *is usable;*
(2) $M(Ro(L(V;U)), G(V;U)) \models \psi_{sml}$.

(b) There is a first order formula $\psi^*_{sml}(U, V)$ *in the language of* $M(Ro(X), H)$ *such that for every* $U \in Ro(X)$ *and* $V \in INV(U) - \{\emptyset\}$,
$M(Ro(X), H) \models \psi^*_{sml}[U, V]$ *iff* $M(Ro(L(V;U)), G(V;U)) \models \psi_{sml}$.

Proof: Part (b) follows trivially from 3.15(c) and (d) and 3.14(b).

(a) Let U and V be as in part (a). Let $L = L(V;U)$ and $G = G(V;U)$. It follows easily from 3.14(a) that if $\langle U, V\rangle$ is usable, then $M(Ro(L), G) \models \psi_{sml}$.

Suppose next that $M(Ro(L), G) \models \psi_{sml}$. Let $T \in Ro(L) - \{\emptyset\}$. Let $T^* \in Ro(X)|U$ be such that $T = T^* \cap L$. Let $\gamma \in \Gamma$ be such that $B_\gamma \cap T \neq \emptyset$. For notational simplicity we assume that $\gamma = Id$. We may assume that $\bar{B}_\gamma \subseteq T^*$. Hence since $V \in INV(U)$, there are $V' \subseteq \bar{B}^E$ and $L' \subseteq \bar{B}^E$ such that $V \cap \bar{B}_\gamma = V' \times \bar{B}^F$ and $T \cap \bar{B}_\gamma = L \cap \bar{B}_\gamma = L' \times \bar{B}^F$. Since $M(Ro(L), G) \models \psi_{sml}$ there are $R, S \in Ro(L)|(T \cap B_\gamma) - \{\emptyset\}$ such that $R \prec^G S$.

Suppose by contradiction that $|p_{E,F}(S)| > 1$. Let $S_1, S_2 \in Ro(L)|S - \{\emptyset\}$ be such that $p_{E,F}(S_1) \cap p_{E,F}(S_2) = \emptyset$. Since $R \prec S$, there are $g_i \in Sp_G(S)$, $i = 1, 2$, such that $g_i(R) \subseteq S_i$. Clearly, $g_i \in Sp_G(L \cap B_\gamma)$. By the definition of G, there are $h_i \in H$ such that $g_i = h_i|L$. So $h_i(L' \times \bar{B}^F) = g_i(L' \times \bar{B}^F) = L' \times \bar{B}^F$, and $h_i|(L' \times S^F) = g_i|(L' \times S^F) = Id$. But $h_2^{-1} \circ h_1(S_1) = S_2$. The properties of $h_2^{-1} \circ h_1$ contradict Proposition 3.14(b). So $|p_{E,F}(S)| = 1$.

Let $S^* \in Ro(X)$ be such that $S = S^* \cap L$, and let $(y, z) \in S$. Let $r > 0$ be such that $X \cap (B^E(y, r) \times B^F(z, r)) \subseteq S^*$. It thus follows that $T \cap (B^E(y, r) \times B^F(z, r)) = \{y\} \times B^F(z, r)$. □

If $\langle U, V\rangle$ is a usable pair, then $\langle L(V;U), G(V;U)\rangle$ is rather well-behaved. We shall indicate this situation later, but for the time being we restrict ourselves to the case that F is finite dimensional. Under this assumption we obtain the following conclusion.

Corollary 3.19 *Suppose that* F *is finite dimensional. Let* $\langle U, V\rangle$ *be a usable pair. Then* $\langle L(V;U), G(V;U)\rangle \in K_{RCM}$, *and* $Dc(L(V;U), G(V;U))$ *is a dense open subset of* $L(V;U)$.

Proof: The proof is trivial. In fact, $Dc(L(V;U), G(V;U))$ is a manifold based on F, and $G(V;U)$ contains all the homeomorphisms of $L(V;U)$ which are the identity outside the range of a chart of $L(V;U)$. □

We have proved two facts on expressibility that have yet to be translated into a reconstruction result. These are Proposition 3.15(c),(d) and Lemma 3.18(b). In order to state this reconstruction result, we define a new structure.

Definition 3.20 (a) Let $Usbl(X) \stackrel{def}{=} \{\langle U, V, x\rangle \mid \langle U, V\rangle$ is a usable pair, and $x \in Dc(L(V;U))\}$. Note that H has a natural action on $Usbl(X)$. Let $p_1(\langle U, V, x\rangle) = U$ and $p_2(\langle U, V, x\rangle) = V$.

(b) Let $Mus(X,H) \stackrel{def}{=} \langle Ro(X), Usbl(X), H; p_1, p_2, \leq^{Ro(X)}, \circ, Ap\rangle$, where Ap describes the action of H on $Ro(X) \cup Usbl(X)$.

Note every $x \in X$ is represented in $Mus(X,H)$ by infinitely many objects. That is, if $x \in Dc(L(V;U))$ and $x \in Dc(L(V';U'))$, then both $\langle U,V,x\rangle$ and $\langle U',V',x\rangle$ represent x. We shall later see that the fact that two triples represent the same point is expressible in $Mus(X,H)$.

Corollary 3.21 *For $l = 1,2$ let $\mathcal{F}^l = \langle X^l, \sim^l\rangle$ be thin foliated spaces based on E^l and F^l. Suppose that F^l is finite dimensional. Let H^l be either $H_0(\mathcal{F})$ or $H_1(\mathcal{F})$, and $\varphi : H^1 \cong H^2$. Then φ can be extended to an isomorphism between $Mus(X^1,H^1)$ and $Mus(X^2,H^2)$.*

Proof: We indicate the steps in the proof. All of them are similar to arguments that have already been presented.

We begin by applying 3.11 and 2.10 to H^1 and H^2, thus obtaining an isomorphism between $M(Ro(X^1),H^1)$ and $M(Ro(X^2),H^2)$.

Then we apply 3.15(c) and (d) to capture the set of all $G(V;U)$'s in $M(Ro(X^l),H^l)$.

Next we apply 3.14(c) and 2.10 to the set of $G(V;U)$'s captured above. By this we capture the set of $M(Ro(L(V;U)),G(V;U))$'s, for all $U \in Ro(X)$ and $V \in INV(U) - \{\emptyset\}$.

The set of all usable $M(Ro(L(V;U)),G(V;U))$'s is then captured by using 3.18.

Finally we apply 3.19 and 3.4 to construct an isomorphism between $Mus(X^1,H^1)$ and $Mus(X^2,H^2)$. □

There are still three easy steps before the proof of Theorem 3.10 can be concluded.

Proposition 3.22 *For every $x \in X$ there is a usable pair $\langle U,V\rangle$ such that $x \in Dc(L(V;U))$.*

Proof: Let $x \in X$. Let $\gamma \in \Gamma$ be such that $x \in B_\gamma$. We assume that $\gamma = Id$. Since $\mathcal{F}$ is thin, there is a thin $Y' \subseteq B^E$ such that $B_\gamma = Y' \times B^F$. Suppose that $x = (y,z)$. Let $U' \in Nbr^{Y'}(y)$ and $V' \in Ro(Y')|U'$ be such that $bd^{U'}(V') = \{y\}$. Let $U = U' \times B^F$ and $V = V' \times B^F$. Then $\langle U,V\rangle$ is a usable pair, and $x \in L(V;U)$. □

Proposition 3.23 *(a) There is a first order formula $\psi_\varepsilon(u,W)$ in the language of $Mus(X,H)$ such that for every $\langle U,V,x\rangle \in Usbl(X,H)$ and $W \in Ro(X)$: $Mus(X,H) \models \psi_\varepsilon[\langle U,V,x\rangle, W]$ iff $x \in W$.*

(b) There is a first order formula $\psi_{Eq}(u,v)$ in the language of $Mus(X,H)$ such that for every $\langle U,V,x\rangle, \langle U',V',x'\rangle \in Usbl(X,H)$:
$Mus(X,H) \models \psi_{Eq}[\langle U,V,x\rangle, \langle U',V',x'\rangle]$ iff $x = x'$.

Proof (a) Let $\psi_\varepsilon(u,W) \stackrel{def}{\equiv} (\exists g \in Sp_H(W) \cap Sp_H(p_1(u)))(g(u) \neq u)$. It is easy to check that $\psi_\varepsilon(u,W)$ is as required.

(b) Let $\psi_{Eq}(u,v) \stackrel{def}{\equiv} (\forall W \in Ro(X))(\psi_\varepsilon(u,W) \leftrightarrow \psi_\varepsilon(v,W))$. It is trivial that $\psi_{Eq}(u,v)$ is as required. □

Proof of Theorem 3.10: We prove the special case when F is finite dimensional. Combine Corollary 3.21, Proposition 3.22 and Proposition 3.23. □

Proof of Theorem 3.8: The set $\{\langle \boldsymbol{T}_{\mathcal{L}}, H_i(\boldsymbol{T}_{\mathcal{L}})\rangle \mid \mathcal{L}$ is a parallel family and $i \in \{0,1\}\}$ is contained in the class $\{\langle X, H_i(\mathcal{F})\rangle \mid \mathcal{F} = \langle X, \sim\rangle$ is a thin foliated space based on some E, F, and F is finite dimensional, and $i \in \{0,1\}\}$. □

We shall next indicate how to prove Theorem 3.10 without the extra assumption that F is finite dimensional.

Definition 3.24 Let F be a normed vector space over $\mathbb{R}$ and X be a topological space.

(a) Let $\gamma : \bar{B}^F \to X$ be a map with the following properties: (1) $\gamma : \bar{B}^F \cong Rng(\gamma)$; (2) $\gamma(\bar{B}^F)$ is closed in X; and (3) $\gamma(B^F)$ is open in X. Then γ is called a *normal* F*-map*.

(b) Let Γ be a set of normal F-maps such that $\bigcup\{Rng(\gamma) \mid \gamma \in \Gamma\}$ is a dense subset of X. Then we say that $\langle X, \Gamma\rangle$ is a regionally normed space based on F with atlas Γ.

(c) Let $\langle X, \Gamma\rangle$ be a regionally normed space based on F, and $H \leq H(X)$. We say that H is adequate, if $\{(\gamma \circ h \circ \gamma^{-1}) \cup (Id{\restriction}(X - Rng(\gamma))) \mid h \in H(\bar{B}^F)\} \subseteq H$.

We say that $\langle X, H\rangle$ is an *adequate pair* if there is Γ such that $\langle X, \Gamma\rangle$ is a regionally normed space, and H is an adequate subgroup of $H(X)$.

Proposition 3.25 *If $\langle U, V\rangle$ is usable, then $\langle L(V;U), G(V;U)\rangle$ is adequate.*

Proof: The trivial proof is left to the reader. □

In [12] Theorem 3.43, it was proved that for the class of adequate $\langle X, H\rangle$'s, $Dc(X,H)$ can be captured from $M(Ro(X), H)$. If in the proof of Theorem 3.10 we apply Theorem 3.43 of [12] rather than Theorem 3.4, then the finite dimensionality of F is never used.

4 Linear Orders and Circular Orders

In this section we shall demonstrate how the Reconstruction Theorem for Local Movement Systems can be applied to linear orders and circular orders. We shall not present the strongest known results, but rather try to give a unified treatment of the full automorphism groups of such structures. We shall deal with four closely related types of structures: linear orders, circular orders, so called "Equal Direction Structures" and "Equal Orientation Structures".

The main result in this section is stated in Theorem 4.6. It more or less says that under certain homogeneity assumptions on such structures, their Dedekind completion can be reconstructed from their automorphism group.

The interpretation uses only first order formulas. As a result of this it becomes possible in many cases to distinguish between such automorphism groups by first order sentences.

The results presented in this section are known. Some of the history of these results is described in the introduction. However, I am not aware of any published presentation of a first order reconstruction for the class of circular orders.

Definition 4.1 Let $\langle L, <\rangle$ be a dense linearly ordered set.

(a) Assume that $\langle L, <\rangle$ does not have a maximum. *The circular order* based on $\langle L, <\rangle$ is defined as $CR(\langle L, <\rangle) \stackrel{def}{=} \langle L, Cr\rangle$, where $Cr = \{\langle x, y, z\rangle \in L^3 \mid (x < y < z) \vee (z < x < y) \vee (y < z < x)\}$.

We next define structures associated with $\langle L, <\rangle$ and $CR(\langle L, <\rangle)$ whose automorphism groups contain respectively, the order reversing permutations of $\langle L, <\rangle$ and the orientation reversing permutations of $CR(\langle L, <\rangle)$. Let $\vec{x} \in L^n$ denote $\langle x_1, \ldots, x_n\rangle$.

(b) We define the *"Equal Direction Structure"* based on $\langle L, <\rangle$ as $ED(\langle L, <\rangle) \stackrel{def}{=} \langle L, Ed\rangle$, where $Ed = \{\vec{x} \,\hat{}\, \vec{y} \mid \vec{x}, \vec{y} \in L^2 \text{ and } ((x_1 < x_2) \wedge (y_1 < y_2)) \vee ((x_1 > x_2) \wedge (y_1 > y_2))\}$. Here $\vec{x} \,\hat{}\, \vec{y}$ denotes the concatenation of $\vec{x}$ and $\vec{y}$.

(c) We define the *"Equal Orientation Structure"* based on $\langle L, <\rangle$ as $EO(\langle L, <\rangle) \stackrel{def}{=} \langle L, Eo\rangle$, where $Eo = \{\vec{x} \,\hat{}\, \vec{y} \mid \vec{x}, \vec{y} \in L^3$ and $((Cr(x_1, x_2, x_3) \wedge Cr(y_1, y_2, y_3)) \vee (Cr(x_2, x_1, x_3) \wedge Cr(y_2, y_1, y_3)))\}$.

Note that $EO(\langle L, <\rangle)$ is defined only for L's that do not have a maximum.

When needed, we shall denote Cr, Ed and Eo by $Cr(\langle L, <\rangle)$, $Ed(\langle L, <\rangle)$ and $Eo(\langle L, <\rangle)$ respectively.

Structures of type $\langle L, <\rangle$ and $ED(\langle L, <\rangle)$ in which L has no maximum and no minimum, and structures of the types defined in parts (a) and (c) are called *almost ordered structures*. We also say that they are based on $\langle L, <\rangle$. We shall use L alone to stand for $\langle L, <\rangle$.

(d) Let N be an almost ordered structure. We define N^{s} and N^{r}, the *strict and relaxed versions* of N as follows. If $N = L$ or $N = ED(L)$, then $N^{s} = L$ and $N^{r} = ED(L)$. If $N = CR(L)$ or $N = EO(L)$, then $N^{s} = CR(L)$ and $N^{r} = EO(L)$.

(e) Let $\langle L, <\rangle$ be a linearly ordered set. We denote the reverse ordering of $<$ by $<^*$, and $\langle L, <\rangle^*$ denotes $\langle L, <^*\rangle$. We abbreviate $\langle L, <\rangle^*$ by L^*.

Example: Let C denote the circle. Let $Cw = \{\vec{x} \in C^3 \mid x_1, x_2, x_3$ are clockwise oriented$\}$. Then $\langle C, Cw\rangle$ is isomorphic to $CR(\langle [0,1), <\rangle)$. However, $\langle C, Cw\rangle$ is not isomorphic to $CR(\langle \mathbb{R}, <\rangle)$.

On the other hand, $CR(\mathbb{Q}) \cong CR(\mathbb{Q}^{\geq 0})$.

We denote $\langle L, < \rangle$ by L, and from now on we assume that L is a dense linear ordering. If L has a minimum, then $min(L)$ denotes that minimum. If L has a maximum, then $max(L)$ denotes that maximum.

Definition 4.2 Let L be a linearly ordered set.

(a) We say that L and $ED(L)$ are *adequately homogeneous* (or for short, we say that they are *adequate*) if for every $\vec{x}, \vec{y} \in L^2$: if $x_1 < x_2$ and $y_1 < y_2$, then there is $g \in Aut(L)$ such that $g(\vec{x}) = \vec{y}$.

(b) We say that $CR(L)$ and $EO(L)$ are *adequately homogeneous* (or for short, we say that they are *adequate*) if for every $\vec{x}, \vec{y} \in L^3$: if $Cr(x_1, x_2, x_3)$ and $Cr(y_1, y_2, y_3)$, then there is $g \in Aut(CR(L))$ such that $g(\vec{x}) = \vec{y}$.

Our definitions are set in such a way that every adequate structure is almost ordered.

Definition 4.3 (a) For distinct $x, y \in L$, let $(x, y) = \{z \in L \mid x < z < y\}$, $(x, \infty) = \{z \in L \mid x < z\}$, $(-\infty, x) = \{z \in L \mid z < x\}$ and $(x, y)^{Cr} = \{z \in L \mid Cr(x, z, y)\}$. Nonempty sets of the first three types are called *intervals* of L, nonempty sets of the second and third types are called *rays* of L, and nonempty sets of the forth type are called *circular intervals* of L.

(b) Let $\tau(L)$ be the topology on L generated by all intervals of L, and $\tau^{Cr}(L)$ be the topology on L generated by all circular intervals of L. If $N = \langle L, < \rangle$ or $N = ED(L)$, then the set of N-*intervals* is the set of intervals of L, and the set of *proper* N-*intervals* is the set of N-intervals which are not rays. Similarly, if $N = CR(L)$ or $N = EO(L)$, then the set of N-intervals is the set of circular intervals of L, and every N-interval is a proper N-interval.

If $N = \langle L, < \rangle$ or $N = ED(L)$, then $\tau^N \stackrel{def}{=} \tau(L)$. If $N = CR(L)$ or $N = EO(L)$, then $\tau^N \stackrel{def}{=} \tau^{CR}(L)$. We shall regard N as a topological space equipping it with the topology τ^N.

Note that if N is almost ordered, then every proper open subset U of N can be uniquely represented as the union of a set $\mathcal{U}$ of pairwise disjoint N-intervals. Also, $U \in Ro(N)$ iff no two distinct members of $\mathcal{U}$ have a common end point. In particular, every N-interval is a member of $Ro(N) - \{0, 1\}$.

We next set the conventions for the Dedekind completion of an almost ordered structure. Recall that we assume that L is a dense linear ordering.

(c) Let $\bar{\bar{L}}$ denote the full Dedekind completion of L. That is, $\bar{\bar{L}}$ includes a minimum and a maximum. Let $\bar{L}$ denote $\bar{\bar{L}} - \{min(\bar{\bar{L}}), max(\bar{\bar{L}})\}$, and $\bar{L}^{Cr}$ denote $\bar{\bar{L}} - \{max(\bar{\bar{L}})\}$.

(i) If $N = \langle L, < \rangle$, then $\bar{N}$ is $\bar{L}$.

(ii) If $N = ED(L)$, then $\bar{N} \stackrel{def}{=} ED(\bar{L})$.

Assume that L has no maximum.

(iii) If $N = CR(L)$, then $\bar{N} \overset{def}{=} CR(\bar{L}^{Cr})$.
(iv) If $N = EO(L)$, then $\bar{N} \overset{def}{=} EO(\bar{L}^{Cr})$.
Recall that if $M = \langle A, \ldots \rangle$ is a structure, then A is denoted by $|M|$.

Note that if N is almost ordered, then $\bar{N}$ is almost ordered.

Proposition 4.4 *(a)* $Aut(L) \leq Aut(ED(L)) \leq H(\langle L, \tau(L)\rangle)$. *Also,* $Aut(ED(L)) = Aut(L) \cup \{g \mid g : L \cong L^*\}$, *and the index of* $Aut(L)$ *in* $Aut(ED(L))$ *is* ≤ 2.

(b) $Aut(CR(L)) \leq Aut(EO(L)) \leq H(\langle L, \tau^{Cr}(L)\rangle)$. *Also,* $Aut(EO(L)) = Aut(CR(L)) \cup \{g \mid g : CR(L) \cong CR(L^*)\}$, *and the index of* $Aut(EO(L))$ *in* $Aut(CR(L))$ *is* ≤ 2.

(c) Let N *be an almost ordered structure. For every* $g \in Aut(N)$ *there is a unique* $\bar{g} \in Aut(\bar{N})$ *such that* $\bar{g}$ *extends* g. *Let* $Aut^{-}(N) = \{\bar{g} \mid g \in Aut(N)\}$.

The function γ^N *defined by:*

$$U \mapsto int^{\bar{N}}(cl^{\bar{N}}(U)),\ U \in Ro(N);\ \ g \mapsto \bar{g},\ g \in Aut(N)$$

is an isomorphism between $M(Ro(N), Aut(N))$ *and* $M(Ro(\bar{N}), Aut^{-}(N))$. *We identify* $Aut(N)$ *with* $Aut^{-}(N)$.

(d) Let N *be an adequate structure of the form* $\langle L, < \rangle$ *or* $CR(L)$. *Let* $\vec{x}, \vec{y} \in |N|^n$ *be such that: (i) if* $N = \langle L, < \rangle$, *then for every* $1 \leq i < j \leq n$, $x_i < x_j$ *and* $y_i < y_j$; *(ii) if* $N = CR(L)$, *then for every* $1 \leq i < j < k \leq n$, $\langle x_i, x_j, x_k\rangle, \langle y_i, y_j, y_k\rangle \in Cr$. *Then there is* $g \in Aut(N)$ *such that for every* $i = 1 \ldots n$, $g(x_i) = y_i$.

If in addition, $u, v \in |N|$ *and* $\vec{x}, \vec{y} \in (u,v)^N$, *then such a* g *exists in* $Sp_{Aut(N)}((u,v)^N)$.

(e) If N *is adequate, then* $\langle Ro(N), Aut(N)\rangle$ *is a local movement system.*

Proof: Trivial. □

Remark: Part (d) in the above proposition is the only fact that we need to use in order to construct automorphisms. It will be used at several points in the proof of the main theorem, not always with an explicit reference.

Corollary 4.5 *For* $l = 1, 2$ *let* N_l *be an adequate almost ordered structure. Suppose that* $\varphi : Aut(N_1) \cong Aut(N_2)$. *Then there is* $\psi : M(Ro(\bar{N}_1), Aut(N_1)) \cong M(Ro(\bar{N}_2), Aut(N_2))$ *such that* ψ *extends* φ.

We now state the main result of this section.

Theorem 4.6 *For* $l = 1, 2$ *let* N_l *be an adequate almost ordered structure.*

(a) Suppose that $\varphi : Aut(N_1) \cong Aut(N_2)$. *Then there is* $\alpha : \bar{N}_1^{\mathrm{r}} \cong \bar{N}_2^{\mathrm{r}}$ *such that* α *induces* φ.

(b) Suppose that $Aut(N_1)$ *and* $Aut(N_2)$ *are elementarily equivalent. (That is, every first order sentence that holds in one of them, holds in the other.) Then the structures* $Act(\bar{N}_1^{\mathrm{r}}, Aut(N_1))$ *and* $Act(\bar{N}_2^{\mathrm{r}}, Aut(N_2))$ *are also elementarily equivalent.*

Remark: Note that $\bar{N}$ is a locally compact space. So we can use Theorem 3.1 as the main step in the proof of part (a) in the above theorem. However, the proof of Theorem 3.1 uses second order formulas and second order objects to represent points. We shall obtain a stronger type of a reconstruction result by using only first order formulas and first order objects. This extra strength is the content of part (b) of the theorem.

Lemma 4.7 *Let N be adequate. Then*

$$Aut(N^{s}) = \{g_1 g_2 \mid g_1, g_2 \in Aut(N) \text{ and } \mathit{fix}(g_1), \mathit{fix}(g_2) \neq 0\}.$$

Proof: Let $N = \langle L, R\rangle$ and $N^{s} = \langle L, R^{s}\rangle$. Let

$$G = \{g_1 g_2 \mid g_1, g_2 \in Aut(N) \text{ and } \mathit{fix}(g_1), \mathit{fix}(g_2) \neq 0\}.$$

We prove that $G \subseteq Aut(N^{s})$. If $N = N^{s}$, then there is nothing to prove. Suppose otherwise. Let $g \in Aut(N)$ be such that $\mathit{fix}(g) \neq 0$. So there is $\vec{x} \in R^{s}$ such that $g(\vec{x}) = \vec{x}$. Let $\vec{y} \in R^{s}$. Then $\vec{x}\,\hat{}\,\vec{y} \in R$. So $g(\vec{x})\,\hat{}\,g(\vec{y}) \in R$. That is, $\vec{x}\,\hat{}\,g(\vec{y}) \in R$. Hence $g(\vec{y}) \in R^{s}$. Hence $g \in Aut(N^{s})$.

We next prove that for every $g \in Aut(N^{s})$ there are $g_1, g_2 \in Aut(N^{s})$ such that $g = g_1 g_2$ and $\mathit{fix}(g_1), \mathit{fix}(g_2) \neq 0$. Suppose first that $N^{s} = \langle L, <\rangle$. Let x be such that $g(x) \neq x$. We may assume that $g(x) > x$. Let $y > g^2(x)$. Then there is $g_1 \in Aut(L)$ such that $g_1 \restriction (y, \infty) = Id$ and $g_1 \restriction (x, g(x)) = g \restriction (x, g(x))$. Let $g_2 = g_1^{-1} g$. Hence $g_1 g_2 = g$ and $g_2 \restriction (x, g(x)) = Id$.

Suppose next that $N^{s} = CR(L)$. There is an interval $(x, y)^{Cr}$ such that $g((x, y)^{Cr}) \cap (x, y)^{Cr} = \emptyset$. Let $u = g(x)$ and $v = g(y)$. Hence $u, v \notin (x, y)^{Cr}$. We may assume that $v \neq x$. It follows easily that $y, u \in (x, v)^{Cr}$ and that $y \in (x, u)^{Cr}$. Let $w, z \in (v, x)^{Cr}$ be such that $\langle v, w, z\rangle \in Cr$.

It is easy to construct $g_1 \in Aut(N^{s})$ such that $g_1 \restriction (x, y)^{Cr} = g \restriction (x, y)^{Cr}$, $g_1((z, w)^{Cr}) = (z, w)^{Cr}$ and $g_1 \restriction (w, z)^{Cr} = Id$. Let $g_2 = g_1^{-1} g$. Then $(u, v)^{Cr} \subseteq \mathit{fix}(g_2)$, and $g = g_1 g_2$. □

Corollary 4.8 *There is a first order formula $\varphi_s(g)$ such that for every adequate N and $g \in Aut(N)$: $M(Ro(N), Aut(N)) \models \varphi_s[g]$ iff $g \in Aut(N^{s})$.*

Proof: This follows trivially from the previous lemma. □

Let us now see that the set of proper intervals of $\bar{N}$ is definable in $M(Ro(\bar{N}), Aut(N))$ by a first order formula.

Let B be a complete atomless BA and $G \leq Aut(B)$. Let $a \in B$. We say that a is *flexible* in $\langle B, G\rangle$, if for every nonzero $b, c \leq a$ there is $g \in Sp_G(a)$ such that $g(b) \cdot c \neq 0$.

Note that there is a first order formula $\varphi_{Flx}(x)$ in the language of $M(B, G)$ that expresses the fact that x is flexible in $\langle B, G\rangle$.

Proposition 4.9 *Let N be an adequate almost ordered structure and $U \in Ro(\bar{N}) - \{0,1\}$. Then:*

(a) U is an $\bar{N}$-interval iff U is flexible in $\langle Ro(\bar{N}), Aut(N)\rangle$.

(b) The formula $\varphi_{Invl}(U) \stackrel{def}{\equiv} (U \in Ro(\bar{N}) - \{0,1\}) \wedge \varphi_{Flx}(U)$ defines the set of $\bar{N}$-intervals in every adequate almost ordered structure N.

(c) The formula

$$\varphi_{Pinvl}(U) \stackrel{def}{\equiv} \varphi_{Invl}(U) \wedge (\exists g \in Aut(N^s))((U \not\leq g(U)) \wedge (g(U) \not\leq U))$$

defines the set of proper intervals in every adequate almost ordered structure N.

Proof: The proof is trivial. □

There is now a choice of ways of representing the points of $\bar{N}$ by objects from $M(Ro(\bar{N}), Aut(N))$. We shall represent these points by pairs of $\bar{N}$-intervals with a common endpoint. This representation has the advantage of working for all four types of almost ordered structures using exactly the same formulas.

Definition 4.10 Let N be an adequate almost ordered structure. We shall use a simplified notation for $\bar{N}$-intervals. Let $x, y \in \bar{N}$ be distinct.

(i) If N is of the form $\langle L, <\rangle$ or $CR(L)$, then (x,y) denotes respectively $\{z \in \bar{L} \mid x < z < y\}$ and $\{z \in \bar{N} \mid \langle x,z,y\rangle \in Cr(\bar{L}^{Cr})\}$.

(ii) If for some L, $N = ED(L)$, then choose $L^N \in \{L, L^*\}$, and define (x,y) to be $\{z \in \bar{L}^N \mid x < z < y\}$.

(iii) If for some L, $N = EO(L)$, then choose $L^N \in \{L, L^*\}$ such that $N = EO(L^N)$. We may assume that $\bar{N} = EO(\bar{L}^{Cr})$. Now, define (x,y) to be $\{z \in \bar{N} \mid \langle x,z,y\rangle \in Cr((\overline{L^N})^{Cr})\}$.

Note that in (ii) and (iii) one needs to assume the existence of a global choice function. But we ignore this little inaccuracy.

Let $x \in |\bar{N}|$ and U, V be proper $\bar{N}$-intervals. We say that $\langle U, V\rangle$ *represents* x if there are distinct $y, z \in \bar{N} - \{x\}$ such that either $U = (x,y)$ and $V = (x,z)$ or $U = (y,x)$ and $V = (z,x)$.

A pair $\langle U, V\rangle$ of proper $\bar{N}$-intervals is called a *representative* if it represents some $x \in \bar{N}$; x is then denoted by $x_{U,V}$. Let $Rep(N)$ denote the set of representatives in $\bar{N}$.

Let Eqr^N be the following relation on $Rep(N)$.

$$Eqr^N = \{\langle \vec{U}, \vec{V}\rangle \in (Rep(N))^2 \mid x_{\vec{U}} = x_{\vec{V}}\}.$$

Lemma 4.11 *(a) There is a first order formula $\varphi_{Rep}(U,V)$ in the language of $M(Ro(\bar{N}), Aut(N))$ such that for every adequate N and every $U, V \in Ro(N)$: $M(Ro(\bar{N}), Aut(N)) \models \varphi_{Rep}[U,V]$ iff $\langle U,V\rangle \in Rep(N)$.*

(b) There is a first order formula $\varphi_{Eqr}(\vec{U}, \vec{V})$ in the language of $M(Ro(\bar{N}), Aut(N))$ such that for every adequate N and every $\vec{U}, \vec{V} \in Rep(N)$: $M(Ro(\bar{N}), Aut(N)) \models \varphi_{Eqr}[\vec{U}, \vec{V}]$ iff $x_{\vec{U}} = x_{\vec{V}}$.

Proof: Let $-^{Ro}$ denote the difference operation in the Boolean algebra $Ro(\bar{N})$.

(a) Let $\varphi_{Rep}(U,V) \overset{def}{\equiv} \varphi_{Pinvl}(U) \wedge \varphi_{Pinvl}(V) \wedge$
$(((U < V) \wedge \varphi_{Pinvl}(V -^{Ro} U)) \vee ((V < U) \wedge \varphi_{Pinvl}(U -^{Ro} V)))$.

It is easy to see that $\varphi_{Rep}(U,V)$ is as required.

(b) We define the "same side equivalence" of representatives. Let $\vec{U}, \vec{V} \in Rep(N)$. Then $\vec{U}$ *Sseq* $\vec{V}$, if there are $x, u_1, u_2, v_1, v_2 \in |\bar{N}|$ such that either for every $i \in \{1,2\}$, $U_i = (u_i, x)$ and $V_i = (v_i, x)$,
or for every $i \in \{1,2\}$, $U_i = (x, u_i)$ and $V_i = (x, v_i)$.

Let $\varphi_{Sseq}(\vec{U},\vec{V}) \overset{def}{\equiv} \bigwedge_{i,j \in \{1,2\}}((U_i = V_j) \vee \varphi_{Rep}(U_i, V_j))$.

Claim 1: Let N be adequate and $\vec{U}, \vec{V} \in Rep(N)$. Then $\vec{U}$ *Sseq* $\vec{V}$ iff $M(Ro(\bar{N}), Aut(N)) \models \varphi_{Sseq}[\vec{U},\vec{V}]$.

Proof of Claim 1: It is trivial that if $\vec{U}$ *Sseq* $\vec{V}$, then
$M(Ro(\bar{N}), Aut(N)) \models \varphi_{Sseq}[\vec{U},\vec{V}]$.

Let $\vec{U}, \vec{V} \in Rep(\bar{N})$, and assume that $M(Ro(\bar{N}), Aut(N)) \models \varphi_{Sseq}[\vec{U},\vec{V}]$. Then U_1 and U_2 are comparable under inclusion and have precisely one common endpoint x_{U_1,U_2}. Also, V_1 is comparable under inclusion with each U_i and has at least one endpoint in common with each. So V_1 must have x_{U_1,U_2} as an endpoint and must lie on the same side of x_{U_1,U_2} as do U_1 and U_2. This applies likewise to V_2, proving Claim 1.

The equivalences classes of *Sseq* represent every $x \in \bar{N}$ twice.
Both $\langle (y_1, x), (y_2, x)\rangle / Sseq$ and $\langle (x, z_1), (x, z_2)\rangle / Sseq$ represent x. Our next goal is to find a formula that expresses the fact that two representatives represent the same point.

Let $\varphi_{Eqr}(\vec{U},\vec{V}) \overset{def}{\equiv} (\forall g \in Aut(N))(\varphi_{Sseq}(\vec{U}, g(\vec{U})) \rightarrow \varphi_{Sseq}(\vec{V}, g(\vec{V})))$.

If N is adequate, then for every distinct $x, y \in \bar{N}$ there is $g \in Aut(N)$ such that $g(x) = x$ and $g(y) \neq y$. It follows easily from this fact that φ_{Eqr} is as required. □

It now remains to show that the relation of $\bar{N}^{\mathrm{r}}$ (which by Definition 4.1(d) is either *Ed* or *Eo*) is definable by a first order formula in $M(Ro(\bar{N}), Aut(N))$.

For a linear ordering L let $Edt(L) \overset{def}{=} \{\vec{x}\,\hat{}\,\vec{y} \mid \vec{x}, \vec{y} \in L^3$ are 1-1 sequences, and the set $\{\langle x_i, y_i\rangle \mid i = 1,2,3\}$ is an order preserving function$\}$.
Note that $Edt(L)$ is first order definable in $ED(L)$, and $Ed(L)$ is first order definable in $\langle L, Edt(L)\rangle$. So it does not matter whether we capture *Ed* or we capture *Edt*. For an almost ordered N based on L, let the "relaxed uniformized relation" of N be defined as follows: if $N = \langle L, <\rangle$ or $N = ED(L)$, then $Ru(N) = Edt(L)$; and if $N = CR(L)$ or $N = EO(L)$, then $Ru(N) = Eo(L)$. Let the "relaxed uniformized version" of N be defined as follows: $N^{\mathrm{ru}} \overset{def}{=} \langle |N|, Ru(N)\rangle$.

Proposition 4.12 *(a) There is a first order formula $\varphi_{\epsilon}(\vec{U}, V)$ in the language of $M(Ro(\bar{N}), Aut(N))$ such that for every adequate N, $\vec{U} \in Rep(N)$ and $V \in Ro(\bar{N})$,*

$M(Ro(\bar{N}), Aut(N)) \models \varphi_\varepsilon[\vec{U}, V]$ *iff* $x_{\vec{U}} \in V$.

(b) There is a first order formula $\varphi_{Ru}(\vec{U}^1, \vec{U}^2, \vec{U}^3, \vec{V}^1, \vec{V}^2, \vec{V}^3)$ *in the language of* $M(Ro(\bar{N}), Aut(N))$ *such that for every adequate* N *and* $\vec{U}^1, \vec{U}^2, \vec{U}^3, \vec{V}^1, \vec{V}^2, \vec{V}^3 \in Rep(N)$:
$M(Ro(\bar{N}), Aut(N)) \models \varphi_{Ru}[\vec{U}_1, \vec{U}_2, \vec{U}_3, \vec{V}_1, \vec{V}_2, \vec{V}_3]$ *iff* $Ru(x_{\vec{U}_1}, x_{\vec{U}_2}, x_{\vec{U}_3}, x_{\vec{V}_1}, x_{\vec{V}_2}, x_{\vec{V}_3})$ *holds in* $\bar{N}^{\mathrm{ru}}$.

Proof: Part (a) is needed in the proof of (b).

(a) It is easy to see that $\varphi_\varepsilon(\vec{U}, V) \stackrel{def}{\equiv} (\exists g \in Sp_{Aut(N)}(V))\neg\varphi_{Eqr}(g(\vec{U}), \vec{U})$ is as required.

(b) We denote members of $Rep(N)$ by the letters r, s. For $l = 1, 2$ let $\vec{r}_l$ denote a member of $(Rep(N))^3$. Let

$$\varphi_{Ru}(\vec{r}_1 \,\hat{}\, \vec{r}_2) \equiv \Big(\bigwedge_{\substack{0 \le i < j \le 3 \\ l = 1,2}} \neg\varphi_{Eqr}(r_{l,i}, r_{l,j})\Big) \wedge$$
$$(\forall \vec{W}_1, \vec{W}_2 \in (Ro(\bar{N}))^3) \Big(\bigwedge_{\substack{0 \le i < j \le 3 \\ l = 1,2}} \varphi_\varepsilon(r_{l,i}, W_{l,i}) \rightarrow$$
$$(\exists g \in Aut(N^s))(\exists \vec{r} \in (Rep(N))^3) \Big(\bigwedge_{0 \le i < j \le 3} (\varphi_\varepsilon(r_i, W_{1,i}) \wedge \varphi_\varepsilon(g(r_i), W_{2,i}))\Big)\Big).$$

It follows easily from Proposition 4.4(d) that φ_{Ru} is as required. □

Proof of Theorem 4.6: (a) The proof of Part (a) follows the pattern of the proof of the Reconstruction Theorem for Local Movement Systems. So we just indicate the main steps in the proof.

For an adequate N, let $M = M^{\mathrm{ru}}(N)$ be defined as follows.
$M = \langle |M|, \le^{Ro}, \circ, Ap, Ru(\bar{N}), \varepsilon\rangle$, where $|M| - |M(Ro(\bar{N}), Aut(N))| \cup |\bar{N}|$, and $\circ$, Ap and ε are defined as follows. $\circ$ is the composition operation on $Aut(N)$, (defined in a trivial way outside $(Aut(N))^2$); Ap is the application function defined on $Aut(N) \times (|\bar{N}| \cup Ro(\bar{N}))$; and ε is the belonging relation defined on $|\bar{N}| \times Ro(\bar{N})$.

By Propositions 4.11 and 4.12, if for $l = 1, 2$, N_l is adequate, then every isomorphism between $M(Ro(\bar{N}_1), Aut(N_1))$ and $M(Ro(\bar{N}_2), Aut(N_2))$ can be extended to an isomorphism between $M^{\mathrm{ru}}(N_1)$ and $M^{\mathrm{ru}}(N_2)$.

Combining this with Corollary 4.5, we conclude that if for $l = 1, 2$, N_l is adequate, then for every $\varphi : Aut(N_1) \cong Aut(N_2)$ there is $\beta : M^{\mathrm{ru}}(N_1) \cong M^{\mathrm{ru}}(N_2)$ such that β extends φ. Let $\alpha = \beta|\bar{N}_1$. Then $\alpha : \bar{N}_1^{\mathrm{ru}} \cong \bar{N}_2^{\mathrm{ru}}$, and α induces φ. By the mutual definability of Ed and Edt, $\alpha : \bar{N}_1^{\mathrm{r}} \cong \bar{N}_2^{\mathrm{r}}$. This proves Part (a).

(b) Let N be adequate. We define the set of "automorphism representatives" of N. $Repa(N) \stackrel{def}{=} \{\langle g_1, g_2\rangle \in (Aut(N))^2 \mid \langle var(g_1), var(g_2)\rangle \in Rep(N)\}$. For $\langle g_1, g_2\rangle \in Repa(N)$, let $x_{g_1,g_2} \stackrel{def}{=} x_{var(g_1),var(g_2)}$.
Let $Eqra(N) \stackrel{def}{=} \{\langle r_1, r_2\rangle \in (Repa(N))^2 \mid x_{r_1} = x_{r_2}\}$.

By 4.11 and 1.19(b), there are formulas $\varphi_{Repa}(x_1, x_2)$ and $\varphi_{Eqra}(\vec{x}_1, \vec{x}_2)$ in the language of groups such that for every adequate N, φ_{Repa} defines in $Aut(N)$ the set $Repa(N)$, and $\varphi_{Eqra}(\vec{x}_1, \vec{x}_2)$ defines in $Aut(N)$ the relation $Eqra(N)$.

Similarly, by 4.12, 4.11(b) and 1.19(b), there are formulas φ_{Rua} and φ_{Apa} in the language of groups, that respectively define in $Aut(N)$ the relation $\{\vec{t} \in (Repa(N))^6 | \langle x_{t_1}, \ldots, x_{t_6} \rangle \in Ru(\bar{N})\}$, and the application function on $Aut(N) \times (Repa(N)/Eqra(N))$.

It now follows easily from the existence of the above formulas that for every formula $\varphi(x_1, \ldots, x_k, h_1, \ldots, h_m)$ in the language of $Act(\bar{N}^{\mathrm{ru}}, Aut(N))$ there is a formula $\varphi^*(f_1, g_1, \ldots, f_k, g_k, h_1, \ldots, h_m)$ in the language of groups such that for every adequate N, $\langle f_1, g_1 \rangle, \ldots \langle f_k, g_k \rangle \in Repa(N)$ and $h_1, \ldots, h_m \in Aut(N)$: $Act(\bar{N}^{\mathrm{ru}}, Aut(N)) \models \varphi[x_{f_1,g_1}, \ldots, x_{f_k,g_k}, h_1, \ldots, h_m]$ iff $Aut(N) \models \varphi^*[f_1, g_1, \ldots, f_k, g_k, h_1, \ldots, h_m]$.

It follows that for every adequate N_1 and N_2: if $Aut(N_1) \equiv Aut(N_2)$, then $Act(\bar{N}_1^{\mathrm{ru}}, Aut(N_1)) \equiv Act(\bar{N}_2^{\mathrm{ru}}, Aut(N_2))$. By the mutual definability of Ed and Edt, this implies that $Act(\bar{N}_1^{\mathrm{r}}, Aut(N_1)) \equiv Act(\bar{N}_2^{\mathrm{r}}, Aut(N_2))$. □

Theorem 4.6 is a powerful tool in proving that the automorphism groups of some adequate almost ordered structures can be characterized in the class of all such automorphism groups by a first order sentence. The following theorem due to Gurevich and Holland [4] is such an example.

Corollary 4.13 *There are first order sentences $\varphi_{\mathbb{R}}, \varphi_{\mathbb{Q}}$ in the language of groups such that for every adequate N: $Aut(N) \models \varphi_{\mathbb{R}}$ iff $N \cong \mathbb{R}$ and $Aut(N) \models \varphi_{\mathbb{Q}}$ iff $N \cong \mathbb{Q}$ or $\bar{N} - N \cong \mathbb{Q}$.*

Proof: The proof is not difficult. It uses the fact that the separability of N, the completeness of N and the countability of an orbit in $Act(\bar{N}^{\mathrm{r}}, Aut(N))$ can be expressed by first order sentences in the language of $Act(\bar{N}^{\mathrm{r}}, Aut(N))$.

□

References

[1] J. L. Alperin, J. Covington and D. Macpherson, *Automorphisms of quotients of symmetric groups*, this volume.

[2] R. Bieri and R. Strebel, *On groups of PL-homeomorphisms of the real line*, Preprint 1985.

[3] J. R. Büchi, *Die Boolesche Partialordnung und die Paarung von Gefügen*, Portugal. Math. **7**(1948), 119–190.

[4] Y. Gurevich and W. C. Holland, *Recognizing the real line*, Trans. American Math. Soc. **265**(1981), 527–534.

[5] M. Giraudet and J. K. Truss, *On distinguishing quotients of ordered permutation groups*, Quat. J. Oxford **45**(1994), 181–209.

[6] W. C. Holland, *Transitive lattice-ordered permutation groups*, Math. Zeit. **87**(1965), 420–433.

[7] S. Koppelberg, Handbook of Boolean Algebras, Edited by J. D. Monk, Vol 1, North Holland, Amsterdam, 1989.

[8] W. Ling, *A classification theorem for manifold automorphism groups*, Preprint 1980.

[9] D. Maharam, *On homogeneous measure algebras*, Proc. Nat. Acad. Sc. **28**(1942), 108–111.

[10] S. H. McCleary, *Groups of homeomorphisms with manageable automorphism groups*, Comm. in Algebra **6**(1978), 497–528.

[11] E. B. Rabinovich, *On linearly ordered sets with 2-transitive groups of automorphisms*, Vesti Akad. Nauk Belaruskai SSR **6**(1975), 10–17.

[12] M. Rubin, *On the reconstruction of topological spaces from their groups of homeomorphisms*, Trans. American Math. Soc. **312**(1989), 487–538.

[13] M. Rubin, *On the reconstruction of Boolean algebras from their automorphism groups*, in *Handbook of Boolean Algebras*, ed. J. D. Monk, Vol 2 Chapter 15, 547–605, North Holland, Amsterdam, 1989.

[14] M. Rubin, *On the automorphism groups of homogeneous and saturated Boolean algebras*, Algebra Universalis **9**(1979), 54–86.

[15] M. Rubin, *The reconstruction of trees from their automorphism groups*, Contemporary Mathematics **151**(1993), American Math. Soc.

[16] M. Rubin and Y. Yomdin, in preparation.

[17] S. Shelah and J. K. Truss, *On distinguishing quotients of symmetric groups*, in preparation.

[18] F. Takens, *Characterization of a differentiable structure by its group of diffeomorphisms*, Bol. Soc. Bras. Mat. **10**(1979), 17–26.

[19] J. K. Truss, *On recovering structures from their automorphism groups*, this volume.

[20] J. V. Whittaker, *On isomorphic groups and homeomorphic spaces*, Ann. Math. **78**(1963), 74–91.

Infinite Jordan Permutation Groups

S. A. Adeleke

University of Western Illinois
Macomb, Illinois, 61455
USA

1 Introduction

A *Jordan group* is a transitive permutation group which contains in a non-trivial way a smaller permutation group, the smaller group fixing every point outside the smaller space. More precisely, a transitive permutation group (G, Ω) is Jordan if (i) G contains a group H, (ii) Ω contains a non- empty, non-singleton set Γ, (iii) H fixes every point in $\Omega\backslash\Gamma$ with (H, Γ) transitive and, (iv) in case $|\Omega\backslash\Gamma|$ is finite, then G is not $(|\Omega\backslash\Gamma| + 1)$-transitive on Ω.

This paper presents a summary of the extensive work on invariant relations in infinite primitive Jordan groups; it indicates some articles close to that work, and gives some new results on the topic. What makes Jordan groups very interesting is that with very few assumptions, one can get far-reaching conclusions as to their invariant relations. The relations are: linear order, linear betweenness relation, circular order, separation relation, semilinear order (that is, the order on a "tree"), C-relation (that is, the relation on maximal chains on a "tree"), D-relation (relation on the "leaves" or points at infinity on a "tree"), Steiner systems, and limits of some of these relations. Those are all the possibilities, according to the work in [1]–[4], [6], [7], [11], [24]–[27], [31]. The contents below also includes some new results on Jordan sets and linear order, and on invariant relations on a group satisfying slightly weaker properties than those of Jordan groups.

One new result below contains as a corollary that a transitive primitive ordered permutation group need not be weakly doubly transitive, the right regular representation of a subgroup of $\mathbb{R}$ or periodic. The second new result shows that if a primitive permutation group has a syzygetic subset, then it has an invariant semilinear order relation, a B-relation, a C-relation or a D-relation.

This summary is limited in several ways. It is limited to invariant relations; so, only a brief mention is given in section 5.8 to the very important work on normal subgroups and other structures of Jordan groups contained in publications like [13]–[15], [17], [26], [28], [29], [30]. It is limited to infinite Jordan groups; its

W. C. Holland (ed.), Ordered Groups and Infinite Permutation Groups, 159–194.

only reference to finite Jordan groups is to areas of very close connection to the infinite case. Thus, the seminal work of Jordan [19] and the pioneering work in [20], [21] are merely mentioned. Even on invariant relations, the paper does not give details of the work in [11], [16], and [32] on when an invariant Steiner system is a projective or affine space. Some survey articles like [27] and [20] address some of the areas of limitation of this article.

There is also an excellent survey on Jordan groups by Macpherson [22]. But its emphasis is different from the one in this paper. The survey in [22] highlights the pleasant interplay of the works on infinite Jordan groups and Model Theory. This paper, however, leaves out Model Theory altogether and focuses on infinite Jordan groups, especially on the evolution of the ideas that led to the final classification of the invariant relations on Jordan groups. Those early stages on bounded groups (generalization of finitary groups) and the rapid sequence of results in the classification of infinite Jordan groups with proper primitive Jordan sets get most attention here. Whenever possible, it presents alternative but elementary proofs. The overall goal is that this article, when read in conjunction with the very short summary in [4, section 4] on the later proofs on Jordan groups not containing proper primitive Jordan sets should give useful insight to the reader.

A word about proofs in infinite Jordan groups. At the heart of almost all the proofs are maximal Jordan sets containing one element and excluding others. The reasons are two: First is the fact that the union of two nondisjoint Jordan sets is Jordan, the union of a chain of Jordan sets is Jordan and the image of a Jordan set under any group element is also a Jordan set. Second is that the primitivity of the group (G, Ω) can be used to produce from one Jordan set a collection of larger Jordan sets. For example, in the case of a primitive but not 2-transitive Jordan group (G, Ω) with a Jordan set Γ, $\alpha \in \Gamma$ and $\beta \in \Omega \backslash \Gamma$, the maximal Jordan set containing α and excluding β exists, is unique, and with all other such maximal Jordan sets plays a role in revealing the invariant relations in (G, Ω). Details about these general features of the proofs are in section 4 below.

The contents is divided into sections as follows: Section 2: definitions and notation; section 3: statement of the classification theorem; section 4: vital lemmas and properties; section 5: bounded permutation groups and comments on other works; section 6: Jordan groups with proper primitive Jordan sets; section 7: general simply primitive Jordan groups; sections 8 and 9: new results.

2 Definitions and Notation

2.1 *Definitions on Permutation Groups*

1. A permutation group (G, Ω) is *transitive* if for every pair α and β in Ω, there is a permutation in G mapping α to β.

2. The group (G, Ω) is *n-transitive* if for every pair $(\alpha_1, \alpha_2, \ldots, \alpha_n)$ and $(\beta_1, \beta_2, \ldots, \beta_n)$ of n-tuples with distinct components in Ω, there is a permutation

in G mapping $(\alpha_1, \alpha_2, \ldots, \alpha_n)$ to $(\beta_1, \beta_2, \ldots, \beta_n)$. If (G, Ω) is n-transitive for every natural number n, then it is said to be *highly transitive.*

3. In a permutation group (G, Ω), a non-empty subset $\Gamma \subseteq \Omega$ is a *block* if for every $g \in G$, either $\Gamma g \cap \Gamma = \phi$ or $\Gamma g = \Gamma$. A permutation group (G, Ω) with no blocks besides Ω and singleton sets is said to be *primitive.* A permutation group (G, Ω) is *n-primitive* with $n \geq 2$ if for any distinct $n-1$ elements $\alpha_1, \alpha_2, \ldots, \alpha_{n-1}$ in Ω, the group $(H, \Omega \backslash \{\alpha_1, \alpha_2, \ldots, \alpha_{n-1}\})$ is primitive where H is the collection of all $g \in G$ such that g fixes each of $\alpha_1, \alpha_2, \ldots, \alpha_{n-1}$. A primitive group which is not 2-transitive is called *simply primitive.*

4. Let (G, Ω) be a transitive permutation group. A non-empty, non-singleton subset, Γ, of Ω is a *Jordan set* if for every α and β in Γ, there is a permutation in G mapping α to β and fixing every point in $\Omega \backslash \Gamma$. The Jordan set Γ is a *primitive Jordan set* if the action on Γ of the subgroup of G fixing every point of $\Omega \backslash \Gamma$ is primitive. In a similar way, n-transitive and n-primitive Jordan sets are defined.

5. The *support of a permutation h*, denoted supp(h), is the set of elements moved by h. The *support of a subgroup H* of permutations, denoted supp(H), is the union of the supports of the permutations in H.

6. As usual, $|\Gamma|$ shall represent the size of set Γ.

7. A Jordan set Γ in (G, Ω) is a *proper Jordan set* if $\Omega \backslash \Gamma$ is infinite, or if finite, G is not $(|\Omega \backslash \Gamma| + 1)$-transitive. Thus a Jordan set is proper if its existence cannot be derived trivially from multiple transitivity of (G, Ω). Proper primitive Jordan sets and proper n-transitive Jordan sets have similar definitions.

8. A transitive group (G, Ω) is a *Jordan group* if Ω contains proper Jordan sets.

9. In a permutation group (G, Ω), if $\sigma \in \Omega$, then G_σ denotes the subgroup of G fixing σ. Similarly, if $\sigma_1, \sigma_2, \ldots, \sigma_n \in \Omega$, then $G_{\sigma_1 \sigma_2 \ldots \sigma_n}$ denotes the subgroup of G fixing each of $\sigma_1, \sigma_2, \ldots,$ and σ_n. In the same way, if $\Pi \subseteq \Omega$, $G_{(\Pi)}$ denotes the subgroup of G fixing each element in Π. Two Jordan sets Σ_1 and Σ_2 are *typical* if $\Sigma_1 \cap \Sigma_2 \neq \phi$, $\Sigma_1 \backslash \Sigma_2 \neq \phi$, and $\Sigma_2 \backslash \Sigma_1 \neq \phi$.

10. In a Jordan group with Jordan set Σ, every set of the form Σg where $g \in G$ is called a *G-translate* of Σ, or a *translate* of Σ *under G.*

11. In any group under consideration, $MPJ(\beta_1, \beta_2, \ldots, \beta_k; \alpha)$ shall denote *the maximal primitive Jordan set containing α and excluding $\beta_1, \beta_2, \ldots, \beta_k$* whenever the set exists. Likewise $MJ(\beta_1, \beta_2, \ldots, \beta_k; \alpha)$ shall mean *the maximal Jordan set containing α and excluding $\beta_1, \beta_2, \ldots, \beta_k$.*

12. Throughout, $A \subset B$ means set A is a *proper subset of set B.*

2.2 *Definitions of Structures*

1. Let $(\Omega, \leq)$ be a dense linear order. Define on Ω :

$$\rho(\alpha;\beta,\gamma) \quad \text{if and only if} \quad \beta \leq \alpha \leq \gamma \quad \text{or} \quad \gamma \leq \alpha \leq \beta.$$

Then ρ is a *dense linear betweenness relation* induced by the dense linear order.

2. A partial order $\leq$ on a non-empty set Σ is an *upper semilinear order* if (i), every pair in Σ has an upper bound in Σ, and (ii), the set of upper bounds of any element of Σ is linearly ordered. Then, $(\Sigma, \leq)$ is called *an upper semilinearly ordered set.* (An upper semilinearly ordered set is like a set of points on an inverted "tree".)

Let α, β be any two incomparable elements in a nonlinear upper semilinearly ordered set $(\Sigma, \leq)$. Define

$$U\{\alpha,\beta\} := \{\sigma \in \Sigma \mid \sigma \geq \alpha \wedge \sigma \geq \beta\}$$
$$T\{\alpha,\beta\} := \{\sigma \in \Sigma \mid \sigma < \omega \forall \omega \in U\{\alpha,\beta\}\}.$$

Then α, $\beta \in T\{\alpha,\beta\}$ since α and β are chosen to be incomparable. Define $\sim$ on $T\{\alpha,\beta\}$:

$$\xi \sim \eta \quad \text{if and only if} \quad (\exists \mu \in T\{\alpha,\beta\})(\xi \leq \mu \wedge \eta \leq \mu).$$

Then it is not difficult to verify that $\sim$ is an equivalence relation on $T\{\alpha,\beta\}$ and that α, β belong to different classes of $\sim$. Define $U\{\alpha,\beta\}$ as a *branch point* in $(\Sigma, \leq)$, and refer to each class of $\sim$ as a *branch* below the branch point $U\{\alpha,\beta\}$.

3. A ternary relation B on a non-empty set Ω is a *general betweenness relation* (and if $B(\alpha;\beta,\gamma)$ holds then we shall say that α *lies between* β *and* γ) if it satisfies the following conditions:

(B1) $B(\alpha;\beta,\gamma) \rightarrow B(\alpha;\gamma,\beta)$
(B2) $B(\alpha;\beta,\gamma) \wedge B(\beta;\alpha,\gamma) \leftrightarrow \alpha = \beta$
(B3) $B(\alpha;\beta,\gamma) \rightarrow B(\alpha;\beta,\delta) \vee B(\alpha;\gamma,\delta)$
(B4) $\neg B(\alpha;\beta,\gamma) \rightarrow (\exists \delta \neq \alpha)(B(\delta;\alpha,\beta) \wedge B(\delta;\alpha,\gamma))$
If, in addition
(B5) $\alpha \neq \beta \rightarrow (\exists \gamma)(\gamma \neq \alpha \wedge \gamma \neq \beta \wedge B(\gamma;\alpha,\beta))$
then we say (Ω, B) is *dense*.

The pair (Ω, B) of a non-empty set Ω and a general betweenness relation B is called a *general betweenness set.*

4. A ternary relation K on a non-empty set Ω is a *C-relation* (or a *chain relation*) if the following conditions hold:

(C1) $K(\alpha;\beta,\gamma) \rightarrow K(\alpha;\gamma,\beta)$
(C2) $K(\alpha;\beta,\gamma) \rightarrow \neg K(\beta;\gamma,\alpha)$

(C3) $(K(\alpha;\beta,\gamma) \wedge \neg K(\delta;\beta,\gamma)) \rightarrow K(\alpha;\delta,\gamma)$
(C4) $(\alpha \neq \beta) \rightarrow (\exists\mu)((\mu \neq \beta) \wedge K(\alpha;\beta,\mu))$
(C5) $(\forall\alpha,\beta)(\exists\eta)(K(\eta;\alpha,\beta))$.
If in addition
$K(\alpha;\beta,\gamma) \rightarrow (\exists\delta)(K(\delta;\beta,\gamma) \wedge K(\alpha;\beta,\delta))$
holds, then (Ω, K) is said to be *dense*.

(Any set possessing a C-relation can be identified with some collection of maximal chains of an upper semilinearly ordered set whose union is the whole semilinearly ordered set.)

5. A quaternary relation D on a non-empty set Ω is a D-relation if it satisfies the following conditions:
(D1) $D(\alpha,\beta;\gamma,\delta) \rightarrow (D(\beta,\alpha;\gamma,\delta) \wedge D(\alpha,\beta;\delta,\gamma) \wedge D(\gamma,\delta;\alpha,\beta))$
(D2) $D(\alpha,\beta;\gamma,\delta) \rightarrow \neg D(\alpha,\gamma;\beta,\delta)$
(D3) $D(\alpha,\beta;\gamma,\delta) \rightarrow (D(\alpha,\beta;\delta,\varepsilon) \vee D(\gamma,\delta;\alpha,\varepsilon))$
(D4) α,β,γ distinct $\rightarrow (\exists\delta)((\gamma \neq \delta) \wedge D(\alpha,\beta;\gamma,\delta))$.
If also
(D5) $D(\alpha,\beta;\gamma,\delta) \rightarrow (\exists\varepsilon)(D(\varepsilon,\beta;\gamma,\delta) \wedge D(\alpha,\varepsilon;\gamma,\delta)$
$\wedge D(\alpha,\beta;\varepsilon,\delta) \wedge D(\alpha,\beta;\gamma,\varepsilon))$
then we say (Ω, D) is *dense*. (A set with a D-relation can be identified with a set of "points at infinity" or "leaves of a tree.")

6. A *circular order* is a ternary relation ρ satisfying the following conditions:
(ρ1) $\rho(\alpha,\beta,\gamma) \rightarrow \rho(\beta,\gamma,\alpha)$
(ρ2) $(\rho(\alpha,\beta,\gamma) \wedge \rho(\beta,\alpha,\gamma)) \leftrightarrow ((\alpha = \beta) \vee (\beta = \gamma) \vee (\gamma = \alpha))$
(ρ3) $\rho(\alpha,\beta,\gamma) \rightarrow (\rho(\alpha,\beta,\delta) \vee \rho(\delta,\beta,\gamma))$
(ρ4) $\rho(\alpha,\beta,\gamma) \vee \rho(\beta,\alpha,\gamma)$

7. Let ρ be a circular order on a non-empty set Λ. A group G of permutations of Λ *preserves the separation relation associated with* ρ on Λ if and only if every element of G either preserves or reverses the circular order ρ. That is,

$$(\forall g \in G)\left[\begin{array}{l}(\forall\alpha,\beta,\gamma \in \Lambda)(\rho(\alpha,\beta,\gamma) \rightarrow \rho(\alpha g,\beta g,\gamma g))\\ \vee(\forall\alpha,\beta,\gamma \in \Lambda)(\rho(\alpha,\beta,\gamma) \rightarrow \rho(\alpha g,\gamma g,\beta g))\end{array}\right].$$

The separation relation associated with ρ is the quaternary relation S defined by:

$$S(\alpha,\beta;\gamma,\delta) \leftrightarrow ((\rho(\alpha,\beta,\gamma) \wedge \rho(\alpha,\delta,\beta)) \vee (\rho(\alpha,\gamma,\beta) \wedge \rho(\delta,\alpha,\beta))).$$

8. Let k be a positive integer greater than 1, and let Λ be a non-empty set. A non-trivial *Steiner k-system* on Λ is a collection of subsets of Λ of the same size greater than k called *blocks* such that given any k distinct elements in Λ, there is a unique block containing them.

9. Let $(\Omega, \leq)$ be a linearly ordered set. Denote by $\bar{\Omega}$ the Dedekind completion of Ω (see [18, section 1.8] for details). A subset Γ of $\bar{\Omega}$ is *coterminal* in $\bar{\Omega}$ if for every $\omega \in \bar{\Omega}$, there exist $\gamma_1, \gamma_2 \in \Gamma$ such that $\gamma_1 \leq \omega \leq \gamma_2$.

10. Let (G, Ω) be a transitive permutation group preserving a linear order $\leq$ on Ω. Then, (G, Ω) is said to be *weakly doubly transitive* [18, section 4.3] if for any given $\alpha < \beta < \gamma$ in Ω, there exists $g \in G_\alpha$ such that $\beta g \geq \gamma$, and there exists $h \in G_\gamma$ such that $\beta h \leq \alpha$. The group (G, Ω) is said to be *periodic* if there exists an order-preserving permutation z of $\bar{\Omega}$ preserving $\leq$ such that

(i) $\omega z > \omega, \quad \forall \omega \in \bar{\Omega}$

(ii) there exists $\alpha \in \Omega$ such that $\{\alpha z^n \mid n \in \mathbb{Z}\}$ is coterminal in $\bar{\Omega}$, and

(iii) the set of all permutations of $\bar{\Omega}$ that commute with each element of G (that is, the centraliser of G) is precisely $\langle z \rangle$, the group generated by z.

3 The Final Classification Theorem

The full proof of the classification theorem is in [2]–[4].

Theorem 3.1 *Let (G, Ω) be an infinite primitive Jordan group.*

If (G, Ω) is simply primitive, then G leaves invariant a linear order, a semilinear order or a C-relation on Ω.

If (G, Ω) is 2-transitive not 2-primitive, then G leaves invariant a linear betweenness relation, a general betweenness relation, a C-relation, a D-relation or a Steiner system.

If (G, Ω) is 2-primitive not 3-transitive, then G preserves a circular order, a D-relation, a limit of general betweenness relations or of D-relations.

If (G, Ω) is 3-transitive but not 3-primitive, then G preserves a separation relation, a D-relation, a Steiner system or a limit of Steiner systems.

If (G, Ω) is 3-primitive, then it is highly transitive or it preserves a Steiner system or a limit of Steiner systems.

Remark 3.2. In sections 5, 6 and 7 below, we set out chronologically the development of the proof of the classification theorem. But we omit the sketch proofs on 2-transitive Jordan groups with no proper primitive Jordan sets for two reasons. Firstly, [4, section 4] contains very short sketch proofs of the omitted part. Secondly, the account and sketch proofs below should help the reader with the intuition necessary for a quick understanding of all the full proofs on the topic. A good reference for definitions of limits of betweenness relations, D-relations and Steiner systems is [4, Definitions 2.1.9 and 2.1.10]. The limit cases in Theorem 3.1 occur when (G, Ω) contains no proper primitive Jordan sets.

4 Vital Lemmas and Properties

Let (G, Ω) be a Jordan group, infinite or finite. Let $\mathfrak{p}$ denote any of the following properties of Σ :

Σ is a Jordan set;
Σ is a primitive Jordan set;
Σ is a t-transitive Jordan set for some $t \in \mathbb{N}$;
Σ is a t-primitive Jordan set for some $t \in \mathbb{N}$.

The following lemmas are not surprising and are easy to prove (see [2]).

Lemma 4.1.1 *If Σ is a Jordan set with property $\mathfrak{p}$, then Σg is a Jordan set with property $\mathfrak{p}$ for any $g \in G$.*

Lemma 4.1.2 *If $\{\Sigma_i \mid i \in I\}$ is a chain of Jordan sets with property $\mathfrak{p}$, then $\cup\{\Sigma_i \mid i \in I\}$ is a Jordan set with property $\mathfrak{p}$.*

Lemma 4.1.3 *Let Σ_1 and Σ_2 be Jordan sets with property $\mathfrak{p}$ such that $\Sigma_1 \cap \Sigma_2 \neq \phi$. Then $\Sigma_1 \cup \Sigma_2$ is a Jordan set with property $\mathfrak{p}$.*

The next Lemma is a simple consequence of the three Lemmas above as the sketch proof shows.

Lemma 4.1.4 *If (G, Ω) is primitive and contains a proper Jordan set with property $\mathfrak{p}$, then Ω has property $\mathfrak{p}$.*

Sketch proof. A maximal Jordan set Σ with property $\mathfrak{p}$ exists by Lemma 4.1.2 and must be Ω. Otherwise by the primitivity of G, there is $g \in G$ such that $\Sigma g \cap \Sigma \neq \phi$ and $\Sigma g \backslash \Sigma \neq \phi$, which will imply that $\Sigma g \cup \Sigma$ is a Jordan set with property $\mathfrak{p}$ strictly larger than Σ, a contradiction to maximality of Σ.

One of the few results on finite Jordan groups which is extended *repeatedly* in the proofs on the infinite case is the following.

Lemma 4.1.5 [27, Lemma 2, section 3] *Suppose (G, Ω) is a finite Jordan group. Let Γ_1 be a non-trivial Jordan subset of Ω which is not a block of imprimitivity for G, let Γ_2 be minimal amongst Jordan subsets containing Γ_1 properly, and let $\Delta := \Gamma_2 \backslash \Gamma_1$.*

(a) Then Δ is a block of imprimitivity for $G_{\{\Gamma_2\}}$ in Γ_2, and $|\Delta| < |\Gamma_1|$.

(b) If $|\Delta| = 1$ then Γ_1 is an improper Jordan subset for G in Ω. If $|\Delta| > 1$ then Γ_1 is a proper Jordan set and Δ is a Jordan subset for $G_{\{\Gamma_1\}}$ in $\Omega \backslash \Gamma_1$.

(c) If ρ is the $G_{\{\Gamma_2\}}$-congruence on Γ_2 having Δ as a block of imprimitivity then $G_{\{\Gamma_2\}}$ is 2-transitive on Γ_2/ρ; moreover, if Γ is any Jordan set contained in Γ_2 then, provided that Γ is not contained in a single ρ-class, Γ is a union of ρ-classes and Γ/ρ is a Jordan subset for $G_{\{\Gamma_2\}}$ in Γ_2/ρ.

4.1.6 *Remark on Lemma 4.1.5.* (a) The Lemma leads to a chain of Jordan sets $\Gamma_1 \subset \Gamma_2 \subset \cdots \subset \Gamma_k$ where for each i, $\Gamma_{i+1} \backslash \Gamma_i$ is a block of the unique maximal congruence of $G_{(\Omega \backslash \Gamma_{i+1})}$ on Γ_{i+1}, and $\cup \Gamma_i = \Gamma_k = \Omega$. The sets $\Gamma_{i+1} \backslash \Gamma_i$ are called *minimal increments.* If (G, Ω) is primitive, it follows immediately that $|\Gamma_k \backslash \Gamma_{k-1}| = 1$ and thus (G, Ω) is 2-transitive.

(b) One extension of Lemma 4.1.5 to the infinite case defines Γ_1 as a maximal Jordan set with respect to containing one element and excluding a finite number of others. The size of the finite number depends on the context. For example, suppose (G, Ω) is k-transitive not $(k+1)$-transitive, and Γ is a proper Jordan set. Let $\beta_1, \beta_2, \ldots, \beta_k$ be k distinct elements in $\Omega \backslash \Gamma$ with $\alpha \in \Gamma$. Then define Γ_1 as the maximum Jordan set containing α and excluding $\beta_1, \beta_2, \ldots, \beta_k$. That way, $\Gamma_1 \supseteq \Gamma$, and every Jordan set in $\Omega \backslash \{\beta_2, \ldots, \beta_k\}$ larger than Γ_1 must contain β_1. It may well be that every Jordan set in $\Omega \backslash \{\beta_2, \ldots, \beta_k\}$ larger than Γ_1 must contain other elements λ besides β_1. The collection of β_1 together with all such elements λ (if any) now plays the role in the infinite case that the minimal increments in finite Jordan groups play (compare with Remark (a) above). They are like "new" minimal increments in the infinite case; and they point the way to invariant relations in (G, Ω).

5 Infinite Bounded Permutation Groups and Comments on Other Works

The starting point of the whole theory was the work of Jordan. One of his important Theorems was the following Theorem of 1875 in [19].

Theorem 5.1 *There is a function f such that whenever a finite primitive permutation group (G, Ω) contains a permutation $h \in G$ with $h \neq 1$, $|supp(h)| = m$ and $|\Omega| > f(m)$, then G contains the alternating group (see Definition 2.1.5 for support of a permutation).*

Next comes the following generalization of Theorem 5.1 to infinite groups by Wielandt [31].

Theorem 5.2 *If (G, Ω) is an infinite primitive permutation group and G contains some permutation with finite support, then G contains the alternating group.*

Comment on proof of Theorem 5.2.

Wielandt first produced a finite Jordan set in (G, Ω), then used a variant of the result in Lemma 4.1.5 to show that the subgroup of G consisting of maps with finite support is highly transitive and thence concluded that G contains the alternating group. How he produced the Jordan set went like this. Let G_0 be *any* finite subgroup with finite support. Expand G_0 to a larger finite subgroup H of G such that the support of each element in H is finite, and any block of H that contains one point in supp(G_0) contains supp(G_0) as a subset. If Σ is the minimal block of H containing supp(G_0), then (H^{Σ}, Σ) is primitive. (Here, H^{Σ} is the action of H on Σ). Moreover $\langle G_0^{H^{\Sigma}} \rangle := \langle h^{-1} G_0 h | h \in H^{\Sigma} \rangle$ is a normal subgroup of H^{Σ}. Hence it is transitive on Σ. Thus Σ is a finite Jordan set.

Definition 5.3. Let Ω be infinite. A permutation g of Ω is said to be *finitary* if the support of g is finite.

Denote by $SF(\Omega)$ the group of all finitary permutations of Ω.

Theorem 5.4 [26, Theorem 3] *Let (G, Ω) be an infinite group of finitary permutations which is transitive but not primitive and which has a maximal congruence. Then each class of the maximal congruence is finite, and*

$$W' \leq G \leq W$$

where $W = H\ Wr\ SFp$ is the wreath product of H and SFp, W' = commutator group of W, p = number of congruence classes of the maximal congruence (= an infinite cardinal), H corresponds to the action on a congruence class, and SFp is the symmetric finitary group on the congruence classes permuting them as blocks.

One possible generalization of finitary permutations to higher cardinals is as follows.

Definition 5.5. Let (G, Ω) be an infinite permutation group, and m be an infinite cardinal number. Say G is *m-bounded* if $(\forall g \in G)(|\mathrm{supp}(g)| \leq m)$.

Assume the ZFC axioms of set theory. Let m^+ be the successor cardinal of m, and $n := |\Omega|$.

The following problem naturally arises as an extension of Wielandt's Theorem 5.2.

Problem 5.6. Assuming $n > m \leq \aleph_0$, classify all primitive m-bounded permutation groups (G, Ω) with $|\Omega| = n$.

Its solution contains more possibilities than the finitary case in Wielandt's Theorem.

Theorem 5.7 (in [1]) *Let (G, Ω) be an infinite primitive m-bounded permutation group. Then (G, Ω) is a Jordan group containing primitive Jordan sets. Moreover, it is highly transitive or leaves invariant one of these relations: a dense linear order, a dense upper semilinear order or a dense C-relation.* (A Sketch proof is in section 5.9 below).

5.8 *Remarks on related work.*

5.8.1. *Droste's work on partial orders.* Peter Neumann and Adeleke were quite aware, at the early stages, of highly transitive examples of m-bounded groups built on the product topology, and also of a simply primitive example (automorphism Group of $\mathbb{Q} \times \omega_1$). But the question lingered as to the existence or non-existence of other examples that are not highly transitive. That was until December 1984 when they constructed the example on the automorphism group of the semilinear order; they were quite unaware of the work of Manfred Droste on partial orders [13] which was published in 1985. In that work, Droste gave an impressive classification and characterization of partially ordered structures using their automorphism groups as tools. His groups certainly satisfy the properties of what are called Jordan groups but he made no explicit mention of that fact.

5.8.2. *Cameron's work on Combinatorics.* Still on the surprising connections to other works, we turn to the articles of Peter Cameron on combinatorics. In his deep work on counting finite structures [8], Cameron constructed in 1983 an infinite group (H, Λ) with the following properties:

(i) the number of orbits of H on k-element subsets in Ω equals the number of rooted binary trees with k end-vertices,

(ii) (H, Λ) is 3-homogeneous, 2-transitive but not 2-primitive, and

(iii) (H, Λ) is universal in the sense that every 3-homogeneous and 2-transitive but not 2-primitive group of countable degree is isomorphic to a subgroup of (H, Λ) or to a subgroup of the group of order-preserving or -reversing permutations of $\mathbb{Q}$.

He went further. He constructed a group (J, Ω), a one-point extension of (H, Λ) which is 5-homogeneous, 3-transitive but not 3-primitive such that the number of boron trees (all vertices have valency 1 or 3) with k end-vertices equals the number of orbits of J on k-element subsets of Ω. He then asked specifically what the connection of J to the infinite trivalent tree might be. Related to all these was the work of Covington [12] (see also [10]) on N-free graphs. Cameron's question was answered by Neumann and Adeleke in a joint seminar in 1985 and was reported by Cameron in [10]. The answer was that (J, Ω) was actually the automorphism group of the set of points at infinity ("leaves") of a general betweenness relation induced by a semilinear order. The names B, C, D relations came much later. Cameron's work has connections with several areas including chemistry.

5.8.3. *Application to Philosophy of Mathematics.* The work on semilinear order [1]–[3] had an application also to the philosophy of mathematics [5] answering one of Frege's questions. The question pertained to the independence of Frege's axioms in constructing real numbers. The automorphism group of a semilinear order provided the answer.

5.8.4. *Geometric Jordan Groups.* Notice that the classification Theorem 3.1 in one case stopped at a Steiner system. A finer classification of this case would be greatly valued. What conditions make the Steiner system an affine space or a projective space need to be explored. For finite Jordan groups, Kantor [20] and Neumann [27] showed that the Steiner system has to be affine or projective or a space of one of the Mathieu groups or connected with the linear space one way or another. But they used the classification of finite simple groups in pinning down the detailed structure of the Steiner system.

For infinite Jordan groups with enough cofinite Jordan sets (that is, Jordan sets with finite complements), Evans [16], without the classification of finite simple groups, showed that the invariant relation is connected to a linear space or is highly transitive. Zilber's work [32] went further, showing that in finite Jordan groups with dimension six or greater, the relation has to be linked to a projective or an affine space. That, of course, was without using the classification of finite simple groups. Earlier, with the use of the classification of finite simple groups, Cherlin and Zilber [11, Theorem 2.1], and Neumann [27, section 4] independently proved the existence of a link between infinite Jordan groups having cofinite Jordan sets and groups of the linear space.

5.8.5. *Other related work.* Other publications close to the contents summarized in this paper include the work of McDonough [24] on the classification of Jordan groups with perfect Jordan sets. Droste, Holland and Macpherson [15] gave a description of the normal subgroups of the automorphism groups of semilinear orders. Droste also did extensive work on normal subgroups of automorphism groups of linear orders (see [14], for example). Wiegold [30], Giorgetta [17], and Segal [29] worked on finitary permutation groups in a way connected to the contents of this paper.

5.9 *Sketch proof of Theorem 5.7*

The following lemma connects bounded groups to Jordan groups.

Lemma 5.9.1 *An infinite transitive m-bounded group (G, Ω) contains a proper Jordan set. If (G, Ω) is primitive, then it contains a primitive Jordan set.*

Sketch proof. We will assume that (G, Ω) is primitive since the reasoning for the ordinarily transitive case is similar. Let $\alpha \in \Omega$. Put $\Gamma_0 := \{\alpha\}$, $H_0 := \{1\}$. Expand recursively and in a natural way (H_i, Γ_i) $i \in \mathbb{N}$ such that

(a) $\Gamma_0 \subset \Gamma_1 \subset \Gamma_2 \subset \ldots$

(b) $H_0 < H_1 < H_2 < \ldots$

(c) every block of H_{i+1} that contains α contains Γ_i, and

(d) $\Gamma_{i+1} = \Gamma_i \cup \bigcup\{\text{supp}(g) | g \in H_{i+1}\}$.

For details of (c), pick any β, γ, $\delta \in \Gamma_i \backslash \{\alpha\}$ with $\alpha \neq \beta$, $\gamma \neq \delta$. By primitivity of (G, Ω), there exist sequences $g_1, g_2, \ldots, g_k \in G$, and $\eta_0, \eta_1, \ldots, \eta_k \in \Omega$ such that $\eta_0 = \gamma$, $\eta_k = \delta$ and $\{\alpha, \beta\}g_p = \{\gamma_{p-1}, \gamma_p\}$ for $p = 1, 2, \ldots, k$. For each such triple β, γ, δ, pick a corresponding sequence $g_1, g_2, \ldots, g_k$ with the above properties and denote by $\Pi(\beta, \gamma, \delta)$ the set of $g_1, g_2, \ldots, g_k$ picked. Define

$$H_{i+1} = H_i \cup \bigcup\{\Pi(\beta, \gamma, \delta) \mid \beta, \gamma, \delta \in \Omega, \beta \neq \alpha, \gamma \neq \delta\}.$$

Now put $\Sigma := \bigcup_{i=0}^{\infty} \Gamma_i$ and $H = \bigcup_{i=0}^{\infty} H_i$. Then H is primitive on Σ and so Σ is a primitive Jordan set in Ω with $|\Sigma| \leq m$. Since $|\Sigma| < m^+$, the successor cardinal to m, which is less than or equal to n, then Σ is a proper primitive Jordan set.

5.9.2. *Simply primitive case of Theorem 5.7.*

Assume (G, Ω) is simply primitive and m-bounded.

By Lemma 5.9.1, Ω contains a proper primitive Jordan set Γ. Let $\alpha \in \Gamma$ and $\beta \in \Omega \backslash \Gamma$. Define $\Lambda := MPJ(\beta; \alpha)$—that is, the maximal primitive Jordan set in Ω containing α and excluding β. Its existence follows from Lemma 4.1.2. If

$$(\exists \mu, \nu \in \Omega)(\mu \neq \nu)(\forall g \in G)(\nu \in \Lambda g \rightarrow \mu \in \Lambda g), \tag{5.9.3}$$

then define $\nu > \mu$ for such pairs. It is easy to see that $\leq$ is G-invariant, reflexive and transitive. Primitivity of (G, Ω) implies that $\leq$ is antisymmetric and also that $\leq$ satisfies condition 2.2.2(i) of an upper semilinear order in section 2. The use of the stabiliser $G_{(\Omega\backslash\Lambda_x)}$—as defined in 2.1.9—shows that $\leq$ satisfies condition 2.2.2(ii) as well. Denseness of $\leq$ holds because no Jordan set can have a maximal element. Thus (G, Ω) preserves a dense upper semilinear order (which is possibly a linear order) on Ω if (5.9.3) holds.

Suppose condition (5.9.3) fails but the following holds:

(5.9.4) $$(\forall g \in G)(\Lambda g \supseteq \Lambda \quad \text{or} \quad \Lambda g \subseteq \Lambda \quad \text{or} \quad \Lambda g \cap \Lambda = \phi).$$

Then define for any α, β, γ in Ω,

$$C(\alpha; \beta, \gamma) \quad \text{if and only if} \quad (\exists h \in G)(\beta, \gamma \in \Lambda h \wedge \alpha \notin \Lambda h).$$

It follows easily that the defined relation C above satisfies conditions (C1)-(C4) of a C-relation in (2.2.5). Axiom (C5) holds since otherwise the relation $\sim$ defined by $\omega_1 \sim \omega_2$ if and only if

$$(\exists \xi \in \Omega)(C(\xi; \omega_1, \omega_2))$$

is a proper congruence of (G, Ω). And denseness follows because Λ and its translates under G are primitive.

Lastly, we claim that if (5.9.3) fails, then (5.9.4) holds. We shall supply some details here since the method of proof in [1] is different but more general. Suppose, with hope of a contradiction, that (5.9.3) and (5.9.4) both fail to hold. Then

(5.9.5) $$(\exists g \in G)(\Lambda g \cap \Lambda \neq \phi, \Lambda g\backslash\Lambda \neq \phi \quad \text{and} \quad \Lambda\backslash\Lambda g \neq \phi).$$

In other words, Λ and Λg are typical. It will be considerably helpful to the reader to draw a diagram here. By definition of Λ, $\Lambda g = MPJ(\beta g; \alpha g)$. Since $\Lambda g \cup \Lambda$ is then a primitive Jordan set larger than Λ and Λg, then $\beta \in \Lambda g\backslash\Lambda$ and $\beta g \in \Lambda\backslash\Lambda g$. Furthermore, $(\beta g)g \in \Lambda g$ since $\beta g \in \Lambda$. Pick $h \in G_{(\Omega\backslash\Lambda g)}$ such that $\beta g^2 h = \beta$. Since $\beta g \notin \Lambda g$, then $\beta g h = \beta g$, and so $\beta(gh)^2 = \beta ghgh = \beta ggh = \beta g^2 h = \beta$. Moreover $\Lambda gh = \Lambda g$ by definition of h. Thus, because we can change g to gh, we can assume without loss of generality that $\beta g^2 = \beta$. Then since $\Lambda = MPJ(\beta; \beta g)$ and $\beta g^2 = \beta$, it follows that $\Lambda g^2 = MPJ(\beta g^2; \beta g^3) = MPJ(\beta; \beta g) = \Lambda$.

Define

$$\Delta(\Lambda) := \{\xi \in \Omega | \Lambda = MPJ(\xi; \alpha)\}.$$

That is, Λ is maximal with respect to excluding every element in $\Delta(\Lambda)$. Put another way, every primitive Jordan set properly containing Λ must contain $\Delta(\Lambda)$. (Note that the definition of $\Delta(\Lambda)$ and the arguments in this paragraph can be extended to Jordan groups beyond bounded ones. We put it here because of our view that the reader who understands the concept in this simpler setting can extend it easily to other cases.) Of course $\beta \in \Delta$. Then, in a way, the set Δ corresponds to the minimal increment described in Remark 4.1.6(a) on *finite* Jordan groups. Just as in the finite case, it is easy to show that $\Delta(\Lambda)$ is a block for $G_{(\Omega\setminus\Sigma)}$ where $\Sigma := \Lambda \cup \Lambda g$. But, by Lemma 4.1.3, Σ is a primitive Jordan set. Thus $\Delta(\Lambda)$ must be precisely $\{\beta\}$.

This fact about $\Delta(\Lambda)$ implies in the following way that $\Sigma = \Omega$ and (G, Ω) is 2-transitive.

Let $\mu \in \Sigma\setminus\{\beta\}$. If $\mu \in \Lambda$, then $MPJ(\beta;\mu) = \Lambda$ with $|\Delta(MPJ(\beta;\mu)| = 1$. Suppose $\mu \in \Lambda g\setminus(\Lambda \cup \{\beta\})$. Since $\Delta(\Lambda) = \{\beta\}$, $\mu \notin \Delta(\Lambda)$. So, if we pick $z \in G_{(\Omega\setminus\Lambda g)}$ such that $\beta z = \mu$, then $\Lambda z \supseteq \Lambda \cup \{\beta\}$ and $\Lambda z \subset \Sigma$. Pick $x \in G_{(\Omega\setminus\Lambda z)}$ such that $\beta g x = \beta$. Then $\mu x = \mu$, $\Lambda g x \subset \Sigma$ and $\Lambda g x = MPJ(\beta;\mu)$. Then from $|\Delta(\Lambda)| = 1$, we deduce $|\Delta(\Lambda g x)| = 1$ and so $|\Delta(MPJ(\beta;\mu))| = 1$. Thus for any $\mu \in \Sigma\setminus\{\beta\}$, $|\Delta(MPJ(\beta;\mu)| = 1$.

Suppose now that $\Omega \neq \Sigma$ and that, unlike μ, some element ν exists in $\Omega\setminus\Sigma$. Since we are assuming that (5.9.3) fails for Λ, then $MPJ(\beta;\nu)$ exists. Let $\Gamma := MPJ(\beta;\nu)$. From $\nu \notin \Lambda$, we deduce that Γ and Λ are disjoint both being maximal primitive Jordan sets with respect to excluding β but with $\nu \in \Gamma\setminus\Lambda$. Furthermore, $\Gamma \cap \Lambda g = \phi$ otherwise $\Gamma \cup \Lambda g$ will be a primitive Jordan set larger than Λg and excluding βg and then Λg will cease to be $MPJ(\beta g;\beta)$. Thus, $\Gamma \cap \Sigma = \phi$. By use of maps in $G_{(\Omega\setminus\Sigma)}$, we see that $\Delta(\Gamma) \supseteq \Sigma$. That is, $|\Delta(MPJ(\beta;\nu))| > 1$: a result different from that of $\Delta(MPJ(\beta;\mu))$ for any $\mu \in \Sigma\setminus\{\beta\}$. We conclude then that $\nu \notin \mu G_\beta$ for any $\mu \in \Sigma\setminus\{\beta\}$. Thus, $\Sigma G_\beta = \Sigma$. And for any $w \in G_{(\Omega\setminus\Lambda g)}$ with $\beta w \neq \beta$, we see that $\Sigma w = \Sigma$. So, $\Sigma\langle G_\beta, w\rangle = \Sigma$. But then, $G = \langle G_\beta, w\rangle$ by primitivity of G. Hence $\Sigma G = \Sigma$ and $\Sigma = \Omega-$ contradicting our supposition of existence of ν. We deduce therefore that $\Sigma = \Omega$.

From $\Sigma = \Omega$, we now deduce that (G, Ω) is 2-transitive. Let ξ be any element in $\Lambda\setminus(\Lambda g \cup \{\beta g\})$. Then $\xi g \in \Lambda g\setminus(\Lambda g^2 \cup \{\beta g^2\}) = \Lambda g\setminus(\Lambda \cup \{\beta\})$ since $\Lambda g^2 = \Lambda$ and $\beta g^2 = \beta$. Exactly as done for μ when assumed to be in $\Lambda g\setminus(\Lambda \cup \{\beta\})$, we can construct $x' \in G$ such that $(\xi g)x' = \xi g$, $(\beta g)x' = \beta$. Then $\xi g x' = \xi g$ and $(\beta g)g x'^{-1} = \beta x'^{-1} = \beta g$. Thus $g x'^{-1} \in G_{(\beta g)}$ and $\xi(g x'^{-1}) \in \Lambda g$. Therefore, $\xi G_{(\beta g)} \cap \Lambda g \neq \phi$. Hence ξ lies in the orbit of $G_{(\beta g)}$ containing Λg. Since ξ is arbitrary in $\Lambda\setminus(\Lambda g \cup \{\beta g\}$, then (G, Ω) (which is the same as (G, Σ)) is 2-transitive. This contradicts our hypothesis. Thus (5.9.4) must hold if (5.9.3) fails, and so Theorem 5.7 holds when (G, Ω) is simply primitive.

5.9.6. *The 2-transitive not 3-transitive case.* If (G, Ω) is 2-transitive but not 3-transitive, then G has an invariant C-relation. Its proof is similar but simpler than the proof of the simply primitive case. The proof defines Λ as the maximal 2-transitive Jordan set containing α and excluding β and shows that (5.9.4) always holds as follows. If (5.9.4) fails, define $\Sigma = \Lambda \cup \Lambda g$ as before. Then Σ is 2-transitive. Pick $\mu \in \Lambda\setminus\Lambda g$ and $\nu \in \Lambda g\setminus\Lambda$. Then by use of $G_{(\Omega\setminus\Lambda)(\mu)}$ followed by use of $G_{(\Omega\setminus\Lambda g)(\nu)}$, we see that $G_{(\Omega\setminus\Sigma)(\mu)(\nu)}$ is transitive on $\Sigma\setminus\{\mu, \nu\}$. Thus Σ is 3-transitive which gives a contradiction.

5.9.7. (G,Ω) *t-transitive for* $t \geq 3$. Consider a maximal proper t-transitive Jordan set Λ. Since G is t-transitive with $t \geq 3$, then if $\lambda_1, \lambda_2 \in \Lambda$, $\lambda_1 \neq \lambda_2$, and $\lambda_3 \in \Omega\backslash\Lambda$, there is $g \in G$ such that $\lambda_1 g = \lambda_1$, $\lambda_2 g = \lambda_3$ and $\lambda_3 g = \lambda_2$. Thus, the situation in (5.9.5) always occurs. This will imply that $\Lambda \cup \Lambda g$ is a $(t+1)$-transitive Jordan set just as in the last case. (In the last part of the argument, pick $\mu \in \Lambda\backslash\Lambda g$, $\nu \in \Lambda g\backslash\Lambda$ as before and pick any other $t-2$ elements in Λ which leave out at least one element in $\Lambda \cap \Lambda g$). Thus G is $(t+1)$-transitive by Lemma 4.1.4. By induction, (G,Ω) is highly transitive.

5.10 *Transitive not primitive m-bounded groups*

Lemma 5.10.1 *Suppose (G,Ω) is a transitive m-bounded group. If ρ is a proper congruence, then $|\rho| \leq m$.*

Reason: If $g \in G$ and ψ is a block with $|\psi| > m$, then $|\text{supp}(g)| \leq m$ and so $\psi\backslash(\text{supp}(g)) \neq \phi$. Therefore $\psi \cap \psi g \neq \phi$; so $\psi g = \psi \forall g \in G$. Therefore $\psi = \Omega$.

The following are also not difficult to prove.

Lemma 5.10.2 [1, Lemma 2.2] *Suppose that G is transitive and m-bounded and that there is no maximal proper congruence on Ω. Then*

(i) $n = m^+$;

(ii) there is a properly increasing chain $\rho_0 < \rho_1 < \ldots < \rho_\xi < \ldots$ of proper congruences indexed by ν, the initial ordinal of n, and $\cup_{\xi<\nu}\rho_\xi$ is the universal relation;

(iii) if $\Delta \subseteq \Omega$ and $|\Delta| \leq m$ then there exists $\mu < \nu$ such that all members of Δ are equivalent modulo ρ_μ;

(iv) if ρ is any proper congruence then $\rho \leq \rho_\eta$ for some $\eta < \nu$.

Definition 5.10.3 Let (G,Ω) be a transitive m-bounded permutation group. Say (G,Ω) is *almost primitive* if (G,Ω) has a maximal proper congruence $\rho_{\max}$ with each class of the congruence finite.

Lemma 5.10.4 [1, Lemma 2.3] *If G is a transitive, almost primitive, m-bounded group of permutations of Ω then the maximal proper congruence $\rho_{\max}$ is unique.*

Remark 5.10.5. By Lemma 5.9.1, a transitive m-bounded group is a Jordan group.

6 Infinite Primitive Jordan Groups with Proper Primitive Jordan Sets

Bounded groups discussed in Theorem 5.7 constitute a type of infinite primitive Jordan groups with proper primitive Jordan sets as in Theorem 5.7. A natural problem is then to classify all primitive Jordan groups that contain proper primitive Jordan sets. This question is more complex than that of m-bounded groups. For we can no longer establish existence of proper t-transitive Jordan sets in t-transitive Jordan groups for $t \geq 2$ as is done in the proof of Lemma 5.9.1. But the final result is this.

Theorem 6.1 (see [2]) *Let (G, Ω) be an infinite primitive Jordan group that contains a proper primitive Jordan set. Then (G, Ω) either is highly transitive or leaves invariant one of the following relations on Ω :*

(a) a dense linear order, a dense semilinear order or a dense C-relation if (G, Ω) is simply primitive;

(b) a dense linear betweenness relation, a dense general betweenness relation, or a dense C-relation if (G, Ω) is 2-transitive not 2-primitive;

(c) a dense circular order or a dense D-relation if (G, Ω) is 2-primitive not 3-transitive;

(d) a dense separation relation or a dense D-relation if (G, Ω) is 3-transitive not 3-primitive.

In the case when (G, Ω) is 3-primitive then it is highly transitive.

Remark 6.2. The proof of Theorem 6.1 in [2] uses Theorem 6.3 below by Cameron. We will roughly indicate below an alternative proof of Theorem 6.1 not using Theorem 6.3. The aim is to present the reader with an elementary proof which when combined with the version in [2] produces good understanding.

Theorem 6.3 (Cameron) [8]) *Let (G, Ω) be an infinite highly homogeneous group which is t-transitive not $(t+1)$-transitive. Then $t \leq 3$ and G is a group of order-preserving or reversing transformations of a linear or circular order on Ω.*

6.4 *Sketch proof of Theorem 6.1*

6.4.1. (G, Ω) *simply primitive.* This is proved by the arguments for simply primitive m bounded groups in (5.9.2). Before we pass on to the next case, let us make one observation.

Lemma 6.4.2 *Let (G, Ω) be an infinite simply primitive permutation group with a proper primitive Jordan set. Suppose Ω contains a maximal primitive Jordan set Λ which does not induce a G-invariant linear order through property (5.9.3). Then no two proper primitive Jordan sets Γ_1, Γ_2, are typical. That is, the following is not possible:*

$$\Gamma_1 \cap \Gamma_2 \neq \emptyset, \Gamma_1 \backslash \Gamma_2 \neq \emptyset \quad \text{and} \quad \Gamma_2 \backslash \Gamma_1 \neq \emptyset.$$

Sketch proof of Lemma 6.4.2. Let $\Lambda := MPJ(\beta; \alpha)$ as in the proof of Lemma 5.9.2. Assume that Λ induces an invariant *nonlinear* upper semilinear order on Ω through property (5.9.3). It is not hard to see that nonlinearity of the semilinear order implies that if Γ is a Jordan set in Ω containing an element ξ, then Γ also contains an element η incomparable to ξ. So, the map in $G_{(\Omega \backslash \Gamma)}$ that moves ξ to η also moves all elements below ξ. Hence Γ contains all elements below ξ. Thus every Jordan set is a branch below a branch point (as defined in (2.2.2)) in the semilinear order or a union of branches below branch points. And every *primitive* Jordan set in Ω is a branch or a union of an *increasing* chain of branches. Since no two branches are typical, we conclude then that no two primitive Jordan sets are typical. Hence the lemma holds if G preserves a nonlinear semilinear order on Ω. A similar argument proves the lemma when G preserves a C-relation on Ω.

6.4.3. (G,Ω) *2-transitive not 2-primitive.* If (G,Ω) has a proper 2-transitive Jordan set, then the arguments in section 5.9.6 will suffice to show that G preserves a C-relation on Ω.

Let us assume then that (G,Ω) has no proper 2- transitive Jordan sets. Then the proof for this case is a small step beyond that of simply primitive m-bounded groups in section 5.9.2. Let $\Lambda := MPJ(\beta;\alpha)$. Since (G,Ω) is not 2-primitive, then $\Lambda \neq \Omega\backslash\{\beta\}$. If condition (5.9.4) holds, then G preserves a dense C-relation on Ω. If condition (5.9.4) fails, then as before, $\Lambda\cup\Lambda g = \Omega$ for some $g \in G$.

In that case, for every $\alpha,\beta,\gamma\in\Omega$, define $B(\alpha;\beta,\gamma)$ if and only if

$$(\alpha=\gamma) \quad \text{or} \quad (\alpha=\beta) \quad \text{or} \quad (MPJ(\alpha;\beta)\cap MPJ(\alpha;\gamma)=\emptyset).$$

It follows easily that the defined relation B satisfies axioms (B1), (B3), and (B5) of a general betweenness relation in 2.2.3. As for axiom (B2), suppose $B(\alpha;\beta,\gamma)$ with α,β and γ distinct. Then $\beta\notin MPJ(\alpha;\gamma)$. Because the defined set $\Delta(MPJ(\alpha;\gamma)) := \{\xi | MPJ(\xi;\gamma) = MPJ(\alpha;\gamma)\}$ equals the singleton set $\{\alpha\}$ (in a way similar to the procedure in section 5.9.2), then from $\beta\notin MPJ(\alpha;\gamma)$, we deduce that $MPJ(\beta;\gamma)$ is larger than $MPJ(\alpha;\gamma)$. Thus $\alpha\in MPJ(\beta;\gamma)$. Therefore $\neg B(\beta;\alpha,\gamma)$. So axiom (B2) holds as well. We can deduce more. The arguments show that every maximal primitive Jordan set is contained in a larger maximal primitive Jordan set. Equivalently by 2-transitivity, every maximal primitive Jordan set contains a smaller maximal primitive Jordan set. So, to prove (B4), suppose $\neg B(\alpha;\beta,\gamma)$. Then $\gamma\in MPJ(\alpha;\beta)$. Put $\Gamma := MPJ(\alpha;\beta)$. Then Γ contains a smaller maximal primitive Jordan set Σ where Σ is a maximal primitive Jordan set $MPJ(\mu;\nu)$ in Ω containing some element ν and excluding some μ. Hence $(G_{(\Omega\backslash\Gamma)},\Gamma)$ is a simply primitive Jordan group, simply primitive because we have ruled out the case of proper 2-transitive Jordan sets. Whichever of the types of invariant relation in (5.9.2) that Σ induces on Γ, it is easy to see from the definitions of the relations that there exists a translate $\Sigma x\subset\Gamma$, $x\in G_{(\Omega\backslash\Gamma)}$ such that γ, $\beta\in\Sigma x$. Thus Γ contains properly some $MPJ(\delta;\beta)$ with $\gamma\in MPJ(\delta;\beta)$ and $\delta\in\Gamma$. In that case, $B(\delta;\alpha,\beta)$ and $B(\delta;\alpha,\gamma)$ and so (B4) holds. Consequently, G preserves a general betweenness relation B which may reduce to the special case of a linear betweenness relation.

6.4.4. (G,Ω) *2-primitive not 3-transitive.* Let $\delta\in\Omega$. Then $(G_\delta,\ \Omega\backslash\{\delta\})$ is simply primitive.

We first show that on $\Omega\backslash\{\delta\}$, no maximal primitive Jordan set in $\Omega\backslash\{\delta\}$ induces a nonlinear upper semilinear order that is invariant under G_δ. Dugald Macpherson

[23] in a preprint also proved the nonexistence of any G_δ-invariant nonlinear semilinear order on $\Omega\backslash\{\delta\}$ if (G, Ω) is 2-primitive but his proof is not the same as the one below. He showed further that there is no one-point extension of a primitive Jordan group (H, Γ) if H preserves a general betweenness relation on Γ, or preserves a D-relation on Γ. But again his proofs are not the same as those in this paper.

Suppose for a contradiction that there is a maximal primitive Jordan set in $\Omega\backslash\{\delta\}$ which induces a nonlinear upper semilinear order on $\Omega\backslash\{\delta\}$ invariant under G_δ. Define for every $\omega_1, \omega_2 \in \Omega\backslash\{\delta\}$, $\omega_1 \sim \omega_2$ if and only if there is a primitive Jordan set Γ in Ω containing δ and excluding ω_1 and ω_2. Clearly, $\sim$ is reflexive, symmetric and G_δ-invariant. To check if $\sim$ is transitive, suppose $\omega_1 \sim \omega_2$ and $\omega_2 \sim \omega_3$. Then there exist primitive Jordan sets Γ_1 and Γ_2 in Ω such that $\delta \in \Gamma_1 \cap \Gamma_2$, $\omega_2 \notin \Gamma_1 \cup \Gamma_2$, $\omega_1 \notin \Gamma_2$ and $\omega_3 \notin \Gamma_2$. In other words, Γ_1 and Γ_2 are non-disjoint primitive Jordan sets in $\Omega\backslash\{\omega_2\}$ with its nonlinear semilinear order. By Lemma 6.4.2, either $\Gamma_1 \subseteq \Gamma_2$ or $\Gamma_2 \subseteq \Gamma_1$. In either case, the smaller one contains δ and excludes ω_1 and ω_3. Thus, $\omega_1 \sim \omega_3$ and $\sim$ is transitive. So, $\sim$ is a congruence which must then be universal by primitivity of $(G_\delta, \Omega\backslash\{\delta\})$. But that means that if $\omega \in \Omega\backslash\{\delta\}$, then for any $\omega_* \in \Omega\backslash\{\delta\}$ with $\omega_* < \delta$ in the nonlinear semilinear order on $\Omega\backslash\{\omega\}$ there is a primitive Jordan set Γ containing δ and excluding ω_* and ω. Since Γ is Jordan, with $\delta > \omega_*$, $\delta \in \Gamma$, $\omega_* \notin \Gamma$, then each element of Γ is an upper bound of ω_*. In that case, Γ contains some λ such that $\delta > \lambda > \omega_*$. Let Σ be the primitive Jordan set guaranteed by $<$ on $\Omega\backslash\{\omega\}$ which contains λ and omits δ and ω. Then $\omega_* \in \Sigma\backslash\Gamma$, $\lambda \in \Sigma \cap \Gamma$, and $\delta \in \Gamma\backslash\Sigma$. Then Σ and Γ will form a typical pair both excluding ω. This contradicts Lemma 6.4.2. Hence G_δ does not preserve a nonlinear semilinear order on $\Omega\backslash\{\delta\}$.

Suppose G_δ preserves a C-relation on $\Omega\backslash\{\delta\}$. With the same method as in Section 5.9.2 (now applied to $(G_\delta, \Omega\backslash\{\delta\})$, let $\Lambda := MPJ(\delta, \beta; \alpha)$. Also let $\xi_1, \xi_2 \in \Omega\backslash(\Lambda \cup \{\beta\})$ with $\xi_1 \neq \xi_2$. The method of the sketch proof of Lemma 6.4.2 shows that a union of an increasing chain of translates of Λ equals $\Omega\backslash\{\xi_1\}$, and a union of another chain of translates of Λ equals $\Omega\backslash\{\xi_2\}$. It then follows that there are two translates of Λ, one picked from each chain, which are typical. As a consequence, Λ and Λh are typical for some $h \in G$. By Lemma 6.4.2, $\Lambda \cup \Lambda h = \Omega$. So in case of a C-relation on $\Omega\backslash\{\delta\}$, Ω contains typical pairs of translates of Λ with the union of any such pair being Ω. This property on typical pairs prompts the following definition. For $\alpha, \beta, \gamma, \delta \in \Omega$, define

$$\begin{aligned}&D(\alpha, \beta; \gamma, \delta) \quad \text{if and only if}\\ &\quad (\exists x \in G)((\alpha, \beta \in \Lambda x) \wedge (\gamma, \delta \notin \Lambda x)) \quad \text{or} \quad (\exists y \in G)((\alpha, \beta \notin \Lambda y) \wedge (\gamma, \delta \in \Lambda y)).\end{aligned}$$

Then, relation D above satisfies all the axioms of a D-relation contained in 2.2.6 with a little use of Lemma 6.4.2. It is a dense D-relation because of the primitivity of Λ.

Lastly, suppose G_δ preserves a linear order $<$ on $\Omega\backslash\{\delta\}$ through property

(5.9.3). In that case, define for every distinct γ, ξ, η in Ω,

$$\begin{aligned}&\rho(\gamma,\xi,\eta) \quad \text{if and only if}\\&(\gamma=\xi) \quad \text{or} \quad (\xi=\eta) \quad \text{or} \quad (\eta=\gamma) \quad \text{or} \quad (\forall x\in G)(\eta x=\delta)\rightarrow(\gamma x<\xi x)).\end{aligned}$$

Then, it is clear that ρ is G-invariant and satisfies axioms ($\rho2$) and ($\rho4$) in 2.2.7. As for axiom ($\rho1$), suppose in hope of contradiction that $\rho(\alpha,\beta,\gamma)$ and $\neg\rho(\beta,\gamma,\alpha)$. Then α,β,γ are distinct and there exist $x,y\in G$ such that

$$(\gamma x=\delta)\wedge(\alpha y=\delta)\wedge\rho(\alpha x,\beta x,\delta)\wedge\neg\rho(\beta y,\gamma y,\delta).$$

By linearity *of* $<$, we can rewrite condition on ρ as $\rho(\alpha x,\beta x,\delta)$ and $\rho(\gamma y,\beta y,\delta)$. From condition (5.9.3), we see that the primitive Jordan set Λ is an initial segment of the linearly ordered set $(\Omega\backslash\{\delta\},\leq)$. By usual arguments (see first paragraph of proof of Theorem 9.3.2(i) below), G_δ is order 2-transitive on $\Omega\backslash\{\delta\}$. Thus, there is $z\in G_\delta$ such that $\alpha xz=\gamma y$ and $\beta xz=\beta y$. Then $\beta(xzy^{-1})=\beta$, $\alpha(xzy^{-1})=\gamma$ and $\gamma(xzy^{-1})=\delta zy^{-1}=\delta y^{-1}=\alpha$. Therefore, the map xzy^{-1} fixes β but interchanges γ and α. Thus it does not preserve the linear order on $\Omega\backslash\{\beta\}$, a contradiction. So, ρ satisfies condition ($\rho1$) also. Now, to condition ($\rho3$);

$$\rho(\alpha,\beta,\gamma)\wedge\neg\rho(\delta,\beta,\gamma)\rightarrow\rho(\alpha,\beta,\gamma)\wedge\rho(\beta,\delta,\gamma)$$

using the definition of ρ, then the linearity of $<$, and then the G-invariance of ρ;

$$\rightarrow\rho(\gamma,\alpha,\beta)\wedge\rho(\delta,\gamma,\beta)$$

using ($\rho1$)

$$\rightarrow\rho(\delta,\alpha,\beta)$$

using the definition of ρ, followed by the definition of $<$ on $\Omega\backslash\{\delta\}$ through 5.9.3, and then the G-invariance of ρ.

So, ($\rho3$) holds. We conclude therefore that ρ is a G-invariant circular order.

6.4.5. (G,Ω) *3-transitive not 3-primitive.* If $\delta\in\Omega$, then G_δ is 2-transitive not 2-primitive on $\Omega\backslash\{\delta\}$. From 6.4.3, G_δ preserves on $\Omega\backslash\{\delta\}$ a betweenness relation, or a C-relation.

Suppose it is a C-relation. By 3-transitivity of G, Λ and Λg are typical for some $g\in G$. Then, arguments similar to those on the C-relation in 6.4.4, with Lemma 6.4.2 suitably used, imply the existence of a G-invariant D-relation on Ω.

Suppose the invariant relation on $\Omega\backslash\{\delta\}$ is not a C-relation but a nonlinear betweenness relation. We shall show that this cannot occur. Let β_1,β_2 be distinct elements of $\Omega\backslash\{\delta\}$. Then, of course, an invariant nonlinear betweenness relation occurs on $\Omega\backslash\{\beta_1\}$ too. Since $(G_{\beta_1},\Omega\backslash\{\beta_1\})$ is not 2-primitive, and is 2-transitive, then there is some maximal primitive Jordan set $\Gamma:=MPJ(\beta_1,\beta_2;\lambda)$ with respect to excluding β_1 and β_2 which is disjoint from $MPJ(\beta_1,\beta_2;\delta)$. So, Γ is a subset of

$\Omega\backslash\{\delta\}$. (The reader will benefit from two diagrams here; one for the betweenness relation on $\Omega\backslash\{\delta\}$, another for the betweenness relation on $\Omega\backslash\{\beta_1\}$.) Whatever the location of Γ in the betweenness relation on $\Omega\backslash\{\delta\}$, one thing is clear. There is a primitive Jordan set $\Pi \subset \Omega\backslash\{\delta\}$ disjoint from Γ which contains one of β_1, β_2 and excludes the other. Without loss of generality, assume $\beta_1 \in \Pi$ and $\beta_2 \notin \Pi$. By use of $G_{(\Omega\backslash\Pi)}$, we deduce that $\Gamma = MPJ(\omega, \beta_2; \lambda)$ for all $\omega \in \Pi$. But this contradicts the defined set $\Delta(MPJ(\beta_1, \beta_2; \lambda)) := \{\xi | MPJ(\xi, \beta_2; \lambda) = MPJ(\beta_1, \beta_2; \lambda)\}$ being a singleton set as described in 6.4.3. Thus G_δ cannot preserve a nonlinear betweenness relation on $\Omega\backslash\{\delta\}$.

The only other possibility on $\Omega\backslash\{\delta\}$ is a linear betweenness relation invariant under G_δ. This leads to a G-invariant separation relation on Ω proved in [2] using Cameron's Theorem. But there is an alternative proof not using Cameron's Theorem. It is messy though. The proof, like others so far, relies on earlier arguments; in this case, on the arguments for the 2-transitive not 2-primitive possibility. Let δ, ω, α be distinct in Ω. Let $\Lambda := MPJ(\delta, \omega; \alpha)$. Since we are assuming that no G_δ-invariant C-relation exists on $\Omega\backslash\{\delta\}$, then from the first two paragraphs of section 6.4.3 now applied to $(G_\delta), \Omega\backslash\{\delta\})$, we note that $\Lambda \cup \Lambda g = \Omega\backslash\{\delta\}$ for some $g \in G_\delta$. It follows that Λ is not a bounded interval with respect to any of the two dense linear orders associated with the linear betweenness relation on $\Omega\backslash\{\delta\}$ assumed invariant under G_δ. So, for one of the two linear orders, $<$ (say), Λ is an initial segment. Define $H_1 := \{g \in G_\delta \mid \Lambda g \subseteq \Lambda \text{ or } \Lambda g \supseteq \Lambda\}$. Then clearly, $(H_1, \Omega\backslash\{\delta\})$ becomes a simply primitive group. Moreover $<$ satisfies:

$$(6.4.6) \qquad \mu < \nu \quad \text{if and only if} \quad (\forall h \in H_1)((\nu \in \Lambda h) \rightarrow (\mu \in \Lambda h)).$$

Next, observe in the following arguments, that $MPJ(\omega, \alpha; \delta)$ consists of δ, the elements less than α in the linear order $<$, and the elements greater than ω. For if ξ is any element in $\Omega\backslash\{\delta\}$ less than α, then $MPJ(\delta, \alpha; \xi)$ is the set of all elements lower than α. The result on Δ being singleton in 6.4.3 now shows that $MPJ(\delta, \alpha; \xi) \neq MPJ(\omega, \alpha; \xi)$ but rather $MPJ(\delta, \alpha; \xi) \subset MPJ(\omega, \alpha; \xi)$. Thus $\delta \in MPJ(\omega, \alpha; \xi)$. Similarly, if $\eta > \omega$, then $\delta \in MPJ(\omega, \alpha; \eta)$. So, indeed,

$$MPJ(\omega, \alpha; \delta) = \{\gamma \in \Omega | \gamma < \alpha \quad \text{or} \quad \gamma > \omega\} \cup \{\delta\}.$$

This of course extends to MPJ $(\beta_1, \beta_2; \delta)$ for any two distinct β_1, β_2 in $\Omega\backslash\{\delta\}$. All these show that if we define

$$H_2 := \{g \in G_\omega | \Lambda g \subseteq \Lambda \quad \text{or} \quad \Lambda g \supseteq \Lambda\} \quad \text{and} \quad H := \langle H_1, H_2 \rangle,$$

then we can through lengthy but very elementary steps show that (H, Ω) is a 2-primitive group preserving the circular order associated with $<$. With that proven, if we pick $x \in G_\delta \backslash H_1$, then x reverses $<$ while keeping δ fixed. In fact, x reverses the circular order associated with $<$. Moreover, $G_\delta = \langle H_1, x \rangle$ and consequently

$G = \langle H_1, H_2, x\rangle = \langle H, x\rangle$. Therefore G preserves the circular order or reverses it. That is, G preserves a separation relation on Ω if G_δ preserves a linear betweenness relation on $\Omega\backslash\{\delta\}$.

6.4.7. (G, Ω) *3-primitive not 4-transitive.* This case really does not occur. The arguments should, by now, be familiar. The group G_δ is 2-primitive not 3-transitive on $\Omega\backslash\{\delta\}$. So, employ the notation in 6.4.4.

Suppose G_δ preserves a D-relation on $\Omega\backslash\{\delta\}$. Then, for any $\xi_1, \xi_2, \xi_3, \xi_4$ distinct in Ω, $MPJ(\xi_1, \xi_2, \xi_3; \xi_4)$ exists since the invariant relation on $\Omega\backslash\{\xi_1, \xi_2\}$ is not a semilinear order but a C-relation. Let $\Lambda := MPJ(\delta, \beta, \xi; \alpha)$. Every primitive Jordan set in $\Omega\backslash\{\delta, \beta\}$ is either a $G_{\delta\beta}$-translate of Λ (that is, Λx for some $x \in G_{\delta\beta}$), or is a union of an increasing chain of such translates (using the method of proof of Lemma 6.4.2). Let $\{\Lambda_i | i \in I\}$ be an increasing chain of primitive Jordan sets, each containing α, whose union is $\Omega\backslash\{\delta, \beta\}$. Without loss of generality, assume that each i in I has a successor i^+. Let $\lambda_i \in \Lambda_{i^+}\backslash\Lambda_i$. From the definition of the D-relation in 6.4.4, if $i < j$ in I and D_1, D_2 are the D-relations on $\Omega\backslash\{\delta\}$ and $\Omega\backslash\{\beta\}$ respectively, then

$$D_1(\alpha, \lambda_i; \lambda_j, \beta) \wedge D_2(\alpha, \lambda_i; \lambda_j, \delta). \tag{6.4.8}$$

(A diagram will be useful here).

Let $i_0 \in I$ and define

$$\Gamma_1 := MPJ(\alpha, \delta, \lambda_{i_0}; \beta), \quad \Gamma_2 := MPJ(\alpha, \beta, \lambda_{i_0}; \delta).$$

From (6.4.8), $\lambda_i \notin \Gamma_1 \cup \Gamma_2$ if $i < i_0$. Let $\mu \in \Gamma_1\backslash\{\beta\}$. Since $\cup\{\Lambda_i | i \in I\} = \Omega\backslash\{\delta, \beta\}$, then

$$(\exists j_0 \in I)(\forall j > j_0)(D_1(\alpha, \mu; \lambda_j, \beta)).$$

In that case, $\lambda_j \in \Gamma_1$, $\forall j > j_0$. Similarly, we can deduce that for some $k_0 \in I$, $\lambda_k \in \Gamma_2$, $\forall k > k_0$. Hence $\lambda_p \in \Gamma_1 \cap \Gamma_2$ for all $p > \max\{j_0, k_0\}$. Thus Γ_1, Γ_2 are typical primitive Jordan sets in $\Omega\backslash\{\alpha, \lambda_{i_0}\}$. This contradicts Lemma 6.4.2. So, G_δ cannot preserve a D-relation on $\Omega\backslash\{\delta\}$.

If on the other hand G_δ preserves a circular order on $\Omega\backslash\{\delta\}$, then $G_{\delta\beta}$ preserves a linear order $<$ on $\Omega\backslash\{\delta, \beta\}$. Let $\alpha < \eta$ in $\Omega\backslash\{\delta, \beta\}$. Then the arguments in section 6.4.5 on the circular order show that

$$\Pi_1 := MPJ(\delta, \alpha, \eta; \beta) = \{\beta\} \cup \{\xi \mid \xi < \alpha \quad \text{or} \quad \xi > \eta\}$$
$$\Pi_2 := MPJ(\beta, \alpha, \eta; \delta) = \{\delta\} \cup \{\xi \mid \xi < \alpha \quad \text{or} \quad \xi > \eta\}$$

Thus Π_1, Π_2 is a typical pair with $\Pi_1\backslash\Pi_2 = \{\beta\}$. So, $\Pi_1 \cup \Pi_2$ is a proper 2-transitive Jordan set excluding α and η. Then $(G_{\delta\beta}, \Omega\backslash\{\delta, \beta\})$ becomes 2-transitive by Lemma 4.1.4. This contradiction shows that G_δ cannot preserve a circular order on $\Omega\backslash\{\delta\}$ either.

6.4.9. (G, Ω) *4-transitive not 4-primitive.* This case also is not possible.

Suppose (G, Ω) contains proper 2-transitive Jordan sets. Then, as indicated in the first statement in Section 6.4.3, the *arguments* in Section 5.9.6 are enough to show that G_{δ_μ} preserves a C-relation on $\Omega\backslash\{\delta, \mu\}$ for any distinct δ and μ in Ω. Then the second paragraph in Section 6.4.5 shows that G_δ preserves a D-relation on $\Omega\backslash\{\delta\}$.

In that case, replace maximal primitive Jordan sets in the arguments in section 6.4.7 by maximal 2-transitive Jordan sets. Then $\Gamma_1 \cup \Gamma_2$ will be a proper 3-transitive Jordan set at the end of the arguments making $G_{\alpha\lambda_{i_0}}$ 3-transitive and thus (G, Ω) 5-transitive, a contradiction.

Suppose then that (G, Ω) does not contain 2-transitive Jordan sets. If G_δ preserves a D-relation on $\Omega\backslash\{\delta\}$, use arguments in 6.4.7 to get a contradiction. If, however, G_δ preserves a separation relation on $\Omega\backslash\{\delta\}$, use the part of the arguments in section 6.4.7 on circular orders to get a contradiction.

6.4.10. *Conclusion to proof of Theorem 6.1.* From sections 6.4.7 and 6.4.9, we see that if (G, Ω) is 3-primitive then it is 4-primitive. When this result is in turn applied to $(G_\delta, \Omega\backslash\{\delta\})$, the latter is shown to be 4-primitive, and so (G, Ω) is proved 5-primitive. Such an inductive argument shows that if (G, Ω) is 3-primitive, then it is n-primitive for all n, and so is highly transitive.

7 General Infinite Primitive Jordan Groups

(No assumption of presence of proper primitive Jordan sets.)

7.1. *Sketch proof of Theorem 3.1 for Simply Primitive Jordan Groups*

Define $\Lambda := MJ(\beta; \alpha)$, the maximal Jordan set in Ω containing α and excluding β. If Λ satisfies (5.9.3) or (5.9.4), then the arguments in section (5.9.2) show that Λ induces an invariant dense semilinear order or an invariant C-relation, not necessarily dense.

Suppose, however, that Λ violates (5.9.3) and (5.9.4). If $|\Delta(\Lambda)| = 1$, then the arguments of section (5.9.2) again yield a contradiction. Hence if Λ violates (5.9.3) and (5.9.4) then $|\Delta(\Lambda)| > 1$.

From the property that $|\Delta(\Lambda)| \neq 1$, we now show that $\Delta(\Lambda)$ and $\Delta(\Lambda)x$ cannot be typical for any $x \in G$. Suppose $\beta \in \Delta(\Lambda) \cap \Delta(\Lambda)x$ and $\gamma \in \Delta(\Lambda)x\backslash\Delta(\Lambda)$. Since $\Delta(\Lambda)x = \Delta(\Lambda x)$, then $\gamma \in \Delta(\Lambda x)\backslash\Delta(\Lambda)$. So $\Delta(\Lambda x) \neq \Delta(\Lambda)$ and therefore Λ and Λx are different maximal Jordan sets with respect to excluding the same β. Consequently $\Lambda x \cap \Lambda = \phi$. If $\gamma \notin \Lambda$, then since $\gamma \notin \Delta(\Lambda)$, we have $MJ(\gamma; \alpha) \supset \Lambda$, and so $\Delta(\Lambda) \subset MJ(\gamma; \alpha)$; in particular $\beta \in MJ(\gamma; \alpha)$. Put $\Pi := MJ(\gamma; \alpha)$. Then Π and Λx are both maximal Jordan sets with respect to excluding the same γ, one containing Λ and the other disjoint from Λ. So $\Lambda x \cap \Pi = \phi$. Since $\beta \in \Pi$ and $\beta \in \Delta(\Lambda x)$, then the use of $G_{(\Omega\backslash\Pi)}$ yields $\Delta(\Lambda x) \supseteq \Pi$. But $\Delta(\Lambda) \subset \Pi$. Hence $\Delta(\Lambda x) \supset \Delta(\Lambda)$. Suppose on the other hand that $\gamma \in \Lambda$. Consider the typical pair Λ and Λg with $\beta \in \Lambda g\backslash\Lambda$ and $\beta g \in \Lambda\backslash\Lambda g$. Pick $z \in G_{(\Omega\backslash\Lambda)}$ such that $\beta g z = \gamma$. Then $\Lambda g z = MJ(\gamma;\beta)$ with $\Lambda g z$ and Λ being typical. Now $\Lambda g z$ and Λx are both maximal Jordan sets with respect to excluding γ with $\beta \in \Lambda g z$ and $\beta \notin \Lambda x$. Therefore, $\Lambda g z \cap \Lambda x = \phi$. This with the earlier deduction of $\Lambda \cap \Lambda x = \phi$ shows that $\Lambda x \cap (\Lambda \cup \Lambda g z) = \phi$. Since $\beta \in \Delta(\Lambda x)$, then by using elements in G which fix pointwise $\Omega\backslash(\Lambda \cup \Lambda g z)$, we see that $\Delta(\Lambda x) \supseteq \Lambda \cup \Lambda g z$. But $\Delta(\Lambda) \subseteq (\Lambda g z\backslash\Lambda)$ since $(\Lambda g z \cup \Lambda) \supset \Lambda$. Hence $\Delta(\Lambda) \subset (\Lambda \cup \Lambda g z) \subseteq \Delta(\Lambda x)$. Therefore, $\Delta(\Lambda) \subseteq \Delta(\Lambda x)$. Hence, in any case, $\Delta(\Lambda x) \supset \Delta(\Lambda)$ if $\Delta(\Lambda x) \cap \Delta(\Lambda) \neq \phi$ and $\Delta(\Lambda x)\backslash\Delta(\Lambda) \neq \phi$. Thus $\Delta(\Lambda)$ satisfies a property similar to property (5.9.4) on Λ as follows:

$$(7.1.1) \qquad (\forall g \in G)(\Delta(\Lambda)g \cap \Delta(\Lambda) = \phi) \vee (\Delta(\Lambda)g \subseteq \Delta(\Lambda)) \vee (\Delta(\Lambda)g \supseteq \Delta(\Lambda)).$$

And because $\Delta(\Lambda) \subset \Lambda g$ and Λ violates (5.9.3), then $\Delta(\Lambda)$ also violates a property similar to (5.9.3). Therefore if Λ violates (5.9.3) and (5.9.4), then $\Delta(\Lambda)$ induces a C-relation (not necessarily dense) on Ω which is invariant under G.

7.2. *On general Theorem 3.1 when (G, Ω) is 2-transitive*

This paper gives no sketch proof of the general Theorem 3.1 for 2-transitive Jordan groups mainly because a very short section (section 4 of [4]) does so already. The sketch proof in [4] focuses on the situation when no primitive Jordan sets exist in the Jordan group. It should pose no difficulty to a reader who has gone over the sketch proofs in this paper. Even the full proofs in [4] themselves should not be difficult since they are natural extensions of the reasoning described in the earlier sections above. Unexpected examples occur in the general case, of course; see for instance [6].

8 On Invariant Relations in Primitive Permutation Groups

8.1. *Introduction*

From an inverted tree, three other relations can be defined besides the order on the points on the tree (see[2], [3], [4] or [6]. Firstly, a betweenness relation exists between the points on the tree. Secondly, apart from the directly upward and directly downward directions on the tree, there are other directions. An abstraction of the properties of the directions is called a D-relation. Thirdly, the relation induced on the maximal chains on the tree is called a C-relation. Precise definitions of these relations are given in section 2.2 above.

This section addresses the conjecture in [3] on when a simply primitive infinite permutation group possesses invariant relations connected with an inverted tree. Specifically, the work here states that the conjecture is true if "betweenness relation" is relaxed to "B-relation". A B-relation is like a betweenness relation on an inverted tree except that two incomparable elements in the tree may have none of their common upper bounds located between the two elements.

Definition 8.2. A ternary relation satisfying (B1), (B2) and (B3) in (2.2.3) above is called a *B-relation.*

Definition 8.3 Let (G, Ω) be a permutation group and let Σ be a subset of Ω satisfying $|\Sigma| \leq 2, |\Omega \backslash \Sigma| \geq 2$. Then Σ is said to be *syzygetic* with respect to G if

$$(\forall g \in G)(((\Sigma \cap \Sigma g) = \phi) \vee (\Sigma \cup \Sigma g = \Omega) \vee (\Sigma \subseteq \Sigma g) \vee (\Sigma g \subseteq \Sigma)).$$

It is said to be *hypersyzygetic* with respect to G if

$$(\forall g \in G)(((\Sigma \cap \Sigma g) = \phi) \vee (\Sigma \subseteq \Sigma g) \vee (\Sigma g \subseteq \Sigma)).$$

Conjecture 8.4 *The conjecture [3, conjecture 34.8] states the following:*
Let G be a primitive permutation group on Ω. If there exists a subset of Ω that is syzygetic with respect to G, then Ω carries a G-invariant semilinear order relation, betweenness relation, C-relation or D-relation.

Observation 8.5. Note that if Σ is a subset of Ω which is syzygetic with respect G, then $\Omega \backslash \Sigma$ is also syzygetic with respect to G by DeMorgan's laws of set theory.

Observation 8.6. If α and β are distinct elements of Ω and if there is some translate Σg of Σ containing β and excluding α, then by the syzygetic property of Σ, the collection of all translates of Σ containing β and excluding α is a chain. We define the union of the chain as $M\Sigma(\alpha; \beta)$ where 'M' is to connote some 'maximality'. More precisely, we define

$$M\Sigma(\alpha; \beta) := \cup\{\Sigma g \mid g \in G, \beta \in \Sigma g, \alpha \notin \Sigma g.\}$$

We repeat that $M\Sigma(\alpha; \beta)$ fails to exist if no translate of Σ contains β and excludes α. Similarly, if α, β, and γ are distinct, we define

$$M\Sigma(\alpha, \beta; \gamma) := \cup\{\Sigma g \mid g \in G, \gamma \in \Sigma g, \alpha, \beta \notin \Sigma g\}$$

The set $M\Sigma(\alpha, \beta; \gamma)$ may also fail to exist.

8.7 *Results and Comments*

Lemma 8.7.1 *Let (G, Ω) be a permutation group and let Σ be a subset of Ω which is syzygetic with respect to G. Assume the following:*
(H1) $(\forall \alpha, \beta \in \Omega)((\alpha, \beta, \textit{distinct}) \rightarrow (\exists M\Sigma(\alpha; \beta))(M\Sigma(\alpha; \beta) \neq \Omega \backslash \{\alpha\}))$
(H2) $(\exists \alpha, \beta, \gamma \in \Omega)(\alpha, \beta, \gamma, \textit{distinct})((\exists M\Sigma(\alpha, \beta; \gamma)) \wedge (\nexists M\Sigma(\alpha, \gamma; \beta))$.
Then there is a G-invariant B-relation on Ω.

Remark 8.7.2. The conclusion in the last lemma cannot be strengthened to a betweenness relation (see section (8.8) below).

Lemma 8.7.3 *If (H1) in Lemma 8.7.1 holds and (H2) fails to hold, then*

$$(8.7.3.1) \qquad (\forall \alpha, \beta, \gamma \in \Omega)(\alpha, \beta, \gamma \textit{distinct })(\exists g \in G)((\alpha, \beta \notin \Sigma g) \wedge (\gamma \in \Sigma g))$$

Remark 8.7.4. Assume (G, Ω) is primitive and Σ is a syzygetic subset of Ω with neither Σ nor $\Omega \backslash \Sigma$ hypersyzygetic with respect to G. If condition (8.7.3.1) above holds, then G preserves a D-relation on Ω.

Lemma 8.7.5 *Lemma 8.7.1 above shows that Conjecture 34.8 in* [**3**] *(stated in 8.4 above) holds if "betweenness relation" is changed to "B-relation" in the conjecture.*

8.8 *Proof of Results in section 8.7*

Proof of Lemma 8.7.1. Assume the hypothesis of the Lemma. We first show that a stronger form of (H2) holds: namely that

(8.8.1)
$$(\exists \gamma, \varepsilon, \alpha \in \Omega)(\gamma, \varepsilon, \alpha \text{ distinct })(\nexists M\Sigma(\varepsilon, \gamma; \alpha) \wedge (M\Sigma(\alpha; \varepsilon) \cap M\Sigma(\alpha; \gamma) = \phi)).$$

Let $\Gamma := M\Sigma(\alpha, \beta; \gamma)$, $\Pi := M\Sigma(\alpha; \beta)$, both of which exist by (H1) and (H2). If $\gamma \notin \Pi$, then $M\Sigma(\alpha; \gamma) \cap M\Sigma(\alpha; \beta) = \phi$ and $M\Sigma(\beta, \gamma; \alpha)$ does not exist. So, put $\epsilon := \beta$, and (8.8.1) follows. Suppose on the other hand that $\gamma \in \Pi$. In that case, $\Pi \supset \Gamma$. Let $\varepsilon \in \Omega \backslash (\Pi \cup \{\beta\})$. The element ε exists by (H1). Suppose $\Lambda := M\Sigma(\varepsilon, \gamma; \alpha)$ exists. Then $\varepsilon \notin \Pi \cup \Lambda$ and $\gamma \in \Pi \backslash \Lambda$. Therefore, by the syzygetic property of Σ, and the fact that Π and Λ are chains of translates of Σ, we have $\Lambda \subset \Pi$. Hence $\beta \notin \Lambda$, a contradiction to the assumption that $M\Sigma(\beta, \gamma; \alpha)$ does not exist. Therefore, $M\Sigma(\varepsilon, \gamma; \alpha)$ does not exist. Now let $\Pi' := M\Sigma(\alpha; \varepsilon)$. Then $\Pi' \cap \Pi = \phi$. Thus (8.8.1) holds.

Now, define $B(\alpha; \beta, \gamma)$ if and only if

$$(8.8.2) \qquad \begin{aligned} &(\nexists g \in G)(\alpha \in \Sigma g \wedge \beta, \gamma \notin \Sigma g) \\ &\wedge (\nexists g \in G)(\beta, \gamma \in \Sigma g \wedge \alpha \notin \Sigma g). \end{aligned}$$

Note that if (8.8.1) holds for γ, ϵ and α, then by the definition of $M\Sigma(\alpha; \epsilon)$ and $M\Sigma(\alpha; \gamma)$, we see that $B(\alpha; \gamma, \epsilon)$. Thus the relation defined in (8.8.2) is not empty. Clearly (B1) in (2.2.3) holds. Assume $B(\alpha; \beta, \gamma)$ and $B(\beta; \alpha, \gamma)$. Suppose $\alpha \neq \beta$. By primitivity of G, there exists $g \in G$ such that Σg contains α and excludes β or there exists $g \in G$ such that Σg contains β and excludes α. If the former condition holds, then by the definition of $B(\alpha; \beta, \gamma)$, γ, $\alpha \in \Sigma g$ and $\beta \notin \Sigma g$—contradicting $B(\beta, ; \alpha, \gamma)$. The other case leads similarly to γ, $\beta \in \Sigma g$ and $\alpha \notin \Sigma g$ by $B(\beta; \alpha, \gamma)$—contradicting $B(\alpha; \beta, \gamma)$. So, we can deduce that $B(\alpha; \beta, \gamma)$ and $B(\beta; \alpha, \gamma)$ both hold if and only if $\alpha = \beta$. Therefore condition (B2) holds.

Assume, with hope of a contradiction, that

$$B(\alpha;\beta,\gamma) \wedge \neg B(\alpha;\delta,\gamma) \wedge \neg B(\alpha;\beta,\delta). \tag{8.8.3}$$

From $(8.8.3)_2$ (that is, the second part of (8.8.3)), either Σg contains α and excludes δ and γ for some $g \in G$, or $\Sigma g'$ contains δ and γ and excludes α for some $g' \in G$.

If such a Σg exists, then by $(8.8.3)_1$, $\beta \in \Sigma g$. Thus α, $\beta \in \Sigma g$ and δ, $\gamma \notin \Sigma g$. But then if Σh contains α and excludes β and δ for any h, we have that $\alpha \in \Sigma h \cap \Sigma g$, $\delta \notin \Sigma h \cup \Sigma g$ and $\beta \in \Sigma g \backslash \Sigma h$. Then by syzygetic property of Σ, $\Sigma h \subset \Sigma g$. Since $\gamma \notin \Sigma g$, then $\gamma \notin \Sigma h$. Thus $\alpha \in \Sigma h$, and β, γ, $\delta \notin \Sigma h$. This contradicts $(8.8.3)_1$. So there is no Σh containing α and excluding β and δ. In that case, by $(8.8.3)_3$, there exists some $\Sigma h'$ containing β and δ and excluding α. This implies $\beta \in \Sigma h' \cap \Sigma g$, $\alpha \in \Sigma g \backslash \Sigma h'$, $\delta \in \Sigma h' \backslash \Sigma g$. Then by the syzygetic property of Σ, $\Sigma h' \cup \Sigma g = \Omega$. Hence $\gamma \in \Sigma h'$. Therefore β, $\gamma \in \Sigma h'$ and $\alpha \notin \Sigma h'$—contradicting $(8.8.3)_1$. So, (8.8.3) leads to a contradiction if $(8.8.3)_2$ implies existence of Σg containing α and excluding δ and γ.

Suppose on the other hand that $(8.8.3)_2$ implies existence of $\Sigma g'$ containing δ and γ and excluding α. Then $\beta \notin \Sigma g'$ by $(8.8.3)_1$. Then, just as in the last paragraph, any Σh that contains α and excludes β and δ must be disjoint from $\Sigma g'$ by the syzygetic property of Σ, and thus must exclude β and γ. But this will contradict $(8.8.3)_1$. And any $\Sigma h'$ that contains β and δ and excludes α must contain $\Sigma g'$ properly, and so must contain γ—contradicting $(8.8.3)_1$ again. So, (8.8.3) leads to a contradiction in this case too. With this contradiction, we conclude that (8.8.3) does not hold. So (B3) holds.

We conclude therefore that B as defined above is a B-relation. It is clear from its definition that B is G-invariant.

Reasons for Remark 8.7.2.

Let (Λ, B_0) be the discrete B-set which is regular of branching type (2,3), as established by [3, Theorem 2.9.7]. This means that every point of Λ has branching number 2 and every branch point of negative type (that is, a branch point not in Λ) has branching number 3 (see diagram below).

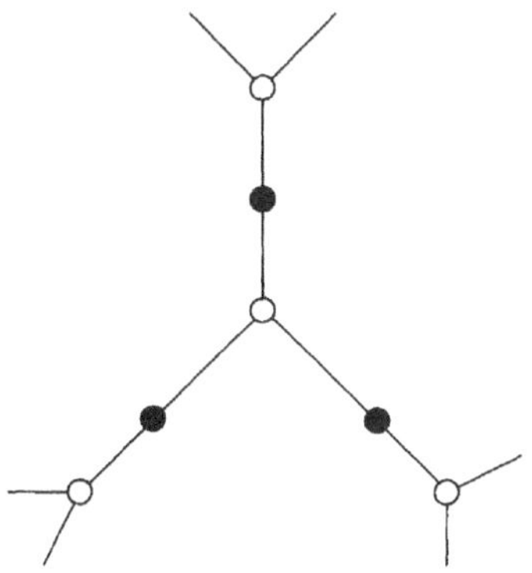

Let $G := \text{Aut}(\Lambda, B_0)$, the group of all maps of Λ preserving the B-relation. From [3, Theorem 29.7], (G, Λ) is primitive. Pick a branch point of negative type and let Σ be a branch at that branch point. It is easy to see that Σ is syzygetic with respect to G. Furthermore, B_0 is not a betweenness relation since the three points in the diagram above demonstrate that (Λ, B) violates axiom (B4).

We now show that *any* G-invariant B-relation on Λ violates axiom (B4). Let α denote the top element in the diagram above; let β be the element at the lower left and γ the third. Suppose B_1 is an arbitrary G-invariant B-relation on Λ. As observed from the action of G preserving B_0, the action of $G_{\{\alpha,\beta,\gamma\}}$ on $\{\alpha, \beta, \gamma\}$ is isomorphic to the symmetric group S_3. Thus, by condition (B2) of the B-relation in Section (2.2.3), none of α, β, γ is between the other two in B_1. Thus $\neg B_1(\alpha; \beta, \gamma)$, and, if there exists $\delta \in \Lambda$ such that $B_1(\delta; \alpha, \beta) \wedge B_1(\delta; \alpha, \gamma)$, then $\delta \neq \beta$ and $\delta \neq \gamma$.

Suppose such a δ exists, and in addition, that $B_0(\alpha; \delta, \beta)$. Extend the diagram of B_0 from α to δ. (This will be away from β). Because the branching number in (Λ, B_0) is 3, then it is easy to see that there exists an element $y \in G$ of order 2 such that

$$\alpha y = \alpha, \quad \beta y = \beta, \quad \delta y \neq \delta \quad \text{and} \quad \delta y^2 = \delta.$$

From $B_1(\delta; \alpha, \beta)$ and condition (B3), we have $B_1(\delta; \alpha, \delta y) \vee B_1(\delta; \beta, \delta y)$. Since B_1 is G-invariant, this means that an application of y to the last property yields

$$(B_1(\delta; \alpha, \delta y) \wedge B_1(\delta y; \alpha, \delta)) \vee (B_1(\delta; \beta, \delta y) \wedge B_1(\delta y; \beta, \delta)).$$

The last statement contradicts condition (B2). Thus $\neg B_0(\alpha; \delta, \beta)$.

Similarly $\neg B_0(\beta; \delta, \alpha)$, $\neg B_0(\alpha; \delta, \gamma)$ and $\neg B_0(\gamma; \delta, \alpha)$ hold for any δ satisfying $B_1(\delta; \alpha, \beta) \wedge B_1(\delta; \alpha, \gamma)$. But there is no $\delta \in \Lambda \backslash \{\alpha, \beta, \gamma\}$ in the diagram of (Λ, B_0) that satisfies the last four conditions simultaneously. So, such a δ does not exist. Thus B_1 violates axiom (B4) and cannot be a G-invariant betweenness relation.

Proof of Lemma 8.7.3.

Suppose (H1) in Lemma 8.7.1 holds but (H2) fails. Then

$$\text{(8.8.4)} \qquad \begin{array}{l} (\forall \alpha, \beta, \gamma \in \Omega)(\alpha, \beta, \gamma \text{ distinct }) \\ (\nexists M\Sigma(\alpha, \beta; \gamma)) \rightarrow ((\nexists M\Sigma(\beta, \gamma; \alpha)) \wedge (\nexists M\Sigma(\gamma, \alpha; \beta))). \end{array}$$

Suppose also that

$$\text{(8.8.5)} \qquad (\exists \alpha_0, \beta_0, \gamma_0 \in \Omega)(\alpha_0, \beta_0, \gamma_0 \text{ distinct })(\nexists M\Sigma(\alpha_0, \beta_0; \gamma_0)).$$

Let $\Gamma := M\Sigma(\alpha_0; \gamma_0)$ and $\varepsilon \in (\Omega \backslash (\Gamma \cup \{\alpha\}))$. Both Γ and ε exist by (H1). By (8.8.5), $\beta_0 \in \Gamma$. Now $M\Sigma(\alpha_0; \varepsilon)$ exists again because of (H1). Furthermore, $M\Sigma(\alpha_0; \varepsilon) \cap \Gamma = \phi$ from the definition of ε and Γ. Hence, $\beta_0, \gamma_0 \notin M\Sigma(\alpha_0; \epsilon)$; so $M\Sigma(\beta_0, \gamma_0; \varepsilon)$ exists. By (8.8.4), we have that $M\Sigma(\varepsilon, \beta_0; \gamma_0)$ exists. Let $\Pi := M\Sigma(\varepsilon, \beta_0; \gamma_0)$. By (8.8.5), $\alpha \in \Pi$. Thus, $\gamma_0 \in \Gamma \cap \Pi$, $\alpha_0 \in \Pi \backslash \Gamma$, $\beta_0 \in \Gamma \backslash \Pi$

and $\varepsilon \notin \Gamma \cup \Pi$. Since Π and Γ are increasing chains of translates of Σ, it follows easily that there exist translates Σ_1, Σ_2, of Σ with $\Sigma_1 \subseteq \Gamma$, $\Sigma_2 \subseteq \Pi$ such that $\gamma_0 \in \Sigma_1 \cap \Sigma_2$, $\alpha_0 \in \Sigma_2 \backslash \Sigma_1$, $\beta_0 \in \Sigma_1 \backslash \Sigma_2$, and $\varepsilon \notin \Sigma_1 \cup \Sigma_2$, a contradiction to the syzygetic property of Σ. This contradiction shows that if (H1) holds and (H2) fails, then (8.8.5) does not hold. Thus, $M\Sigma(\alpha, \beta; \gamma)$ exists for all distinct α, β and γ in Σ.

Proof of Remark 8.7.4. Either (1) use Σ directly and the exact proof of [3, Theorem 34.6] to prove Remark 8.7.4 or (2) note that $\Omega \backslash \Sigma$ is syzygetic if Σ is, and with $\Omega \backslash \Sigma$ replacing Σ, condition 34.6.1 in [3] becomes condition (8.7.3.1) above. Then, Theorem 34.6 proves the result above.

Proof of Lemma 8.7.5.

If Σ or $\Omega \backslash \Sigma$ is hypersyzygetic, then by [3, Theorem 34.4], G preserves an upper semilinear order or a C-relation. So, assume that neither Σ nor $\Omega \backslash \Sigma$ is hypersyzygetic but that Σ is syzygetic.

If (H1) holds and (H2) fails, then by Lemma 8.7.3 and Remark 8.7.4 above, there is a G-invariant D-relation.

If (H1) above fails, then we have two cases:

Case 1: (8.8.6) $(\exists \alpha, \beta \in \Omega)(\alpha, \beta, \text{distinct})(\forall g)(\alpha \in \Sigma g \rightarrow \beta \in \Sigma g)$;

Case 2: $M\Sigma(\alpha; \beta)$ exists for all distinct α and β but $M\Sigma(\alpha_0; \beta_0) = \Omega \backslash \{\alpha_0\}$ for some α_0 and β_0.

If Case 2 holds, then $M\Sigma(\alpha_0; \delta) = \Omega \backslash \{\alpha_0\}$ for all δ distinct from α_0. By transitivity of G, $M\Sigma(\alpha; \beta) = \Omega \backslash \{\alpha\}$ for all distinct α and β. As a consequence

$$(8.8.7) \qquad (\forall \alpha, \beta, \gamma \in \Omega)(\alpha, \beta, \gamma \text{ distinct })(\exists g \in G)(\beta, \gamma \in \Sigma \wedge \alpha \notin \Sigma).$$

Now if (8.8.7) holds, then by [3, Theorem 34.6], G preserves a D-relation on Ω. On the other hand, suppose Case 1 holds. Then by [3, Lemma 34.2], there is a G-invariant partial order $\leq$ on Ω defined by

$$\alpha \geq \beta \quad \text{if and only if (8.8.6) holds.}$$

We shall now show that, in fact, (8.8.6) induces a G-invariant B-relation on Ω. Define $B(\alpha; \beta, \gamma)$ if and only if (8.8.2) above holds. Then B becomes a G-invariant B-relation as shown above in the proof of Lemma 8.7.1. Moreover, if $\alpha > \beta > \gamma$ in the G-invariant partial order defined above, then it follows easily that $B(\beta; \alpha, \gamma)$. Thus, the B-relation is not empty.

With Lemma 8.7.1 above, the proof of Lemma 8.7.5 is complete.

9 Jordan Groups on a Linear Order

9.1.

Let (G, Ω) be a primitive Jordan group with a given G-invariant linear order $\leq$ on Ω. Clearly, (G, Ω) is simply primitive. And true, Theorem 7.1 shows that maximal

Jordan sets or minimal increments of Jordan sets induce either a G-invariant linear order or a nonlinear semilinear order through property (5.9.3), or a C-relation through property (5.9.4) or (7.1.1). But those induced G-invariant relations may be different from the linear order $\leq$ given *a priori* on Ω. This section describes the possible Jordan sets in (G, Ω). It is obvious that in some cases, initial segments and final segments of $(\Omega, \leq)$ are Jordan sets. But there are cases where they are not. In the latter possibility, every Jordan set is open, convex and bounded (that is, bounded below and above), and even then not all open, convex, bounded sets are Jordan sets. The ones that are Jordan sets can be described using a C-relation or a nonlinear semilinear order. Concrete examples of cases with C-relation and semilinear order are given in section (9.4.4) and (9.4.5) below. Theorems 9.3.1 and 9.3.2 (i) are joint work with Peter Neumann.

Remark 9.2. The following example is typical of the groups described in this section. The detailed description is in section (9.4.4) below. Let $\Lambda(\mathbb{Z}, \aleph_0, +)$ be the upper semilinearly ordered set where each maximal chain is isomorphic to $\mathbb{Z}$, and where at each branch point, the set of branches is isomorphic to $\mathbb{Q}$ with its usual order. The notation $\Lambda(\mathbb{Z}, \aleph_0, +)$ is from [3, section 5]. Let Ω be the collection of all *piecewise linear maximal chains* in $\Lambda(\mathbb{Z}, \aleph_0, +)$ using again a term in [3, section 5]. It is not hard to see that when the orders on the sets of branches at branch points are considered together, they induce a linear order on the piecewise linear maximal chains. Moreover, the set of maximal chains with lower sections in any one branch at a branch point is convex in the induced linear order. This is the same order that one gets if one, in a way, projects down the set Ω onto a line. Let G be an automorphism group of Ω that preserves the associated C-relation and the induced linear order on Ω. One ends up having a linearly ordered set $(\Omega, \leq)$ with a C-relation on Ω, and a Jordan group (G, Ω) with G preserving both $\leq$ and the C-relation.

9.3 *Results.*

In the following theorems, let (G, Ω) be a primitive Jordan group with $(G, \Omega) \subseteq (\mathrm{Aut}(\Omega, \leq), \Omega)$ for some given linear order $\leq$ on Ω.

Theorem 9.3.1 *If (G, Ω) is order 3-transitive with respect to $\leq$, then each maximal Jordan set containing one element and excluding another is either an initial segment or a final segment of $(\Omega, \leq)$. Each maximal Jordan set then induces a G-invariant linear order on Ω which is the same as $\leq$ or the reverse of it.*

Remark. Usual arguments show that (G, Ω) in Theorem 9.3.1 is then order n-transitive for all n.

Theorem 9.3.2 *Suppose (G, Ω) is order 2- transitive but not order 3-transitive with respect to $\leq$.*

(i) Then every Jordan set is an open bounded convex set in Ω but not every open bounded convex set is a Jordan set. Every maximal Jordan set violates property

(5.9.3) and satisfies property (5.9.4). So, it induces, as in the proof of Theorem 7.1, a G-invariant relation which is neither linear nor semilinear but rather a C-relation. (ii) Furthermore, G is transitive on the set of nodes of the C-relation. If the number of branches at a node, and therefore at every node, is greater than 2, then it is infinite, and the linear order on Ω induces a dense linear order on the set of branches at each node. On the other hand, if the number of branches at every node is 2, then between any two comparable nodes, there is another.

Theorem 9.3.3 *Suppose (G,Ω) is not order 2-transitive. Then, again, every Jordan set is an open bounded convex set but not vice-versa. Any maximal Jordan set Λ that satisfies property 5.9.3 does not induce through the property an invariant linear order but a nonlinear semilinear order. The other possibility is for a maximal Jordan set to induce a G-invariant C-relation through property (5.9.4).*

Result 9.3.4. An example is indicated in section 9.4.4 below of the C-relation in Theorem 9.3.2 which has an infinite number of branches at each node (that is, branch point), but for which the set of nodes on each maximal chain is isomorphic to $\mathbb{Z}$.

Result 9.3.5. Section 9.4.5 below contains an example of the case in Theorem 9.3.3 where G preserves a nonlinear semilinear order on Ω.

Moreover, the example is neither weakly doubly transitive nor a right regular representation of a subgroup of $\mathbb{R}$, nor periodic. Thus it answers in the negative a primitivity question in [18, page 249].

9.4 *Proofs.*

9.4.1. *Proof of Theorem 9.3.1.* Let Γ be a Jordan set in Ω. Then clearly, Γ is convex. Suppose Γ is bounded with $\alpha \in \Gamma$, $\beta \notin \Gamma$, and $\alpha < \beta$. Since $G_{(\Omega\setminus\Gamma)}$ is transitive on Γ, then α is not a minimal element in Γ. So, Γ contains some δ with $\delta < \alpha$. Let $\mu \in \Omega\setminus\Gamma$ with $\mu < \delta$. By order 3-transitivity of G, there exists $x \in G$ with $(\delta,\alpha,\beta)x = (\mu,\alpha,\beta)$. Thus $\mu \in MJ(\beta;\alpha)$, the maximal Jordan set in Ω containing α and excluding β. Similar arguments show that if $\alpha < \nu < \beta$ then $\nu \in MJ(\beta;\alpha)$. Hence $MJ(\beta;\alpha)$ is the initial segment of $(\Omega,\leq)$ with least upper bound β. We conclude, using similar arguments, that in any case, every maximal Jordan set Λ is an initial segment or a final segment of $(\Omega,\leq)$. It then follows that $\Lambda g \subseteq \Lambda$ or $\Lambda g \supseteq \Lambda$ for every $g \in G$. In addition, Λ satisfies property (5.9.3). Hence Λ induces through property (5.9.3) and in the manner of proof of Theorem 7.1 a G-invariant linear order on Ω. From the description of Λ above, the linear order equals the given order $\leq$ in the hypothesis of this theorem or the reverse of it.

9.4.2. *Proof of Theorem 9.3.2(i).* Let Γ be a Jordan subset in Ω. Then again, Γ is convex with respect to the linear order $\leq$ on Ω. If Γ is unbounded, then by usual arguments, repeated below, G is order 3-transitive. For suppose Γ is unbounded below. Then order 2-transitivity of G implies that for any $\alpha \in \Omega$ the set Σ_α of elements less than α is a Jordan set. Given $\alpha_1 < \alpha_2 < \alpha_3$ and $\alpha_1' < \alpha_2' < \alpha_3'$, we can map (α_2, α_3) to (α'_2, α'_3) by order 2-transitivity and can use the elements of $G_{(\Omega\setminus\Sigma_{\alpha'_2})}$ to adjust our map to take $(\alpha_1, \alpha_2, \alpha_3)$ to $(\alpha'_1, \alpha'_2, \alpha'_3)$. This will mean that G is order 3-transitive, a contradiction to the hypothesis. A similar contradiction results if G has Jordan sets which are unbounded above. So, Γ must be bounded. It follows also easily from order 2-transitivity that $\leq$ is dense. Since $G_{(\Omega\setminus\Gamma)}$ is transitive on Γ then Γ cannot contain a maximal or a minimal element. So Γ is open. Thus Γ is an open convex bounded set with respect to $\leq$, a dense linear order.

Suppose $\alpha \in \Gamma$ and $\beta \notin \Gamma$. Define $\Lambda := MJ(\beta; \alpha)$. From the last paragraph Λ is a bounded interval in $(\Omega, \leq)$. Suppose $\alpha < \beta$. An argument similar to the one below treats the case when $\alpha > \beta$. From boundedness of Λ, there exists $\gamma \in \Omega\setminus\Lambda$ such that $\gamma < \alpha$. By order 2-transitivity of (G, Ω), G contains some element g satisfying $(\gamma, \alpha)g = (\alpha, \beta)$. Thus from $\gamma \notin \Lambda$ and $\alpha \in \Lambda$, we deduce $\alpha \notin \Lambda g$ and $\beta \in \Lambda g$. Thus property (5.9.3) fails to hold for the maximal Jordan set Λ containing Γ.

By the sketch proof in Section 7.1, the two remaining possibilities are property (5.9.4) or (7.1.1) holding. In fact Λ always satisfies property (5.9.4). For if otherwise, then by arguments similar to those in Section 5.9.2, there exists $g \in G$ such that Λ, Λg are typical and g interchanges β and βg—that is, $(\beta, \beta g)g = (\beta g, \beta)$. But such a g cannot occur here where g preserves a linear order. Thus Λ induces a G-invariant C-relation through property (5.9.4).

Proof of Theorem 9.3.2 (ii). The C-relation on Ω means that the elements of Ω behave like some maximal chains in an "inverted tree", or equivalently, in an upper semilinearly ordered set (see [2], [3], [4] or [6]. One of the 4 possible configurations of $C(\alpha; \beta, \gamma)$ in this case is shown.

Definitions. For $\alpha, \beta \in \Omega$ with $\alpha \neq \beta$, define

$$N\{\alpha, \beta\} := \{\gamma \in \Omega \mid \neg C(\gamma; \alpha, \beta)\} \cup \{\alpha, \beta\}$$
$$BR(\alpha; \beta) := \{\gamma \in \Omega | C(\alpha; \beta, \gamma)\}.$$

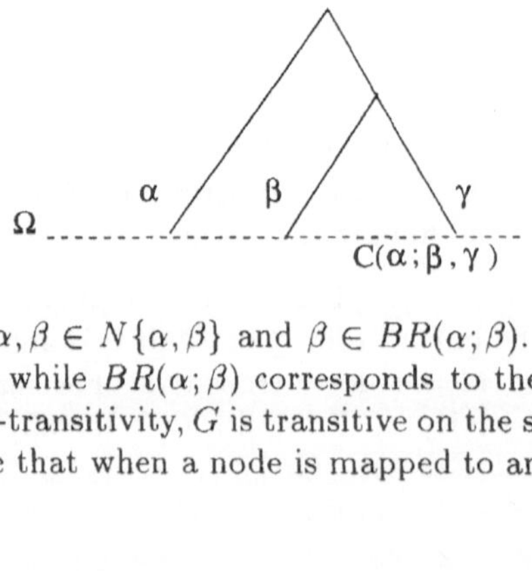

From the definitions, $\alpha, \beta \in N\{\alpha, \beta\}$ and $\beta \in BR(\alpha; \beta)$. We notice that $N\{\alpha, \beta\}$ corresponds to a node while $BR(\alpha; \beta)$ corresponds to the branch at $N\{\alpha, \beta\}$ that contains β. By order 2-transitivity, G is transitive on the set of all nodes of Ω. For a similar reason, we note that when a node is mapped to another node, the branches at the node are mapped in a bijective manner to the branches at the new node. Note also that since, from the proof of Theorem 9.3.2(i), the C-relation is induced by Λ which satisfies property (5.9.4), then $BR(\alpha;\beta)$ is a Jordan set. So, $BR(\alpha;\beta)$ is convex. Consequently, $<$ induces an order on the set of branches at each node.

Suppose the branches at a node are more than 2 in number. In particular, let $\gamma \in N\{\alpha,\beta\}$ with $\alpha < \beta < \gamma$ in the linear order on Ω. By order 2-transitivity of G, we can translate the pair (α,β) to (α,γ). Since $N\{\alpha,\beta\}$ is then fixed and the branches are permuted amongst themselves, we see that there is a branch between α and β. Hence the order on the branches induced by linear order $\leq$ on Ω is dense. By taking (α,β) to (β,γ), we notice that within a node, there is no minimal or maximal branch either. Furthermore, the order on the branches at a node is order 2-homogeneous.

Suppose a node has only two branches. Since G is transitive on the set of all nodes, then every node has two branches. Suppose there is no node between $N\{\alpha,\gamma_1\}$ and $N\{\alpha,\gamma_2\}$ with the latter being lower than the former in the semilinear ordering. Suppose $\alpha < \gamma_2$. (The reader may want to draw a diagram here). Now $BR(\gamma_1;\alpha)$ being Jordan, there exists $z \in G_{(\Omega\setminus BR(\gamma_1;\alpha))}$ such that $\gamma_2 z = \alpha$. In that case $\alpha z < \gamma_2 z = \alpha$. Since $\gamma_2 \not< \alpha$ and $\alpha z < \alpha$, then $N\{\alpha z,\alpha\} \neq N\{\alpha,\gamma_2\}$. We then have two distinct maximal nodes within $BR(\gamma_1;\alpha)$ – *a* contradiction to the upper semilinear ordering on the set of nodes. Hence, between $N\{\alpha,\gamma_1\}$ and $N\{\alpha,\gamma_2\}$, there is another node. This conclusion holds by similar arguments if we assume $\gamma_2 < \alpha$. Hence, the nodes are dense with respect to the semilinear ordering on them.

9.4.3. *Proof of Theorem 9.3.3.* Let $\Lambda := MJ(\beta;\alpha)$. Suppose, with hope of a contradiction, that Λ induces a linear order on Ω through property 5.9.3. Then $\Lambda g \cap \Lambda \neq \phi$ for any $g \in G$. Moreover, from property (5.9.3), any translate of Λ that contains one element of $\Delta(\Lambda)$ contains Λ properly and thus contains all elements of $\Delta(\Lambda)$. Thus $\Delta(\Lambda)$ becomes a block for (G,Ω). Since (G,Ω) is primitive, then $\Delta(\Lambda) = \{\beta\}$. From $\Lambda g \cap \Lambda \neq \phi$ for any $g \in G$, we deduce that Λ is not bounded with respect to the *a priori* order $\leq$ on Ω; otherwise by moving β suitably, some x in G will satisfy $\Lambda x \cap \Lambda = \phi$. Since $\Delta(\Lambda) = \{\beta\}$, we can now assume without loss of generality that Λ is the collection of all elements of Ω less than β. It is then easy to show that G is order 2-transitive on $(\Omega,\leq)$. This contradicts the hypothesis of the Theorem. Thus, no maximal Jordan set can induce an invariant linear order through property 5.9.3. By the arguments at the close of Theorem 9.3.2(i), we then see that every maximal Jordan set must satisfy property 5.9.4 which yields a G-invariant C-relation on Ω. Hence the theorem holds.

9.4.4. *Example on Result 9.3.4.*

Let

$\Lambda :=$

$\{(0n_1q_1n_2q_2...n_{k-1}q_{k-1}n_k | k \in \mathbb{N},\ q_i \in \mathbb{Q}\setminus\{0\}, i = 1,\ldots,k, n_1 > \ldots > n_k \in \mathbb{Z}\}.$

Consider the following order $\leq_s$ on Λ :

$(0 n_1 q_1 n_2 q_2 \ldots n_k -_1 q_k -_1 n_k) \geq_s (0 m_1 p_1 m_2 p_2 \ldots m_r -_1 p_r -_1 m_r)$
if and only if:
(i) $k \leq r$,
(ii) $n_i = m_i$ and $q_i = p_i$ for $i = 1, 2, \ldots, k-1$, and
(iii) $n_k \geq m_k$.

Remarks. $(\Lambda, \leq_s)$ becomes an *inverted tree* with an infinite number of branches at each node, that is, a homogeneous upper semilinear order with the set of elements of Λ on each chain having order type $\mathbb{Z}$. Then Λ equals $\Lambda(\mathbb{Z}, \aleph_0, +)$ using the notation in [3, section 5].

Description of a linear order on some chains of $(\Lambda, \leq_s)$.
Define

$$\Omega := \{(0 n_1 q_1 n_2 q_2 \ldots n_k q_k) \mid k \in \mathbb{N} \cup \{0\}, q_i \in \mathbb{Q} \backslash \{0\}, i = 1, \ldots, k, n_1 > \ldots > n_k \in \mathbb{Z}\}.$$

In [3, section 5], Ω is called the set of piecewise linear maximal chains in Λ. We can view (0) as the vertical maximal chain; $(0 n_1 q_1 n_2 q_2 \ldots n_k q_k)$ as the maximal chain in the q_k^{th} branch at node $(0 n_1 q_1 \ldots n_k)$ which is straight after the stated node (that is, it contains no node of the form $(0 n_1 q_1 \ldots n_k q_k n_{k+1} q_{k+1} n_{k+2})$ or lower). If $\mathbf{w_2} := (0 n_1 q_1 \ldots n_k q_k)$, and $\mathbf{w_1} := (0 m_1 p_1 \ldots m_r p_r)$, define:

$\mathbf{w_2} >_l \mathbf{w_1}$ if and only if
$\mathbf{w_1}$ is an initial segment of $\mathbf{w_2}$ and $q_{r+1} > 0$
or $\mathbf{w_2}$ is an initial segment of $\mathbf{w_1}$ and $p_{k+1} < 0$
or for some $j \geq 1$, $q_i = p_i$ for $i = 1, 2, \ldots, j-1$,
$n_i = m_i$ for $i = 1, 2, \ldots, j$, and $q_j > p_j$;
or for some $j \geq 1$, $q_i = p_i$, $n_i = m_i$ for $i = 1, 2, \ldots, j-1$,
and $n_j > m_j$ and $q_j > 0$;
or for some $j \geq 1$, $q_i = p_i$, $n_i = m_i$ for $i = 1, 2, \ldots, j-1$
and $n_j < m_j$ and $p_j < 0$.

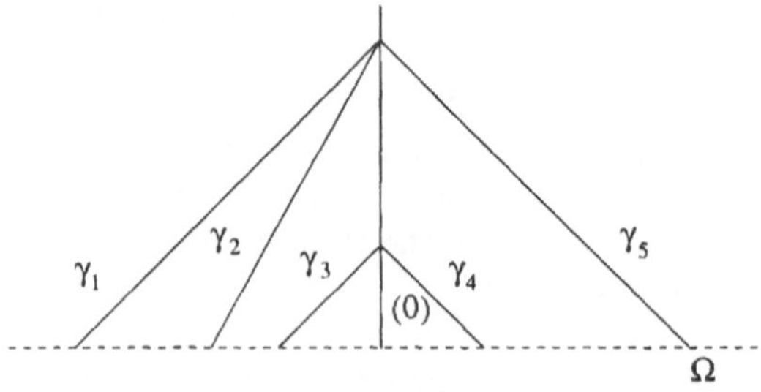

For example, in the diagram, we have $\gamma_1 <_l \gamma_2 <_l \gamma_3 <_l (0) <_l \gamma_4 <_l \gamma_5$. We can verify that relation $\leq$ becomes a dense order 2-homogeneous linear order on

Ω. The relation also induces, on the set of all branches at a node, a linear order isomorphic to the order on $\mathbb{Q}$.

Let G be the subgroup of $\mathrm{Aut}(\Lambda, \leq_s)$ which preserves on each node the linear order amongst the branches at the node and which fixes Ω setwise. One can verify that $G \subseteq \mathrm{Aut}(\Omega, \leq_l)$ and is an order 2-transitive Jordan group which is not order 3-transitive.

9.4.5. *Example on Result 9.3.5.* Consider the case (G, Ω) in Theorem 9.3.2 (ii) where the number of branches at each node is exactly 2 and where the set of nodes on each maximal chain is dense. From this denseness, each branch $BR(\beta; \alpha)$ in Ω is not only a Jordan set but an order 2-transitive Jordan set with respect to the order $\leq$ on the set Ω of maximal chains.

Let Π be the collection of all nodes in the C-relation on Ω. It is not difficult to see that Π is an upper semilinearly ordered set. The order 2-transitivity of the Jordan sets in (G, Ω) now shows that the set Σ of all *nodes* contained in any branch below a given node is a Jordan set in Π. In fact, denseness of Π relative to the upper semilinear order on it implies that Σ is a primitive Jordan set. It thus follows that (G, Π) is a primitive Jordan group. Moreover, the linear order $\leq$ on Ω now induces a linear order $\leq_*$ on Π as follows. To make the definition easier, first note that if $\pi \in \Pi$, the representation $\pi = N\{\omega, \omega'\}$ of π for $\omega, \omega' \in \Omega$ is not unique. Secondly, if π_1, π_2 are comparable relative to the upper semilinear order on Π, then there exist distinct $\omega_1, \omega_2, \omega_3 \in \Omega$ such that $\pi_1 = N\{\omega_1, \omega_2\}$, $\pi_2 = N\{\omega_2, \omega_3\}$ and $\omega_1 < \omega_2 < \omega_3$ or $\omega_1 > \omega_2 > \omega_3$ where $\leq$ is the linear order on Ω. And if π_1, π_2 are incomparable, then there exist distinct $\omega_1, \omega_2, \omega_3, \omega_4$ such that $\pi_1 = N\{\omega_1, \omega_2\}$, $\pi_2 = N\{\omega_3, \omega_4\}$ and $\omega_1 < \omega_2 < \omega_3 < \omega_4$ or $\omega_1 > \omega_2 > \omega_3 > \omega_4$. So, we can state the definition of $\leq_*$ in general as:

$$N\{\alpha_1, \alpha_2\} <_* N\{\beta_1, \beta_2\}$$

if and only if
there exist $\gamma_1, \gamma_2, \ldots, \gamma_m \in \Omega$ such that:

$\gamma_1 < \gamma_2 < \gamma_3 < \ldots < \gamma_m$,
$N\{\alpha_1, \alpha_2\} = N\{\gamma_1, \gamma_2\}$, and
$N\{\beta_1, \beta_2\} = N\{\gamma_{m-1}, \gamma_m\}$.

The form of this definition makes the proof that $<_*$ is a linear order very simple.

Clearly, G is not order 2-transitive with respect to order $\leq_*$ on Π since two comparable elements under $\leq_s$ cannot go into incomparable elements under $\leq_s$. But we shall show even more; namely that (G, Ω) is not weakly doubly transitive (see (2.2.11) above for definition).

Note that $(\Pi, \leq_*)$ contains a subset of the following form:

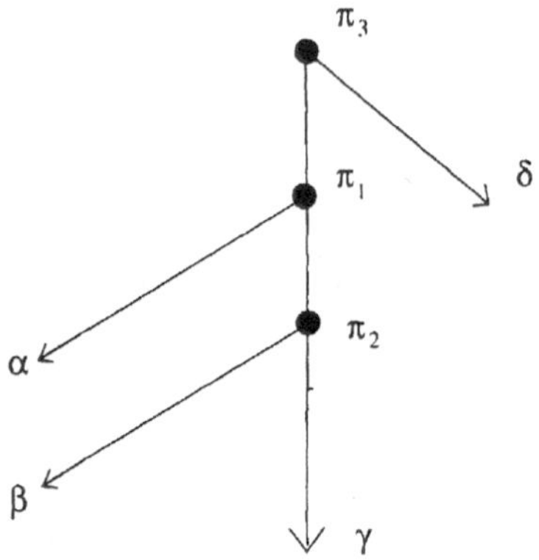

In the diagram, $\alpha, \beta, \gamma, \delta$ refer to maximal chains, and π_1, π_2, π_3 refer to nodes. Note that in the diagram, π_1, π_2, π_3 are comparable nodes under $\leq_s$, and that

$$\alpha < \beta < \gamma < \delta \quad \text{in} \quad (\Omega, \leq)$$
$$\pi_1 = N\{\alpha, \beta\}, \pi_2 = N\{\beta, \gamma\} \quad \text{and} \quad \pi_3 = N\{\gamma, \delta\}.$$

So, $\pi_1 <_* \pi_2 <_* \pi_3$. But there is no element in G that fixes node π_1 and moves π_2 to π_3. In fact, observe that if $\pi \in \Omega$ such that $\pi \geq_* \pi_3$, then $\pi \not\leq_s \pi_1$. Thus for any such π, there is no element in G that fixes node π_1 and moves π_2 to π. Hence, (G, Π) is not weakly doubly transitive using the definition in (2.2.11). It is also easy to see that (G, Ω) is not periodic (see 2.2.11 for definition) since any z satisfying the second condition of periodicity cannot fix setwise each Jordan set in (G, Π), and therefore cannot satisfy the commutativity condition of periodicity. The existence of Jordan sets also shows that (G, Π) is not a regular representation of any subgroup of $\mathbb{R}$. With this, the proof of Result 9.3.5 is complete.

References

[1] S. A. Adeleke and Peter M. Neumann, *Infinite bounded permutation groups*, J. London Math. Soc., to appear.

[2] S. A. Adeleke and Peter M. Neumann, *Primitive permutation groups with primitive Jordan sets*, J. London Math. Soc., to appear.

[3] S. A. Adeleke and Peter M. Neumann, *Relations related to betweenness: their structure and automorphisms*, preprint.

[4] S. A. Adeleke and Dugald Macpherson, *Classification of infinite primitive Jordan permutation groups*, Proc. London Math. Soc., to appear.

[5] S. A. Adeleke, M. A. E. Dummett and Peter M. Neumann, *On a question of Frege's about right-ordered groups*, Bull. London Math. Soc. **19**(1987), 513–521.

[6] S. A. Adeleke, *Semilinear towers of Steiner systems*, J. Comb. Theory Series (A), to appear.

[7] S. A. Adeleke, *Examples of irregular infinite Jordan groups*, preprint.

[8] Peter J. Cameron, *Transitivity of permutation groups on unordered sets*, Math. Zeitschrift. **148**(1976), 127–139.

[9] Peter J. Cameron, *Orbits of permutation groups on unordered sets, IV: homogeneity and transitivity*, J. London Math. Soc. (2) **27**(1983), 238–247.

[10] Peter J. Cameron, *Some tree-like objects*, Quart. J. Math. Oxford **38**(1987), 155–183.

[11] G. Cherlin, L. Harrington, and A. H. Lachlan, $\aleph_0$*-categorical,* $\aleph_0$*-stable structures*, Annals of Pure and Applied Logic **28**(1985), 103-135.

[12] J. A. Covington, *A universal structure for N-free graphs*, Proc. London Math. Soc. (3) **58**(1989), 1-16.

[13] Manfred Droste, *Structure of partially ordered sets with transitive automorphism groups*, Memoirs American Math. Soc. **57**(1985), No. 334.

[14] Manfred Droste, *The normal subgroup lattice of 2-transitive automorphism groups of linearly ordered sets*, Order **2**(1985), 291–319.

[15] M. Droste, W. C. Holland and H. D. Macpherson, *Automorphism groups of semilinear orders, I, II*, Proc. London Math. Soc. (3) **58**(1989), 454–478; 479–494.

[16] David M. Evans, *Homogeneous geometries*, Proc. London Math. Soc. (3) **32**(1986), 305–327.

[17] D. Giorgetta, Dr. Phil. Dissertation, E. T. H., Zürich.

[18] A. M. W. Glass, *Ordered Permutation Groups*, London Math. Society Lecture Notes Series **55**, Cambridge University Press, Cambridge, 1981.

[19] Camille Jordan, *Sur la limite du degré des groupes primitifs que contiennent une substitution donnée*, J. Reine und Angew. Math. (Crelle's J.) **79**(1875), 248–258 (=Oeuvres de Camille Jordan Vol. I (Paris 1961), 485-495).

[20] W. M. Kantor, *Homogeneous designs and geometric lattices*, J. Comb. Theory Series A **8**(1985), 64–77.

[21] William M. Kantor, *Jordan groups*, J. Algebra **12**(1969), 471–493.

[22] H. D. Macpherson, *A survey of Jordan groups*, in *Automorphism Groups of First Order Structures*, eds. R. W. Kaye and H. D. Macpherson, Oxford University Press, Oxford (1994), 73-110.

[23] H. D. Macpherson, *Extensions of Jordan groups*, preprint.

[24] T. P. McDonough, *An infinite version of Marggraff's theorem*, Quart. J. Math. Oxford (2) **32**(1981), 173–179.

[25] Peter M. Neumann, *The lawlessness of groups of finitary permutations*, Archiv Math. (Basel) **26**(1975), 561–566.

[26] Peter M. Neumann, *The structure of finitary permutation groups*, Archiv Math. (Basel) **27**(1976), 3–17.

[27] Peter M. Neumann, *Some primitive permutation groups*, Proc. London Math. Soc. (3) **50**(1985), 265–281.

[28] Peter M. Neumann, *The classification of some infinite Jordan groups*, preprint.

[29] D. Segal, *Normal subgroups of finitary permutation groups*, Math. Z. **140** (1974), 81–85.

[30] James Wiegold, *Groups of finitary permutations*, Arch. Math. **25**(1974), 466–469.

[31] H. Wielandt, *Unendliche Permutationsgruppen*, Lecture notes, Tübingen, 1960 (reprint: York University, Toronto, 1967).

[32] B. I. Zilber, *Finite homogeneous geometries*, in *Proc. of the 6th Easter Conference on Model Theory*, eds. B. I. Dahn and H. Wolter, Humboldt Universität zu Berlin (1988), 186–208.

The Separation Theorem for Group Actions

Cheryl E. Praeger

University of Western Australia
Nedlands, W.A., 6907
Australia

1 Introduction

Let G be a group of permutations of a set Ω, and let Γ and Δ be (not necessarily distinct) subsets of Ω. Under what conditions on G, or on Γ and Δ is it possible to *separate* Γ from Δ by an element of G, that is to have $\Gamma^g \cap \Delta = \emptyset$ for some $g \in G$? In 1976 Peter M. Neumann [25, Lemma 2.3] proved that every finite subset can be separated from itself in this way provided that all the G-orbits in Ω are infinite.

Theorem 1.1 (Separation Theorem) *If G is a group of permutations of a set Ω such that all G-orbits in Ω are infinite, then for each finite subset Γ of Ω there is an element $g \in G$ such that $\Gamma^g \cap \Gamma = \emptyset$.*

This fundamental result for group actions was shown in [25] to be equivalent to an earlier result of B. H. Neumann [24, Lemma 4.1] about covering an abstract group by a finite number of cosets of proper subgroups.

This chapter presents an account of the Separation Theorem for group actions. We shall discuss various improvements and generalisations of Theorem 1.1 which have been proved in an attempt to understand its full power. We shall also survey several diverse applications to problems about group actions, abstract group structure, and even an unsolved problem about combinatorial designs.

The contrapositive of the assertion made by Theorem 1.1 is the following. If there is a finite subset $\Gamma \subseteq \Omega$ such that, for all $g \in G$, $|\Gamma^g \setminus \Gamma| < |\Gamma|$, then G has at least one finite orbit in Ω. It is in this form that the Separation Theorem is usually applied, and it is from this form that questions most readily arise. How large is the finite orbit? Is its size dependent on the size of the finite set Γ? Is any similar assertion possible when Γ is infinite? It is from questions such as these that many of the variations, improvements and generalisations of the Separation Theorem have arisen.

For a finite subset Γ of Ω, the *movement* of Γ under the action of a group G of

W. C. Holland (ed.), Ordered Groups and Infinite Permutation Groups, 195–219.

permutations of Ω is defined as

$$\text{move}(\Gamma) := \max_{g\in G} |\Gamma^g \setminus \Gamma|,$$

and Γ will be said to have *restricted movement* if $\text{move}(\Gamma) < |\Gamma|$. Thus an equivalent version of the Separation Theorem is:

Theorem 1.1′ *Let G be a group of permutations of a set Ω. If there exists a finite subset Γ of Ω with restricted movement, then G has at least one finite orbit in Ω.*

Sections 2 and 5 are concerned with specifying upper bounds on the length of the smallest G-orbit in terms of parameters such as the size and movement of Γ.

Under certain conditions we can extend the definition of restricted movement to infinite subsets: if Γ is any subset of Ω such that, for all $g \in G$, $\Gamma^g \setminus \Gamma$ is finite, and $\{|\Gamma^g \setminus \Gamma| \mid g \in G\}$ is bounded, then we define the *movement* $\text{move}(\Gamma)$ of Γ, as above, as the maximum of $|\Gamma^g \setminus \Gamma|$ over all $g \in G$, and we say that Γ has *restricted movement* if $\text{move}(\Gamma) < |\Gamma|$. Of course if $\text{move}(\Gamma)$ is defined for an infinite subset Γ, then $\text{move}(\Gamma)$ will be finite and hence less than $|\Gamma|$, so Γ will have restricted movement. If Ω is infinite then every cofinite subset of Ω has restricted movement, so there are many infinite subsets with restricted movement. More generally, if Σ is a *G-invariant subset* of Ω (that is, if $\Sigma^g \subseteq \Sigma$ for all $g \in G$) then $\text{move}(\Sigma) = 0$ and Σ has restricted movement. Section 3 reports on a study in [8] of the structure of subsets with restricted movement: it is shown that a subset with restricted movement m differs from a G-invariant subset by a set of size bounded above by a function of m. In Section 4 we apply this result, together with a result from Section 2, to obtain a version of Theorem 1.1′ for arbitary subsets with restricted movement.

Consider the situation in which, for a permutation group G on Ω, a large number of subsets of Ω have movement bounded above by a positive integer m. For example, if all subsets of Ω have finite movement at most m then both the number of nontrivial G-orbits and their lengths are bounded above by functions of m. Similarly if, for some integers m and k such that $0 < m < k \leq |\Omega|$, all k-element subsets of Ω have movement at most m, then the number of nontrivial G-orbits and their lengths are bounded above by functions of m and k. These results are discussed in detail in Section 6.

The Separation Theorem has been used in some surprising ways. In Section 7 we give a brief account of the way it was used in solving a problem about abstract groups. For k a positive integer, a group G is said to have the *small squaring property $SS(k)$* if, for all k-element subsets K of G, the set of all products gh, for g, h (not necessarily distinct) elements of K, has size strictly less than k^2. An abelian group G has the small squaring property $SS(k)$ for all k, $2 \leq k \leq |G|$. A version of the Separation Theorem was used to show that an infinite group with property $SS(k)$ for some k is not far (in a certain sense) from being abelian.

Not surprisingly for such a fundamental result as the Separation Theorem, special cases of it have been proved independently for particular kinds of group actions. The most commonly used quantitative version of the Separation Theorem [2] (stated in Section 2 as Theorem 2.1) was also proved in [18, Lemma 7] in the case of the symmetric group on a set Σ of size n acting on the set Ω of m-element subsets of Σ. This version has been applied to problems in both Graph Theory and the theory of Combinatorial Designs, and we shall discuss these applications in Section 9. Our account in Section 9 makes use also of a probabilistic version of the Separation Theorem which will be presented in Section 8.

2 Quantitative versions of the Separation Theorem

This section contains a discussion of various versions of the Separation Theorem which specify upper bounds on the size of the finite orbit in the conclusion of Theorem 1.1′. The first, and most useful result, was proved by B. J. Birch, R. G. Burns, S. Oates Macdonald and P. M. Neumann [2, Theorem 2] in 1976 and gives an upper bound in terms of the size of the finite set Γ. Their most general result is proved for the separation of possibly different finite subsets Γ, Δ. (In the statement we use the notation $\mathrm{move}(\Gamma)$ for the movement of a finite subset Γ defined in Section 1.)

Theorem 2.1 [2, Theorem 2] *Let G be a group of permutations of a set Ω, and suppose that there are finite subsets Γ, Δ of Ω such that $\Gamma^g \cap \Delta \neq \emptyset$ for all $g \in G$. Then there exists a G-orbit in Ω of length at most $|\Gamma| \cdot |\Delta|$. In particular, if $\mathrm{move}(\Gamma) < |\Gamma|$, then some G-orbit in Ω has length at most $|\Gamma|^2$.*

Theorem 2.1 clearly gives more precise information than the Separation Theorem, and the Separation Theorem follows from Theorem 2.1. However there is no known proof of Theorem 2.1 which does not rely on the Separation Theorem.

Question 1 *Is there a proof of Theorem 2.1 which does not use the Separation Theorem 1.1?*

The upper bound in Theorem 2.1 is the best possible bound in the case of separating *different* subsets Γ and Δ. For example, if $|\Omega| = km$ and $G = S_k \mathrm{\ wr\ } S_m$ is the subgroup of the symmetric group $\mathrm{Sym}(\Omega)$ preserving some partition $\mathcal{P}$ of Ω into m subsets of size k, and if Γ is one of the subsets of $\mathcal{P}$ while Δ consists of exactly one point from each subset of $\mathcal{P}$, then $\Gamma^g \cap \Delta \neq \emptyset$ for all $g \in G$. However, if $\Gamma = \Delta$, then it is implicit in the proof of Theorem 2.1 that the bound $|\Gamma|^2$ is not sharp, and this was formally recorded for transitive group actions in [8, Proposition D]. Moreover if G is intransitive on Ω then a better bound, namely $(|\Gamma| - 1)^2$, was obtained in [29, Corollary 1.5]. (In Section 5 we discuss in more detail the case of intransitive group actions.) Although the bound $|\Gamma|^2$ in Theorem 2.1 is not sharp for transitive groups G, it is not far from best possible, as the following family of examples shows.

Example 2.2 Let $k-1$ be a prime power, let Ω be the set of points of the Desarguesian projective plane $\mathrm{PG}_2(k-1)$, and let Γ be the set of points on a line of $\mathrm{PG}_2(k-1)$. Then $G := \mathrm{PGL}(3, k-1)$ is transitive on Ω, $|\Omega| = k^2 - k + 1$, and Γ is a k-element subset with $\mathrm{move}(\Gamma) < k$.

The set $E := \{\Gamma^g \mid g \in G\}$ forms the set of (hyper)edges of an intersecting hypergraph having all edges of size k, that is a k-uniform hypergraph. According to Z. Füredi (in [14, p.158]) the conjecture that the union of the edges of such a hypergraph contains at most $k^2 - k + 1$ points was made by P. Erdős and by B. Bollobás, and was first proved by L. Lovász in 1975. A second proof was given in 1981 by Füredi [14, Corollary 4] using the theory of fractional matchings and fractional transversals in hypergraphs. Füredi's proof also shows that any example attaining the bound $k^2 - k + 1$ comes from a projective plane of order $k-1$. A proof of these facts involving only elementary counting arguments, and using the language of tactical configurations can be found in [27, Corollary 1].

Theorem 2.3 *Let G be a group of permutations of a set Ω, and suppose that Γ is a k-element subset of Ω with* $\mathrm{move}(\Gamma) < k$. *Then some G-orbit in Ω has length at most $k^2 - k + 1$. Furthermore, if the shortest G-orbit Ω' intersecting Γ nontrivially has length $k^2 - k + 1$, then $\Gamma \subseteq \Omega'$ and the G-translates of Γ are the lines of a projective plane with point set Ω'.*

Extra information about the action of the group G on Ω, and the corresponding bounds on the minimal length of G-orbits in Theorem 2.1, can be obtained by taking into account the value of the following parameter describing the separation of two subsets.

For a permutation group G on Ω, if Γ and Δ are subsets of Ω such that $\Gamma^g \cap \Delta$ is finite for some $g \in G$, we define the *separation defect* of Γ and Δ under the action of G as

$$\mathrm{sepdef}(\Gamma, \Delta) := \min_{g \in G} |\Gamma^g \cap \Delta|,$$

so that Γ and Δ can be separated by G if and only if $\mathrm{sepdef}(\Gamma, \Delta) = 0$.

Theorem 2.4 [30, Theorem 4] *Let G be a group of permutations of a set Ω and suppose that, for some pair of finite subsets Γ, Δ of Ω,* $\mathrm{sepdef}(\Gamma, \Delta) = m > 0$. *Then there exists a G-orbit in Ω of length at most $|\Gamma| \cdot |\Delta|/m$.*

As in the case of Theorem 2.1, the bound in Theorem 2.4 is sharp in the case of separating different subsets Γ and Δ whenever $m = \mathrm{sepdef}(\Gamma, \Delta)$ divides $|\Gamma| \cdot |\Delta|$. I am grateful to Avinoam Mann for pointing out the following generalisation of my earlier class of examples meeting the bound of Theorem 2.4. Write $m = kl$ where k divides $|\Gamma|$ and l divides $|\Delta|$. Let Ω be a set of size $|\Gamma| \cdot |\Delta|/m$ and let $G \cong S_{|\Gamma|/k} \,\mathrm{wr}\, S_{|\Delta|/l}$ be the subgroup of the symmetric group $\mathrm{Sym}(\Omega)$ which preserves a partition $\mathcal{P}$ of Ω into $|\Delta|/l$ subsets of size $|\Gamma|/k$. If Γ is the union of k parts of $\mathcal{P}$ and Δ consists of l points of each of the parts of $\mathcal{P}$, then each image of

Γ under an element of G meets Δ in exactly m points so that sepdef $(\Gamma, \Delta) = m$. From the proof of Theorem 2.4 in [30, Section 5], it follows that, in any example for which the bound $|\Gamma| \cdot |\Delta|/m$ is attained, Γ, Δ must both be subsets of a G-orbit of length $|\Gamma| \cdot |\Delta|/m$, and the cardinality of $\Gamma^g \cap \Delta$ must be equal to m for all $g \in G$. It would be interesting to know if the examples above are the only examples of transitive groups G of degree $|\Gamma| \cdot |\Delta|/m$ for which there are subsets Γ, Δ such that sepdef $(\Gamma, \Delta) = m$.

However, the upper bound provided by Theorem 2.4 in the case where $\Gamma = \Delta$, namely $|\Gamma|^2/m$, is not best possible. This can be seen when $m = 1$ by applying Theorem 2.3. A better upper bound which can be attained for certain arbitrarily large values of the relevant parameters, can be deduced from a result of Frankl and Füredi [12, Theorem 5.4] about intersecting hypergraphs. Alternative proofs have been given in [9] (using the theory of Bose Mesner algebras of association schemes) and in [27, Corollary 2] and [30, Theorem 1] (using elementary counting arguments). The result is the following. (A 2-(v, k, λ) *design* consists of a set Ω of v points together with a collection of k-element subsets of Ω, called blocks, such that each pair of points is contained in exactly λ blocks; such a design is said to be *symmetric* if there are exactly v blocks.)

Theorem 2.5 *Let G be a group of permutations of a set Ω and let m and k be positive integers with $m < k$. If there exists a k-element subset Γ of Ω such that* move$(\Gamma) \leq m$, *then Γ intersects nontrivially a G-orbit in Ω of length at most $(k^2 - m)/(k - m)$. Moreover if the shortest G-orbit Ω' intersecting Γ nontrivially has length $(k^2 - m)/(k - m)$, then $\Gamma \subseteq \Omega'$ and the G-translates of Γ form the blocks of a symmetric 2-$(\frac{k^2-m}{k-m}, k, k - m)$ design with point set Ω'.*

It is quite straightforward using the techniques of [2] to obtain $k^2/(k - m)$ as an upper bound in Theorem 2.5, and this was obtained in [8, Theorem 2.1] in the case where G is transitive on Ω. If $k = m + 1$ we recover Theorem 2.3. Also if $m = 1$ and the shortest G-orbit Ω' meeting Γ has length $k + 1$ (the bound of Theorem 2.5) then we have the trivial case in which Γ is the whole orbit Ω' with one point removed. For the other cases, that is, for $2 \leq m \leq k - 2$, in the case of equality in Theorem 2.5 the fact that $k - m$ divides $k^2 - m$ (or equivalently $k - m$ divides $m^2 - m$) provides a nontrivial restriction on the parameters.

Note that every symmetric 2-(v, k, λ) design has the property that each pair of distinct blocks has precisely λ points in common (see for example [11, p. 57]). Thus in Theorem 2.5, if the transitive action of G on Ω' is equivalent to the action of G on the points of some (in fact, any) point-transitive symmetric 2-(v, k, λ) design, and if Γ is a subset of Ω' which forms a block of this design, then Γ has restricted movement equal to $k - \lambda$ and the upper bound on $|\Omega'|$ is attained. Typical examples of such designs are the designs from projective geometries in Example 2.6 below. Other examples of point-transitive symmetric designs are the Paley designs, defined in [11, p. 97], which have $k = 2m - 1$ where $4m - 1$ is an odd prime power (see Example 3.5 in the next section).

Example 2.6 Let q be a power of a prime, and n an integer, $n \geq 2$. Let Ω be the set of points of the projective geometry $\mathrm{PG}_n(q)$, let $G = \mathrm{PGL}(n+1, q)$ (or any subgroup transitive on Ω), and let Γ be the set of points on a hyperplane. Then $k := |\Gamma| = (q^n - 1)/(q-1)$, $m := \mathrm{move}(\Gamma) = q^{n-1}$, and $|\Omega| = (q^{n+1} - 1)/(q-1) = (k^2 - m)/(k-m)$.

3 Subsets with restricted movement: the closest invariant set

In [8] a study was made of the structure of (possibly infinite) subsets Γ of a set Ω which have *restricted movement* under the action of some permutation group G on Ω, that is, subsets Γ such that $\Gamma^g \setminus \Gamma$ is finite and of bounded size for all $g \in G$ so that $\mathrm{move}(\Gamma)$ is defined, and such that $\mathrm{move}(\Gamma) < |\Gamma|$. The aim in [8] was to show that such a set Γ must have symmetric difference with some G-invariant set Σ of size bounded above by a function of $\mathrm{move}(\Gamma)$.

The suggestion that this might be the case came from a result of Leonid Brailovsky [3] that a subset Γ with (restricted movement and with) $\mathrm{move}(\Gamma) = 1$ must be equal to some G-invariant subset with either one point added or one point removed. It follows from the definitions that a subset Γ is G-invariant if and only if $\mathrm{move}(\Gamma) = 0$, and this is the case if and only if Γ has symmetric difference of size 0 with some G-invariant set, namely with Γ itself. In general, if Γ has size greater than m and if Γ can be obtained from a G-invariant subset Σ by adding and/or removing a total of at most m points, then Γ has restricted movement, $\mathrm{move}(\Gamma) \leq m$, and the symmetric difference $\Gamma\Delta\Sigma$ has size at most m: we shall call such subsets Γ *generic subsets with restricted movement at most* m.

Certainly not all of the subsets Γ with restricted movement at most m (that is $\mathrm{move}(\Gamma) \leq m$) are generic. For example, when m is a prime power, it follows from Example 2.2 that a line Γ in the Desarguesian projective plane $\mathrm{PG}_2(m)$ has $\mathrm{move}(\Gamma) = m < |\Gamma| = m+1$, and the closest $\mathrm{PGL}(3, m)$-invariant subset to Γ is the empty set $\emptyset$. In [7] it was shown first that, if $\mathrm{move}(\Gamma) = 2$, then Γ has symmetric difference at most 3 from some G-invariant subset. Then the bound 3 was shown to be best possible and in all cases meeting this bound the collection $\{\Gamma^g |\ g \in G\}$ was shown to come from one of a handful of possible geometrical configurations. These results prompted the general study, and the best general result obtained in [8] is the following. (For a real number x, $\lceil x \rceil$ denotes the least integer greater than or equal to x. Also, e is the base of natural logarithms.)

Theorem 3.1 [8, Theorems A, B, and Corollary 2.3] *Let G be a group of permutations of a set Ω and suppose that $\Gamma \subseteq \Omega$ has restricted movement with* $\mathrm{move}(\Gamma) = m$. *Then*

(a) there exists a G-invariant subset Σ of Ω such that $|\Gamma\Delta\Sigma| < 2em\lceil(\ln 2m)\rceil$, and

(b) Γ has a proper, nontrivial intersection with at most $2m-1$ of the G-orbits.

(c) Moreover, if G is transitive on Ω then either $|\Gamma| \leq 2m-1$ or $|\Omega \setminus \Gamma| \leq 2m-1$.

The bounds in parts (b) and (c) are sharp for some values of m as Examples 3.2 and 3.3, respectively, show.

Example 3.2 Let $m := 2^r \geq 2$, and let $G := Z_2^{r+1}$. Then G has exactly $2m-1 = 2^{r+1}-1$ subgroups $H_1, \ldots, H_{2m-1}$ of index 2. Let $\Omega_i := \{H_i g \mid g \in G\}$ be the set of (two) H_i-cosets, for each i, and set $\Omega := \bigcup_{1\leq i\leq 2m-1} \Omega_i$. Then G acts faithfully on Ω by right multiplication with $2m-1$ orbits $\Omega_1, \ldots, \Omega_{2m-1}$ of length 2. Note that each nontrivial element $g \in G$ lies in exactly $m-1 = 2^r - 1$ of the H_i and therefore permutes nontrivially the remaining $m = 2^r$ of the Ω_i. Let Γ consist of exactly one point from each of the Ω_i. Then $|\Gamma| = 2m-1$ and $\text{move}(\Gamma) = m$, so Γ has restricted movement, and Γ has a proper nontrivial intersection with each of the $2m-1$ orbits.

Example 3.3 An *Hadamard design* $\mathcal{D}$ is a $2-(4m-1, 2m-1, m-1)$ design, that is $\mathcal{D}$ consists of a set Ω of $4m-1$ points and a set of $(2m-1)$-element subsets of Ω, called blocks, such that each pair of points lies in exactly $m-1$ blocks (see [11, p97] for this definition and for the proofs of various assertions made in the following discussion). An Hadamard design $\mathcal{D}$ is a symmetric design, that is the number of blocks is equal to the number $4m-1$ of points. Also each pair of distinct blocks of $\mathcal{D}$ intersects in exactly $m-1$ points. Let G be the automorphism group of $\mathcal{D}$, so elements of G map blocks to blocks, and let Γ be a block of $\mathcal{D}$. If $g \in G$, then $\Gamma^g \setminus \Gamma$ is either the empty set or has size m. Thus Γ has restricted movement, $|\Gamma| = 2m-1$, and $\text{move}(\Gamma) = m$ under the action of G on Ω. If G were transitive on Ω, then the closest G-invariant set to Γ would be the empty set and $|\Gamma \Delta \emptyset| = 2m-1$. Such designs exist. For example, the Paley designs defined in [11, p97] are examples of Hadamard designs with transitive automorphism groups and they exist whenever $4m-1$ is an odd prime power. Also the design of points and hyperplanes of the d-dimensional projective space $\text{PG}_d(2)$ over $\text{GF}(2)$ is an Hadamard design with transitive automorphism group $\text{PGL}(d+1, 2)$ and with $m = 2^{d-1}$.

However, the bound in Theorem 3.1(a) is not best possible. Better bounds were obtained in [8] for small values of m, namely for $m \leq 31$, combining the information from Theorem 3.1(b) with the following technical result gave a better bound than that in Theorem 3.1(a).

Theorem 3.4 [8, Theorem C] *Let G be a group of permutations of a set Ω and suppose that $\Gamma \subseteq \Omega$ has restricted movement with* $\text{move}(\Gamma) = m$. *Let t be the number of G-orbits with which Γ has a proper nontrivial intersection. Then there exists a G-invariant subset Σ of Ω such that*

$$|\Gamma \Delta \Sigma| \leq 2m(3/2)^{\lceil \log_2 t \rceil} - 2^{\lceil \log_2 t \rceil}.$$

When $t = 1$ the inequality in Theorem 3.4 becomes $|\Gamma \Delta \Sigma| \leq 2m-1$, and this bound can sometimes be attained as Example 3.3 showed. If Γ has restricted movement m and the number t of G-orbits meeting Γ in a non-empty, proper subset is bounded (independently of m), then Theorem 3.4 gives an upper bound for $|\Gamma \Delta \Sigma|$ which is linear in m. At the time of writing the first draft of this chapter, it was not known whether this restriction on t was necessary in order to obtain a linear upper bound on $|\Gamma \Delta \Sigma|$. In 1992, Peter Neumann, in a letter to Leonid Brailovsky, asked the following question.

Question 2 (P. M. Neumann) *Under the conditions of Theorem* 3.1*(a), is it true that* $|\Gamma \Delta \Sigma| \leq 2m - 1$*?*

In a letter to me, dated Match 19, 1995, Peter Neumann showed that Question 2 has an affirmative answer, so that we now have:

Theorem 3.5 (P. M. Neumann) *Let G be a group of permutations of a set Ω and suppose that $\Gamma \subseteq \Omega$ has restricted movement with* $\mathrm{move}(\Gamma) = m$. *Then there is a G-invariant subset Σ of Ω such that* $|\Gamma \Delta \Sigma| \leq 2m - 1$.

4 Infinite subsets with restricted movement

The results of Section 3 make it possible to prove a version of Theorem 2.1 for an infinite subset with restricted movement. To see how this is done we review some of the ideas behind the investigation in [8] and the proof of Theorem 3.1.

Recall from the discussion in Section 3 that a subset Γ obtained from a G-invariant subset by adding or removing a total of m points will have finite movement at most m; in particular if $|\Gamma| > m$ then Γ will have restricted movement (that is, $\mathrm{move}(\Gamma) < |\Gamma|$) and subsets Γ obtained in this way are called *generic subsets with restricted movement at most m*.

Suppose that G is a group of permutations of a set Ω and that Γ is a (not necessarily finite) subset of Ω with restricted movement. Let $m := \mathrm{move}(\Gamma)$, so m is finite and $|\Gamma| > m$. Observe that, if some G-orbit Ω' in Ω is contained in Γ, then from the definition of restricted movement it is clear that $\Gamma \setminus \Omega'$ also has finite movement equal to m. Suppose that we remove from Γ all G-orbits which are completely contained in Γ. The resulting set Γ' has movement m, and either Γ' has restricted movement or $|\Gamma'| = m$. If $|\Gamma'| = m$, then the original set Γ was a generic subset with restricted movement m.

So we shall suppose that $|\Gamma| > m$, $\mathrm{move}(\Gamma) = m$, and for all G-orbits Ω', $\Gamma \cap \Omega'$ is a proper subset of Ω'. Now if $\Gamma \cap \Omega' = \emptyset$ for some Ω' then Γ still has movement m under the permutation group induced by G on $\Omega \setminus \Omega'$. So we may assume that $\Gamma \cap \Omega' \neq \emptyset$ for all G-orbits Ω'.

Suppose next that, for some G-orbit Ω', $\Gamma' := \Gamma \cap \Omega'$ is infinite. It follows from the definition of movement that $\mathrm{move}(\Gamma') \leq m$, and since Γ' is infinite, that Γ' has restricted movement. Then by a result of Peter Neumann [8, Theorem 2.4], $\Omega' \setminus \Gamma'$ is finite (in fact of size at most m by [8, Corollary 2.5]). It follows from [8, Lemma 3.1] that the set $(\Gamma \cup \Omega') \setminus \Gamma'$, obtained from Γ by replacing Γ' with $\Omega' \setminus \Gamma'$, has movement equal to m. Thus, by making such a substitution for all infinite orbit intersections of Γ (if any) we obtain a subset Γ'' with movement m which intersects each G-orbit Ω' in a finite, non-empty, proper subset of Ω'. Again, either Γ'' has restricted movement or Γ'' has size m, and in the latter case the original set Γ must have been a generic subset with restricted movement m.

So we now suppose that $|\Gamma| > m$, $\text{move}(\Gamma) = m$, and for all G-orbits Ω', $\Gamma' := \Gamma \cap \Omega'$ is finite, is non-empty, and is a proper subset of Ω'. In a similar manner, if for some G-orbit Ω' the cardinality of Γ' is greater than that of $\Omega' \setminus \Gamma'$, then by [8, Lemma 3.1] the set $(\Gamma \cup \Omega') \setminus \Gamma'$ obtained from Γ by replacing Γ' with $\Omega' \setminus \Gamma'$ has movement equal to m. By making such a substitution for all orbit intersections $\Gamma \cap \Omega'$ for which $|\Gamma \cap \Omega'| > |\Omega' \setminus \Gamma|$ (if any) we obtain a subset Γ'' with movement m which intersects each G-orbit Ω' in a finite, non-empty, proper subset of cardinality at most $|\Omega' \setminus \Gamma''|$. Again, either Γ'' has restricted movement or Γ'' has size m, and in the latter case the original set Γ must have been a generic subset with restricted movement m. Moreover, by [3], all subsets Γ with movement $m = 1 < |\Gamma|$ are generic.

Finally, we assume that $m \geq 2$ and that Γ'' has size greater than m, so that Γ'' has restricted movement. It follows from Theorem 3.5 that $|\Gamma''| \leq 2m - 1$. Set $k := |\Gamma''|$, so $m < k \leq 2m - 1$. Applying Theorem 2.4 to Γ'' we find that the smallest G-orbit intersecting Γ'' nontrivially (and hence intersecting Γ nontrivially) has length at most $(k^2 - m)/(k - m)$. Now using elementary calculus we see that the function $f(x) := (x^2 - m)/(x - m)$ is strictly decreasing on the interval $[m+1, 2m-1]$, and so the maximum value of $f(x)$ on this interval is $f(m+1) = m^2 + m + 1$. On the other hand, if k is known to be significantly larger than m, say $k \geq k_0$, then $f(k)$ will be bounded above by $f(k_0)$. For example, if $k_0 = a \cdot m > m + 1$ then

$$f(k_0) = \frac{a^2m^2 - m}{am - m} < \frac{a^2m}{a - 1}.$$

Thus we have proved the following result (which is essentially [30, Theorem 2] updated by the improved bound of Theorem 3.5).

Theorem 4.1 *Let G be a group of permutations of a set Ω and suppose that $\Gamma \subseteq \Omega$ (a not necessarily finite subset) has restricted movement with* $\text{move}(\Gamma) = m$. *Then either*

(a) Γ is a generic subset with restricted movement m (that is Γ is obtained from a G-invariant subset by adding or removing a total of m points) or

(b) the smallest G-orbit intersecting Γ nontrivially has length at most $m^2 + m + 1$. Moreover $m \geq 2$, and if Γ differs from a G-invariant set by more than $a \cdot m$ points, where $a \geq (m+1)/m$, then the smallest G-orbit meeting Γ nontrivially has length less than $a^2m/(a - 1)$.

5 Separation under intransitive group actions

Suppose now that G is a permutation group on a set Ω which is intransitive on Ω, and that $\Gamma \subseteq \Omega$ has restricted movement with $\text{move}(\Gamma) = m$. If Γ were contained in a G-orbit Ω' then we could consider the transitive action induced by G on Ω', and in this case the results discussed in Section 2 give the best possible upper bounds on the length of Ω'. Thus in this section we shall be interested in the case where Γ intersects nontrivially at least two of the G-orbits in Ω. Suppose that we can write $\Omega = \Sigma_1 \cup \Sigma_2$ where Σ_i is G-invariant, and $\Gamma_i := \Gamma \cap \Sigma_i \neq \emptyset$, for $i = 1, 2$. It is certainly the case that $\text{move}(\Gamma_i) \leq m$ and even a little more is true.

Theorem 5.1 [8, Proposition 4.1] *Let G be a group of permutations of a set Ω and suppose that $\Gamma \subseteq \Omega$ has restricted movement with* $\text{move}(\Gamma) = m$. *Suppose that Ω is the disjoint union of G-invariant sets Σ_1 and Σ_2 such that, for $i = 1, 2$, $\Gamma_i := \Sigma_i \cap \Gamma \neq \emptyset$. Then*

$$\text{move}(\Gamma_1) + \text{move}(\Gamma_2) \leq 3m/2.$$

First we note that the bound in this result can be attained for some values of m. Let $m = 2^r \geq 2$ and let G, Ω and Γ be as in Example 3.2. Choose a non-identity element $g \in G$, let Σ_1 be the union of the G-orbits Ω_i on which g acts nontrivially, and let $\Sigma_2 := \Omega \setminus \Sigma_1$. Then $\text{move}(\Gamma) = \text{move}(\Gamma_1) = m$ and $\text{move}(\Gamma_2) = m/2$.

However, even though $\text{move}(\Gamma_i) \leq \text{move}(\Gamma)$, with strict inequality for at least one i, it is possible for neither of Γ_1, Γ_2 to have restricted movement (see Example 5.2 below). This illustrates the fact that results about separation of subsets by intransitive groups may not be deduced immediately from similar results for transitive groups. The intransitive case must be considered separately.

Example 5.2 Let $m = k - 1$ be a prime power, let Ω be the set of points of the Desarguesian projective plane $PG_2(m)$, let Σ_1 and Γ be the sets of points on distinct lines of the plane, and let G be the setwise stabiliser of Σ_1 in $PGL(3, m)$. Then the G-orbits in Ω are Σ_1 and $\Sigma_2 := \Omega \setminus \Sigma_1$, and we have $|\Sigma_1| = k = m+1$ and $|\Sigma_2| = (k-1)^2 = m^2$. Moreover Γ is a k-element subset with $\text{move}(\Gamma) = m < k$, and setting $\Gamma_i := \Gamma \cap \Sigma_i$ for $i = 1, 2$, we have $|\Gamma_1| = \text{move}(\Gamma_1) = 1$ and $|\Gamma_2| = \text{move}(\Gamma_2) = m$.

Before proceeding with our discussion we note that the proof of Theorem 3.4 is a straightforward induction on the number t of orbits meeting the set Γ; the proof uses Theorem 3.1(c) to start the induction for $t = 1$, and uses Theorem 5.1 for the inductive step.

If Γ is a finite set with restricted movement intersecting exactly t $(1 \leq t \leq |\Gamma|)$ of the G-orbits nontrivially, then the best general separation result available [17,

Theorem 1.4] (or see [29, Corollary 1.5] when $t = 2$) is a generalisation of Theorem 2.1 which gives an upper bound for the length of the smallest G-orbit in terms of both $|\Gamma|$ and t.

Theorem 5.3 [17, Theorem 1.4] *Let G be an intransitive permutation group on a set Ω, and let Γ be a k-element subset of Ω with restricted movement. If Γ intersects nontrivially t of the G-orbits in Ω, where $1 \leq t \leq k$, then the smallest of these orbits has length at most $(k-t+1)^2+t-2$.*

When $t = k$ and $t-1$ is a prime power, then this bound can always be attained.

Example 5.4 Let $k = t = q+1$ for some prime power q, let Ω be the set of lines of the Desarguesian affine plane $AG_2(q)$, and let G be the group of translations of $AG_2(q)$. Then $|\Omega| = q(q+1)$, G is elementary abelian of order q^2, and G has $q+1$ orbits of length q in Ω, namely the $q+1$ parallel classes of lines. Moreover, each non-identity translation in G fixes each line in exactly one of the parallel classes and acts semiregularly on the lines in each of the other parallel classes. Let $\Gamma \subset \Omega$ consist of exactly one line from each parallel class. Then $|\Gamma| = q+1 = k$ and, from our observation about the action of each non-identity translation g on Ω, $|\Gamma^g \cap \Gamma| = 1$, whence $\text{move}(\Gamma) = q < k$. Thus Γ has restricted movement, Γ meets $t = k$ orbits, and each orbit has size $q = (k-t+1)^2+t-2$.

In fact, the examples given above are essentially the only kinds of examples for which $t = k \geq 3$ and the bound of Theorem 5.3 is attained.

Theorem 5.5 [17, Theorem 1.5] *Let G be an intransitive permutation group on a set Ω, and let Γ be a k-element subset of Ω with restricted movement, where $k \geq 3$. Suppose further that Γ intersects nontrivially k of the G-orbits in Ω, and that the smallest of these orbits has length $k-1$ (the bound of Theorem 5.3). Then $k-1$ is a prime power, and, if Ω' is the union of the G-orbits meeting Γ, then Ω' can be identified with the set of lines of an affine translation plane of order $k-1$ in such a way that Γ consists of one line from each parallel class, $G^{\Omega'}$ fixes each parallel class setwise, and $G^{\Omega'}$ contains the group of translations.*

Although the case $t = k \geq 3$ is the only one where ($t \geq 2$ and) we have a classification of the extreme examples, we do have some fairly restrictive information about the case where $t = 2$ and the bound of Theorem 5.3 is attained.

Theorem 5.6 [29, Theorem 1.4] *Let G be an intransitive permutation group on a set Ω, and let Γ be a k-element subset of Ω with restricted movement which intersects nontrivially 2 of the G-orbits, Ω_1 and Ω_2. Then either the smaller of Ω_1 and Ω_2 has length less than $(k-1)^2$, or $|\Gamma \cap \Omega_1|$ say is equal to 1 and $|\Gamma \cap \Omega_2| = k-1$, and either*

(a) $|\Omega_2| = (k-1)^2$ and $|\Omega_1| \leq |G|/(k-1)$, or

(b) $|\Omega_1| = (k-1)^2$, $|\Omega_2| = (k-1)^2+1$, and $|G| \geq |\Omega_1| \cdot |\Omega_2|(k-1)$.

The conditions of parts (a) and (b) hold when $k = 2$, G is a cyclic group of order 2 acting intransitively with two orbits of lengths 1 and 2, and Γ consists of one point from each G-orbit.

Problem 1 *Determine all values of k for which the conditions of Theorem* 5.6(b) *can hold.*

It is not known whether the bound of Theorem 5.3 can be attained for other values of t and k.

Problem 2 *Determine all values of t and k for which the upper bound for the minimal orbit length in Theorem* 5.3 *can be attained, and classify those groups G and subsets* Γ *attaining the bound.*

As we remarked in Section 2, if Γ is a k-element subset of Ω with restricted movement under the action of a group G, then the collection $E := \{\Gamma^g | g \in G\}$ is the set of edges of an intersecting k-uniform hypergraph with vertex set Ω. The work of Füredi and others on fractional transversals of such hypergraphs is relevant to some of the questions discussed in this chapter. A *fractional transversal* of the hypergraph (Ω, E) is a non-negative real-valued function tr on Ω such that $\sum_{\gamma \in \Gamma^g} \mathrm{tr}(\gamma) \geq 1$ for all $\Gamma^g \in E$. Define $|\mathrm{tr}| := \sum_{\alpha \in \Omega} \mathrm{tr}(\alpha)$. Then the *fractional transversal number* τ^* for (Ω, E) is defined to be the minimum of $|\mathrm{tr}|$ over all fractional transversals tr. Let $\Omega_1, \ldots, \Omega_t$ be the G-orbits, and set $k_i := |\Omega_i \cap \Gamma|$. By averaging over the G-orbits we may assume that an optimal fractional transversal tr on (Ω, E) is constant on each of the Ω_i, say tr takes the value a_i on Ω_i. Then we have $\tau^* = \sum_{1 \leq i \leq t} a_i |\Omega_i|$, and $\sum 1 \leq i \leq t a_i k_i = 1$. If the minimum value of $|\Omega_i|/k_i$ occurs for $i = j$, then these equations imply that

$$|\Omega_j|/(k - t + 1) \leq |\Omega_j|/k_j \leq \tau^*.$$

Thus upper bounds on τ^* will yield information about the minimum length of the G-orbits. The inequality above together with the upper bound $k - 1 + k^{-1}$ on τ^* (see [14, Corollary 4]) yields the upper bound of Theorem 2.3 on the minimal length of the G-orbits. For $t \geq 2$ however, the best known bounds on τ^* yield a slightly weaker upper bound on the minimal orbit length than that given in Theorem 5.3. The paper [15] of Füredi contains a survey of some of the relevant results about hypergraphs, and it has a good bibliography.

We end this section with a general discussion of the relation between the orbit length bounds which can be obtained and the distribution of the points of the set Γ amongst the G-orbits.

Suppose that Γ is a k-element subset with restricted movement and that Γ meets exactly t of the G-orbits, $\Omega_1, \ldots, \Omega_t$. Set $\Gamma_i := \Gamma \cap \Omega_i$ and $v_i := |\Gamma_i|$ for each i. Then each $v_i \neq 0$, $\sum v_i = k$, and we may renumber the Ω_i so that $v_1 \leq \ldots \leq v_t$ and hence $v_t \geq k/t$. In the proof of Theorem 5.3 an upper bound on the length of the smallest of the Ω_i is obtained as an expression involving $v_1, \ldots, v_t$. This expression is greatest when $(v_1, \ldots, v_t) = (1, \ldots, 1, k-t+1)$, thus yielding the upper bound in Theorem 5.3.

We give an illustration of how the bound may be improved when the distribution of the points of Γ over the Ω_i is not as unbalanced as it is in the configuration corresponding to the upper bound of Theorem 5.3. Consider the requirement that $v_t \leq k - xk/t$ for some positive real number x. Since $v_t \geq k/t$ we must have $x \leq t-1$. Also, since $v_t = k - \sum_{i<t} v_i \leq k-t+1$, this inequality represents a genuine restriction on v_t only if $k - xk/t \leq k-t+1$, that is, $x \geq t(t-1)/k$. Thus we are interested only in values of x satisfying $t(t-1)/k \leq x \leq t-1$.

Theorem 5.7 [17, Theorem 1.5] *Suppose that G is an intransitive permutation group on a finite set Ω with orbits $\Omega_1, \ldots, \Omega_t$, and that Γ is a k-element subset of Ω with restricted movement and Γ meets Ω_i in v_i points for $i = 1, \ldots, t$, where $0 < v_1 \leq \ldots \leq v_t$. If x is a real number satisfying $t(t-1)/k \leq x \leq t-1$, and if $v_t \leq k - xk/t$, then the smallest of the Ω_i has length at most*

$$k^2(1 - \frac{2x(t-x)}{t^2}) - (t-1)(t-2).$$

To see that this can be quite powerful in certain situations, consider the case where Γ has restricted movement, $k = |\Gamma| \geq 4$, and Γ meets $t = 3$ orbits $\Omega_1, \Omega_2, \Omega_3$; taking $x = 3/2$, Theorem 5.7 says that the smallest of the three Ω_i has length at most $k^2/2 - 2$ provided that each $\Gamma \cap \Omega_i$ contains no more than half the points of Γ. This is much stronger than the bound from Theorem 5.3 which in this case is $(k-2)^2 + 1$.

6 Permutation groups with bounded movement

This section is devoted to a discussion of permutation groups G on a set Ω for which many subsets of Ω have finite movement at most m, for some given positive integer m. In particular, if $\mathrm{move}(\Gamma) \leq m$ for all $\Gamma \subseteq \Omega$, then G is said to have *bounded movement* and the *movement of G* is defined as the maximum of $\mathrm{move}(\Gamma)$ over all subsets Γ. Having bounded movement is a very strong restriction on a group, but it is natural to ask just which permutation groups have bounded movement m. The question was first posed by Carlo Casolo during a discussion we had in 1989. The bounded movement condition was a generalisation to arbitrary group actions of a condition he was working with for the action of a finite soluble group on a module over a finite field.

Suppose now that G is a permutation group on a set Ω and that G has bounded movement equal to m. Quite a lot can be said about G simply by applying results from the previous sections. For example, the group of permutations induced by G on one of its orbits Ω' also has movement at most m, and it follows from Theorem 2.1, on considering subsets of Ω' of size at most $m+1$, that $|\Omega'| \leq (m+1)^2$. Hence all G-orbits are finite of length at most $(m+1)^2$ (and of course better bounds, which are still quadratic in m, follow from Theorems 2.3 and 2.4). Also, by considering a subset of points consisting of one point from each nontrivial G-orbit (that is each orbit of size greater than 1), it follows from Theorem 3.1(b) that G has less than $2m$ nontrivial orbits in Ω. Thus the number of points of Ω moved by G is less than $2m(m+1)^2$. An investigation in [28] of groups with bounded movement, using the Separation Theorem and some careful but elementary arguments, resulted in the following theorem.

Theorem 6.1 [28, Theorem 1] *Let G be a group of permutations of a set Ω with no fixed points in Ω, and suppose that, for some positive integer m, G has bounded movement equal to m. Then*
(a) the number t of G-orbits is at most $2m-1$,
(b) each G-orbit has length at most $3m$, and moreover
(c) $|\Omega| \leq 3m+t-1 \leq 5m-2$.

The bounds in Theorem 6.1(a) and (b) can be attained for certain values of m, and therefore the bound in part (c) can also be attained for certain m when $t=1$.

First we consider the bound of part (a). The family of permutation groups defined in Example 3.2 consists of permutation groups $G = Z_2^{r+1}$ with $2^{r+1}-1$ orbits of length 2 and with the property that each nontrivial element of G has 2^r cycles of length 2. Thus, for any subset Γ of points and any $g \in G$, the set $\Gamma^g \setminus \Gamma$ consists of at most one point from each of the G-orbits on which g acts nontrivially, and consequently $|\Gamma^g \setminus \Gamma| \leq 2^r$. It follows that G has bounded movement $m = 2^r$ and G has $2m-1$ nontrivial orbits. This family of examples is essentially the only one to attain the bound.

Theorem 6.2 [10] *Let G be a group of permutations of a set Ω with no fixed points in Ω, and suppose that, for some positive integer m, G has bounded movement equal to m and has $2m-1$ orbits. Then G is an elementary abelian 2-group, all the G-orbits have length 2, and $m=2^r$ for some non-negative integer r.*

Let G be a group satisfying the conditions of Theorem 6.2, and let the G-orbits in Ω be $\Omega_1, \ldots, \Omega_{2m-1}$, where $m = 2^r$. Then G is a subgroup of the direct product $G_1 \times \ldots \times G_{2m-1}$ where $G_i \cong Z_2$ is the permutation group induced by G on Ω_i for each i. Further, from our discussion before the statement of Theorem 6.2, each element of G must act nontrivially on at most m of the Ω_i. However it is not known precisely which subdirect products of $G_1 \times \ldots \times G_{2m-1}$ have movement equal to m.

Question 3 *What are the possible orders of the groups G in Theorem 6.2?*

Now we turn to the bound in part (b) of Theorem 6.1. It was shown in [28, Theorem 2] that, for a transitive group G on a set Ω of size $3m$ to have movement m, m must have the form $2^a 3^b 5^c$ with a and c at most 1. Moreover [28, Theorem 2] also gave several examples of such groups, namely transitive groups of exponent

3 on a set of size $3m = 3^r \geq 3$, the symmetric group S_3 of degree $3m = 3$, and the groups A_4 and A_5 acting transitively of degree $3m = 6$. It has recently been proved that these are the only examples. In [16] it was conjectured that the only examples were S_3, A_4, A_5 and transitive 3-groups, and it was shown that a minimal counterexample must be a nonabelian simple group and must be primitive on Ω. Then in [23] it was shown that every example which is a 3-group must have exponent 3, and that the only simple primitive example is A_5 of degree 6.

Theorem 6.3 [16, 19] *Let G be a transitive group of permutations of a set Ω of size $3m$ for some positive integer m. Then G has bounded movement equal to m if and only if $m = 3^r \geq 1$ and G has exponent 3, or $m = 1$ and $G = S_3$, or $m = 2$ and $G = A_4$ or A_5.*

The results in Theorems 6.1 amd 6.3 highlight the special role of the prime 3 in this study. For groups with order relatively prime to 3 we obtain different bounds. Two lemmas from [28] give a starting point.

Theorem 6.4 [28, Lemmas 2.1 and 2.2] *Let G be a group of permutations of a set Ω with no fixed points in Ω, and suppose that, for some positive integers m and t, G has bounded movement equal to m and G has t orbits in Ω.*

(a) If G is a 2-group then each G-orbit has length at most $2m$ and $|\Omega| \leq t + 2m - 1 \leq 4m - 2$.

(b) If G is not a 2-group and if p is the least odd prime dividing $|G|$, then each G-orbit has length less than $1+2mp/(p-1)$ and $|\Omega| < t+2mp/(p-1)$. In particular, if $p \geq 5$, then $|\Omega| \leq (9m-3)/2$.

It is easy to show that every permutation group on a set of size at most $2m$ has bounded movement at most m (see [28, Lemma 3.4]). So the groups of interest (in the context of Theorem 6.4) are those which are not 2-groups and which have order relatively prime to 3. In particular it would be interesting to have an analogue of Theorem 6.3 in this case.

Problem 3 *Let m be a positive integer and p an odd prime. Determine all transitive permutation groups G on a set of size $\lceil 2mp/(p-1) \rceil$ which have bounded movement equal to m, and for which p is the least odd prime divisor of $|G|$.*

There certainly are some families of examples as the next lemma shows.

Lemma 6.5 *Let p be an odd prime and let $m := p^{a-1}(p-1)/2$ for some $a \geq 1$.*

(a) A transitive permutation group of exponent p on a set Ω of size $p^a = 2mp/(p-1)$ has bounded movement equal to m.

(b) Let $a = 1$ and let $s = 2^b$ be the highest power of 2 dividing $p-1$. Let F_{sp} denote the Frobenius group of order sp acting transitively on a set Ω of size $p = 2mp/(p-1)$. Then each transitive subgroup G of F_{sp} has bounded movement equal to m.

Proof. (a) Let $g \in G \setminus \{1\}$ and let $\Gamma \subseteq \Omega$. For each cycle C of g of length p, $(\Gamma \cap C)^g \setminus (\Gamma \cap C)$ has size at most $(p-1)/2$. Thus $|\Gamma^g \setminus \Gamma|$ is at most the number of cycles of g of length p times $(p-1)/2$, that is, $|\Gamma^g \setminus \Gamma| \leq (|\Omega|/p) \cdot (p-1)/2 = m$. Thus G has movement at most m. Moreover, since G is transitive, the average number of fixed points of elements of G is 1 and so G contains an element g, say, with no fixed points in Ω; for this element g, a subset Γ consisting of "every second element of every cycle of g" (that is, if a cycle C of g is $C = (\alpha_1, \ldots, \alpha_p)$ then put into Γ the $(p-1)/2$ points $\alpha_2, \alpha_4, \ldots, \alpha_{p-1})$ has size m and is mapped disjointly from itself by g. Thus the movement of G is equal to m.

(b) By part (a), if $g \in G$ has order p, then $|\Gamma^g \setminus \Gamma| \leq m = (p-1)/2$ for all subsets Γ and there is a subset Γ such that $\Gamma^g \setminus \Gamma$ has size equal to m. Suppose now that $g \in G \setminus \{1\}$ has order $o(g)$ a power 2. Then g has one fixed point and $(p-1)/o(g)$ cycles of length $o(g)$ in Ω. For each $\Gamma \subseteq \Omega$, $\Gamma^g \setminus \Gamma$ consists of at most $o(g)/2$ points from each cycle of g of length $o(g)$, and hence has size at most m. Thus G has bounded movement equal to m.

Work on Problem 3 is currently being carried out by Mehdi Khayarty and Sjabon Sedghi, two PhD students of Akbar Hassani in Tehran. Further, Akbar Hassani is developing a full version of Theorem 6.1 for groups G such that the smallest prime divisor p of $|G|$ is at least 5.

It is not clear whether the bound in part (c) of Theorem 6.1 can be attained for large values of t. Put in another way we could ask:

Question 4 *For which positive integers n is there a positive integer t and a permutation group G on a set of size n with t nontrivial orbits such that G has bounded movement equal to $\lceil (n-t+1)/3 \rceil$?*

The paper [28, Lemma 3.6] contains a construction of such a group if 3 divides n and the 3-adic expansion of n (namely $n = \sum_{i \geq 10} a_i 3^i$ with each a_i equal to 0, 1, or 2) satisfies $1 \leq \sum_{i \geq 1} a_i \leq 3$. The number t of nontrivial orbits in the construction is the number of terms of the 3-adic expansion, that is, $t = \sum_{i \geq 1} a_i$, and so the greatest value of t for this family is 3. It is not known whether the bound of Theorem 6.1(c) is best possible for larger values of t.

Recently these results about permutation groups G on Ω with bounded movement were generalised. The assumption of restricted movement for essentially all subsets of Ω was weakened to the requirement that, for some integers m and k with $0 < m < k$, all k-element subsets of Ω should have movement at most m. If the size of Ω is at most $k+m$ then it follows from the definition of movement that this condition is satisfied for every permutation group G on Ω. Thus it is the case where $|\Omega| > k+m$ which is of interest. Surprisingly, in this case, it turns out that the group G must have bounded movement at most m, and so Theorem 6.1 applies.

Theorem 6.6 [30, Theorem 3] *Let G be a group of permutations of a set Ω with no fixed points in Ω, and let m and k be positive integers such that $m < k$ and $|\Omega| > k+m$. Suppose further that* move$(\Gamma) \leq m$ *for all k-element subsets Γ of Ω. Then* move$(\Gamma) \leq m$ *for every subset Γ of Ω (and hence the number t of G-orbits is at most $2m-1$, $|\Omega| \leq 3m+t-1 \leq 5m-2$, and $k \leq 2m+t-2 \leq 4m-3$).*

Moreover, if some G-orbit Ω' has length greater than $k+m$, then $k \leq 2m-1$ and $|\Omega'| \leq 3m$.

Observe that Theorem 2.5 or Theorem 6.6 gives restrictions on the lengths of the smallest G-orbit or of all G-orbits, when we restrict the movement of one or all k-element subsets of points, respectively. It would be interesting to know what information can be deduced about the G-orbits when we restrict the average of the movements of k-element subsets.

Question 5 *What information about the orbits of a permutation group G on a finite set Ω can be deduced from restricting the average of the movements* move(Γ) *of k-element subsets Γ to be at most m, for some $m < k$?*

7 Groups with a small squaring property

A group G is said to satisfy the *small squaring property on k-sets* (where k is an integer, $k \geq 2$) denoted SS(k) if, for each k-element subset K of G, the set K^2 which is defined as $\{ab \mid a, b \in K\}$ has size strictly less than k^2. Clearly any abelian group G will have the the small squaring property on k-sets for every $k \geq 2$, and the properties SS(k) for various k may be viewed as generalisations of the property of being abelian. The first investigation of groups with one of these properties was carried out by G. A. Freiman. In 1981 he classified groups with the property SS(2) (showing that they were precisely abelian groups and Hamiltonian 2-groups), and began a study of groups with the property SS(3). A significant part of the classification of groups with the property SS(3) was carried out by Ja. G. Berkovich, Freiman and the author in [1], and the classification was completed by Patrizia Longobardi and Mercede Maj in [22].

Theorem 7.1 [1, 13] and [22, Theorems A and C] *(a) A group G has the small squaring property* SS(2) *if and only if either G is abelian or $G = E \times Q$ where E is an elementary abelian 2-group and Q is the quaternion group of order 8.*

(b) A group G has the small squaring property SS(3) *if and only if one of the following holds.*

(i) G is abelian;

(ii) $G = A\langle x\rangle$ where A is abelian, $x^2 \in A$, and $a^x = a^{-1}$ for all $a \in A$;

(iii) G is a nonabelian 2-group of exponent 4, and either $\langle x^2 \mid x \in G\rangle$ has order 2, or $G = E \times D$, where E is an elementary abelian 2-group and D is one of two explicitly given groups, one of order 32 and the other of order 64.

A pattern was emerging here: the examples tended to have either a very large abelian normal subgroup or a very large abelian quotient group. The significance of this pattern would become more evident later. In 1989 Peter Neumann showed that a group having the small squaring property SS(k) for some $k \geq 2$ must be finite-by-abelian-by-finite. His proof involved an ingenious and unexpected use of the Separation Theorem.

Theorem 7.2 [20, Theorem 1] and [26] *If a group G has the small squaring property* SS(k) *for some $k \geq 2$, then G has normal subgroups H and N, with $N \subseteq H$, such that H/N is abelian and both G/H and N are finite.*

It was the quantitative version Theorem 2.1 of the Separation Theorem which was used in the proof of this result to construct the normal subgroup H. We will review the ideas of this part of the proof

Suppose that a group G has the property SS(k) where k is some integer, $k \geq 2$, and let S be the set of all elements $g \in G$ such that the number of conjugates of g by elements of G is at most $k^2(k-1)^2$; so S is the union of all "small" conjugacy classes of G. The idea is to consider the conjugation action of G on itself. Under this action S is a G-invariant subset and we may consider the action of G by conjugation on $\Omega := G \setminus S$. All G-orbits in Ω have length strictly greater than $k^2(k-1)^2$ (by the definition of S), and consequently by Theorem 21 any two subsets of Ω of size at most $k(k-1)$ may be separated completely from each other by some element of G.

This fact was used as follows to prove that G can be written as the union of some collection of $k-1$ translates of S. Suppose that this is not the case. Then k distinct elements $y_1, \ldots, y_k$ of G can be found such that $y_i y_j^{-1} \notin S$ for all $i < j$. Then also, since $y_i y_j^{-1}$ is conjugate to $y_j^{-1} y_i$ and since S is closed under taking inverses, $y_i^{-1} y_j \notin S$ for all $i < j$. It follows that, for all $i \neq j$, both $y_i y_j^{-1}$ and $y_i^{-1} y_j$ are elements of Ω. Set $\Gamma := \{y_i y_j^{-1} \mid i \neq j\}$ and $\Delta := \{y_i^{-1} y_j \mid i \neq j\}$, so that Γ and Δ are both subsets of Ω of size at most $k(k-1)$. By Theorem 2.1, there exists an element $g \in G$ such that $\Gamma \cap \Delta^g = \emptyset$. This means that, for all $i \neq j$ and $i' \neq j'$, $y_i y_j^{-1} \neq g^{-1} y_{i'}^{-1} y_{j'} g$, that is, $(y_i g)(y_j g)^{-1} \neq (y_{i'} g)^{-1}(y_{j'} g)$, which in turn is equivalent to $(y_{i'} g)(y_i g) \neq (y_{j'} g)(y_j g)$. The last inequality, for all $i \neq j$ and $i' \neq j'$, implies that the k-element subset $K := \{y_1 g, \ldots, y_k g\}$ satisfies $|K^2| = k^2$, contradicting the fact that G has the small squaring property SS(k).

Thus there are elements $y_1, \ldots, y_{k-1} \in G$ such that $G = \bigcup_{i \geq 1} S y_i$. Let H be the subgroup of G generated by S. Since S is closed under conjugation, H is a normal subgroup of G, and since $S y_i \subseteq H y_i$ for all i, H has index at most $k-1$ in G. The proof of Theorem 7.2 is completed (in [20] or [26]) by showing that the commutator subgroup $N := [H, H]$ has order bounded above by some function of k.

The information given by Theorem 7.2 was refined in [20] by Herzog, Longobardi and Maj to show that examples of groups with the small squaring property SS(k), for some $k \geq 2$, may be roughly divided into two classes, one with a large abelian normal subgroup, and the other with a large abelian quotient group, a division suggested by the classification of groups with the property SS(3) in Theorem 7.1(b). The statement requires the following definitions. A group G is said to be *nearly dihedral* if it contains a normal abelian subgroup H of finite index, on which each element of G acts by conjugation either as the identity automorphism or as the inverting automorphism. The subgroup of a group G generated by the squares of all elements of G will be denoted by $G^{(2)}$.

Theorem 7.3 [20, Theorem 2] *A group G has the small squaring property* SS (k), *for some $k \geq 2$, if and only if either G is nearly dihedral or $G^{(2)}$ is of finite order.*

Now we finish one part of the story of the investigation of groups with a small squaring property. Since the small squaring property SS (k), for some $k \geq 2$, forces a group to be "nearly abelian" in the sense of Theorem 7.2 or Theorem 7.3, is there a somewhat stronger condition of the same nature which forces a group to be abelian? More precisely, if G is abelian then, for all k-element subsets K of G, $|K^2| \leq k(k+1)/2$, and Peter Neumann (see [6]) asked whether the condition $|K^2| \leq \alpha k^2$ for every k-element subset of G (where $k \geq 2$ and α is a fixed real number satisfying $0 < \alpha < 1$) might imply that G is abelian.

Brailovsky [4] first showed that, if $|K^2| \leq k(k+1)/2$ for all k-element subsets K of G, then G is abelian (where $k > 2$ and if G is finite then $|G| > 2k(k^2-1)$). Next he improved this by showing in [5] that, if $|K^2| \leq k(k+1)/2 + (k-3)/2$ for all k-element subsets K of G, then G is abelian (where $k > 2$ and if G is finite then $|G| > 15(k-1)(k^2-3)$). These results led him to define a *squaring bound* for infinite abelian groups to be a function $f(k)$ of the natural numbers such that, if there exists an integer $k \geq 2$ such that $|K^2| \leq f(k)$ for all k-element subsets K of G, then G is abelian. He sought the *best possible squaring bound for infinite abelian groups*, by which he meant a squaring bound $f(k)$ for infinite abelian groups such that the following condition holds:

If $g(k)$ is a function on the natural numbers such that $g(k) > f(k)$ for some natural number k, then $g(k)$ is not a squaring bound for infinite abelian groups.

He was able to construct such a best possible bound [6, Theorem], namely

$$f(k) = \lceil \frac{5k^2 - 3k - 2}{6} \rceil.$$

One consequence of this result is that $f(k) = \lceil \alpha k^2 \rceil$ is a squaring bound for infinite abelian groups if and only if $0 < \alpha < 5/6$. The reader may be interested to know that the proof of the result of [5] makes essential use of the result in [3], which was the principal motivation for the investigation described in Section 3.

8 A probabilistic version of the Separation Theorem

For some purposes an understanding of the Separation Theorem from a probabilistic point of view is helpful. Suppose that G is a permutation group on a finite set Ω and that Γ, Δ are subsets of Ω of size k and n respectively. Consider an experiment in which we randomly select an element g from G in such a way that the probability of selecting any given element of G is $1/|G|$; we construct the image Γ^g of Γ under g and find the size $X(g)$ of the intersection $\Gamma^g \cap \Delta$. Then $X := X(g)$ is a random variable which takes integer values (which of course depend on the element g chosen) in the closed interval $I := [0, \min\{k, n\}]$.

We determine the expected value $E(X)$ and the variance $\mathrm{var}(X)$ of X. In general $E(X)$ depends on the distribution of the points of Γ and Δ over the G-orbits, but if G is transitive on Ω, then $E(X)$, often written $\overline{X}$, depends only on k, n and the size of Ω. The expression for the variance $\mathrm{var}(X)$ (that is, the expected value of $(X-\overline{X})^2$) is more complicated than that for $E(X)$ and it depends on the distribution of the ordered pairs of points of Γ and of Δ among the G-orbits on ordered pairs of points from Ω. When there are few G-orbits on ordered pairs from Ω, the expression for $\mathrm{var}(X)$ is of course simpler, and we give it explicitly, as a function of k, n and $|\Omega|$, in the case where G is 2-transitive on Ω. (Recall that G induces a natural action on $\Omega \times \Omega$ by $(\alpha, \beta)^g := (\alpha^g, \beta^g)$ for $\alpha, \beta \in \Omega$ and $g \in G$. Let $\Sigma^{(1)}, \ldots, \Sigma^{(r)}$ denote the G-orbits in $\Omega \times \Omega$. The expression for $\mathrm{var}(X)$ involves the number of ordered pairs from Γ and Δ which lie in $\Sigma^{(j)}$, for each j.)

Theorem 8.1 [17, Section 7] *Let G be a group of permutations of a finite set Ω, and let Γ, Δ be subsets of Ω of size k and n respectively. Let X be a random variable defined by $X := X(g) = |\Gamma^g \cap \Delta|$, where g is a uniformly distributed random element of G.*

(a) If Ω_j, $j \in J$, is the set of G-orbits in Ω, then the expected value of X is

$$E(X) = \sum_{j \in J} \frac{|\Gamma \cap \Omega_j| \cdot |\Delta \cap \Omega_j|}{|\Omega_j|}.$$

In particular when G is transitive, $E(X) = kn/|\Omega|$.

(b) Let $\Sigma^{(1)}, \ldots, \Sigma^{(r)}$ be the G-orbits in $\Omega \times \Omega$, and, for $j = 1, \ldots r$, let y_j and z_j denote the number of pairs $(\alpha, \beta) \in \Sigma^{(j)}$ such that $\{\alpha, \beta\} \subseteq \Gamma$ and $\{\alpha, \beta\} \subseteq \Delta$ respectively. Then the variance of X is

$$\mathrm{var}(X) = \sum_{1 \leq j \leq r} \frac{y_j \cdot z_j}{|\Sigma^{(j)}|} - E(X)^2.$$

In particular, if G is 2-transitive on Ω, then

$$\mathrm{var}(X) = \frac{k\ n\ (|\Omega| - k)\ (|\Omega| - n)}{|\Omega|^2\ (|\Omega| - 1)}.$$

It is a simple matter to deduce Theorem 2.1 from Theorem 8.1(a), and the arguments used in the proof of Theorem 8.1(a) are essentially those in the proof of Theorem 2.1: if all the G-orbits have length greater than kn then $E(X)$ is strictly less than $(\sum_{j \in J} |\Gamma \cap \Omega_j| \cdot |\Delta \cap \Omega_j|)/kn$ by Theorem 8.1, and this expression is at most $(\sum_{j \in J} |\Gamma \cap \Omega_j|) \cdot (\sum_{j \in J} |\Delta \cap \Omega_j|)/kn = 1$. It follows that $X(g)$ must be zero for some $g \in G$, that is $\Gamma^g \cap \Delta = \emptyset$ for some $g \in G$.

9 Combinatorial applications of the Separation Theorem

The version of the Separation Theorem stated as Theorem 2.1 was proved independently in 1977 by B. Ganter, J. Pelikan and L. Tierlinck in the situation where the set Ω is the collection $\binom{S}{k}$ of all k-element subsets of a v-element set S, and the group G is $\mathrm{Sym}(S)$ in its natural action on Ω. Their motivation came from combinatorial considerations. If $k = 2$ then any subset Γ of Ω may be identified with (the edge set of) a graph with vertex set S; for larger values of k, a subset Γ of Ω satisfying certain extra conditions may be identified with (the block set of) a combinatorial design.

Throughout this section we assume, therefore, that $\Omega = \binom{S}{k}$ and $G = \mathrm{Sym}(S)$ for some set S of size v and some integer k, with $2 \leq k < v$.

In [18] a subset Γ of Ω with restricted movement is said to be *sprawling*. The main aim of the paper [18] was to analyse the behaviour of the function $b(k,v)$ of the integer variables k and v $(0 < k < v)$ where $b(k,v)$ is defined as the least positive integer n such that there exists a sprawling n-element subset $\Gamma \subseteq \binom{S}{k}$, for $|S| = v$. Clearly $b(k,v) = b(v-k,v)$ for all k and v, and $b(1,v) = \lfloor v/2 \rfloor + 1$ for all v. In [18, Lemma 1] it was shown that $b(k,v) \geq \binom{v}{k}^{1/2}$ for all k, v. This is precisely the information given by the last sentence of Theorem 2.1 in the case of $G = \mathrm{Sym}(S)$ and $\Omega = \binom{S}{k}$. The remaining assertions of Theorem 2.1 were obtained for this group action in [18, Lemma 7]. Asymptotic results about the growth of $b(k,v)$ were proved in [18, Theorem 1] but, for example, the question of whether or not $b(k,v)$ is a monotonically increasing function in k (for $k < v/2$) or in v was asked but not answered. In the special case where $k = 2$, so that a subset Γ of Ω may be identified with a graph with vertex set S, it was shown in [18, Theorem 2] that $b(2,v) = v - 1$ and all sprawling graphs with v vertices and $v - 1$ edges were classified.

The problem of separating two subsets Γ and Δ of $\Omega = \binom{S}{k}$, for $2 < k < v$, arose again (see [19], and [31, Section 4.5]) as part of an investigation of construction methods for combinatorial designs. The sets Γ and Δ were the block sets of two t-(v,k,λ) designs with point sets S, and as the starting point for a method of constructing new designs, it was necessary to produce computationally an image Γ^g of Γ, for some $g \in \mathrm{Sym}(S)$, which was disjoint from Δ. The existence of such an image Γ^g is guaranteed by Theorem 2.1 if $|\Gamma| \cdot |\Delta| < \binom{v}{k}$, that is, if v is "large enough" with respect to k. However Theorem 2.1 takes no account of any special structure or special properties of the sets Γ and Δ, and G. B. Khosrovshahi has conjectured that such an image exists for many other nontrivial t-(v,k,λ) designs.

Recall that, for integers $2 \leq t < k < v$ and $\lambda > 0$, a t-(v,k,λ) *design* consists of a set S of v points and a family of k-element subsets of S, called blocks, such that each t-element subset of S is contained in precisely λ blocks. The *trivial* design based on S is the one in which every k-element subset of S is a block, and all other t-designs are said to be nontrivial. Since the set of complements of the blocks of a

t-(v, k, λ) design forms the block set of a t-$(v, v-k, \lambda')$ design, for some λ', we shall assume from now on that $k \leq v/2$.

Conjecture (Khosrovshahi) *Suppose that t, k, v, λ are positive integers $(2 \leq t < k \leq v/2)$ such that a nontrivial t-(v, k, λ) design exists, and such that no t-(v, k, λ') design exists for any $\lambda' < \lambda$. If Γ is the block set of such a design, then another t-(v, k, λ) design exists, with block set Δ say, such that $\Gamma \cap \Delta = \emptyset$.*

A strong version of this conjecture is true for 2-$(v, k, 1)$ designs, namely, for Γ, Δ the block sets of two 2-$(v, k, 1)$ designs on the same point set S, there exists an element $g \in \mathrm{Sym}(S)$ such that $\Gamma^g \cap \Delta = \emptyset$. An algorithm for finding an image Γ^g disjoint from Δ is given in [31, Section 4.5].

Suppose that Γ, Δ are the block sets of two t-(v, k, λ) designs with the same point set S. While not providing proof of the existence of an image Γ^g disjoint from Δ, or giving an estimate of the likelihood of finding such an image computationally, some insight into the distribution of the intersection sizes $X(g) := |\Gamma^g \cap \Delta|$, for $g \in \mathrm{Sym}(S)$, can be gained from computing the expected value and the variance of the random variable $X := X(g)$ (where g is a uniformly distributed random element of G). The tools for doing this were presented in the previous section. When applied to this problem about combinatorial designs they yield the following results.

Theorem 9.1 [21] *Let t, v, k, λ be positive integers such that $2 \leq t < k \leq v/2$. Further let $G = \mathrm{Sym}(S)$, where S is a set of size v, and consider the action of G on the set $\Omega = \binom{S}{k}$. Let Γ, Δ be subsets of Ω which form the block sets of two t-(v, k, λ) designs with point sets S. Let X be a random variable defined by $X := X(g) = |\Gamma^g \cap \Delta|$, where g is a uniformly distributed random element of G. Then the expected value of X is*

$$E(X) = \frac{\lambda^2 \cdot \binom{v}{k}}{\binom{v-t}{k-t}^2}.$$

Moreover, for $j = 0, 1, \ldots, k$, let y_j and z_j denote the number of pairs $(\alpha, \beta) \in \Omega \times \Omega$ with $|\alpha \cap \beta| = j$ such that $\{\alpha, \beta\} \subseteq \Gamma$ and $\{\alpha, \beta\} \subseteq \Delta$ respectively. Then the variance of X is

$$\mathrm{var}(X) = \frac{1}{\binom{v}{k}} \sum_{0 \leq j \leq k} \frac{y_j \cdot z_j}{\binom{k}{j}\binom{v-k}{k-j}} - E(X)^2.$$

The expression for the variance $\mathrm{var}(X)$ is quite complicated and depends in general on the structure of the design and not simply on the parameters t, v, k, λ. However, in the case of 2-$(v, k, 1)$ designs, it turns out that $\mathrm{var}(X)$ depends only on v and k.

Corollary 9.2 [21] *Let v, k be positive integers such that $2 < k \leq v/2$, let S be a set of size v, let $G = \mathrm{Sym}(S)$, and consider the action of G on the set $\Omega = \binom{S}{k}$. Let*

Γ, Δ be subsets of Ω which form the block sets of two 2-$(v, k, 1)$ designs with point sets S. Let X be a random variable defined by $X := X(g) = |\Gamma^g \cap \Delta|$, where g is a uniformly distributed random element of G. Then the expected value of X is

$$E(X) = \frac{\binom{v}{k}}{\binom{v-2}{k-2}^2},$$

and the variance of X is

$$\operatorname{var}(X) = E(X)\big(1 + \frac{k(r-1)^2}{\binom{v-k}{k-1}} + \frac{(v-k)^2(r-k)^2}{k^2\binom{v-k}{k}}\big) - E(X)^2,$$

where $r = (v-1)/(k-1)$.

References

[1] Ja. G. Berkovich, G. A. Freiman and C. E. Praeger, *Small squaring and cubing properties for finite groups*, Bull. Austral. Math. Soc. **44**(1991), 429-450.

[2] B. J. Birch, R. G. Burns, S. O. Macdonald and P. M. Neumann, *On the orbit sizes of permutation groups containing elements separating finite subsets*, Bull. Austral. Math. Soc. **14**(1976), 7-10.

[3] L. Brailovsky, *Structure of quasi-invariant sets*, Arch. Math. (Basel) **59**(1992), 322-326.

[4] L. Brailovsky, *A characterisation of abelian groups*, Proc. Amer. Math. Soc. **117**(1993), 627-629.

[5] L. Brailovsky, *On the small squaring and commutativity*, Bull. London Math. Soc. **25**(1993), 330-336.

[6] L. Brailovsky, *Combinatorial conditions forcing commutativity of an infinite group*, J. Algebra, **165**(1994), 394 - 400.

[7] L. Brailovsky, D. V. Pasechnik and C. E. Praeger, *Classification of 2-quasi-invariant sets*, Ars Combin., to appear.

[8] L. Brailovsky, D. V. Pasechnik and C. E. Praeger, *Subsets close to invariant subsets under group actions*, Proc. Amer. Math. Soc., to appear.

[9] A. R. Calderbank, *Symmetric designs as the solution of an extremal problem in combinatorial set theory*, European J. Combin. **8**(1987), 171-173.

[10] J. R. Cho, P. S. Kim and C. E. Praeger, *The maximum number of orbits of a permutation group with bounded movement*, (in preparation).

[11] P. Dembowski, *Finite Geometries*, Springer-Verlag, New York, (1967).

[12] P. Frankl and Z. Füredi, *Finite projective spaces and intersecting hypergraphs*, Combinatorica **6**(1986), 335-354.

[13] G. A. Freiman, *On two and three-element subsets of groups*, Aeq. Math. **22**(1981), 140-152.

[14] Z. Füredi, *Maximum degree and fractional matchings in uniform hypergraphs*, Combinatorica **1**(1981), 155-162.

[15] Z. Füredi, *Intersecting designs from linear programming and graphs of diameter two*, Discrete Maths **127**(1994), 187-207.

[16] A. Gardiner and C. E. Praeger, *Transitive permutation groups with bounded movement*, J. Algebra **168**(1994), 798-803.

[17] A. Gardiner and C. E. Praeger, *Bounds on orbit sizes for permutation groups admitting subsets with restricted movement*, (preprint, 1995).

[18] B. Ganter, J. Pelikan and L. Tierlinck, *Some sprawling systems of equicardinal sets*, Ars Combin. **4**(1977), 133-142.

[19] A. S. Hedayatt, G. B. Khosrovshahi and D. Majumdar, *A prospect for a general method of constructing t-designs*, Discrete Appd. Math. **42**(1993), 31-50.

[20] M. Herzog, P. Longobardi and M. Maj, *On a combinatorial problem in group theory*, Israel J. Math. **82**(1993), 329-340.

[21] G. B. Khosrovshahi and C. E. Praeger, *On the intersection problem for combinatorial designs*, (in preparation).

[22] P. Longobardi and M. Maj, *The classification of groups with the small squaring property on 3-sets*, Bull. Austral. Math. Soc. **46**(1992), 263-270.

[23] A. Mann and C. E. Praeger, *Transitive permutation groups of minimal movement*, J. Algebra, to appear.

[24] B. H. Neumann, *Groups covered by permutable subsets*, J. London Math. Soc. **29**(1954), 236-248.

[25] P. M. Neumann, *The structure of finitary permutation groups*, Arch. Math. (Basel) **27**(1976), 3-17.

[26] P. M. Neumann, *A combinatorial problem in group theory*, (unpublished manuscript, 1989).

[27] P. M. Neumann and C. E. Praeger, *An inequality for tactical configurations*, Bull. London Math. Soc. (to appear).

[28] C. E. Praeger, *On permutation groups with bounded movement*, J. Algebra **144**(1991), 436-442.

[29] C. E. Praeger, *Restricted movement for intransitive group actions*, Proceedings of Groups Korea 1994, (to appear).

[30] C. E. Praeger, *Movement and separation of subsets of points under group actions*, (preprint, 1995).

[31] A. P. Street and D. J. Street, *Combinatorics of Experimental Design*, Clarendon Press, Oxford, 1987.

Permutation Groups Whose Subgroups Have Just Finitely Many Orbits

Dugald Macpherson

School of Mathematical Sciences
Department of Pure Mathematics
University of Leeds
Leeds LS2 9JT
England

1 Introduction

In this note we answer a question of Peter Neumann, based on some earlier more general questions of R. Zimmer concerning actions of arithmetic groups and Lie groups on manifolds. The main question of Zimmer is the following.

> Which groups can act on an infinite set so that every element of infinite order generates a cyclic subgroup which has only finitely many orbits on the set?

We prove the following theorem.

Theorem 1.1 *The free group F_2 on two generators has a faithful permutation representation on $\mathbb{N}$ in which every non-trivial subgroup has just finitely many orbits. Furthermore, for every non-identity $w \in F_2$, w is a power of some element which induces a single infinite cycle on a cofinite subset of $\mathbb{N}$.*

It is not hard to sharpen the construction of the above group slightly in order to ensure that the action of F_2 is highly transitive; that is, that it is k-transitive for all positive integers k. The details of this strengthening are omitted. I do not know if the permutation representation of F_2 constructed in the proof of Theorem 1.1 is already highly transitive. However, by the remark following Proposition 3.6 below, the group F_2 is oligomorphic; that is, it has finitely many orbits on $\mathbb{N}^k$ for each positive integer k.

The theorem is proved in Section 2. In Section 3 we discuss Zimmer's question in more generality. In particular we explain why our theorem answers a natural extension of Zimmer's question.

W. C. Holland (ed.), Ordered Groups and Infinite Permutation Groups, 221–229.

I am very grateful to Peter Neumann, both for encouraging me to publish this note, and for allowing me to include the material in Section 3. The latter is entirely due to him.

2 Proof of the theorem

In the proof below we shall use letters x, y for the generators of the free group F_2, letters $a, b, c, d, e, f, m, n, p, q$ for elements of the set $\mathbb{N}$ on which F_2 acts (and simultaneously as indices of certain words), and u, v, w for words in x, y, x^{-1}, y^{-1}. We use K, L, M, N, S, T as exponents of certain words. For $a \in \mathbb{N}$ and $w \in F_2$, we denote by a^w the image of a under w.

Proof of Theorem 1.1. Let F_2 be the free group on generators x, y. Let W be the set of reduced words (in x, y, x^{-1}, y^{-1}) in F_2. Define a relation $\sim$ on W by putting $u \sim v$ if there are $w_1, w_2 \in W$ and $M, N \in \mathbb{Z} \setminus \{0\}$ such that $w_1^{-1} u^M w_1 = w_2^{-1} v^N w_2$. Then $\sim$ is an equivalence relation. Let $\{w_i : i \in \mathbb{N}\}$ consist of exactly one element, of least possible length, from each $\sim$-class other than $\{1\}$. By the minimality assumption, no element w_i can be a proper power in F_2; and each word w_i is a reduced word, not conjugate to any shorter word.

To give the required permutation representation of F_2 on $\mathbb{N}$, we specify the actions of x and y on $\mathbb{N}$ point-by-point. We shall build the two permutations x and y, and then verify that the subgroup of Sym($\mathbb{N}$) generated by x and y is indeed the free group F_2 on x, y and that the permutation representation of F_2 has the required properties. The construction of the permutations x and y is in five kinds of steps. At any of these steps, each $w \in F_2$ will have been defined on a finite subset of $\mathbb{N}$. In particular, at any of these steps w, regarded as a finite partial permutation of $\mathbb{N}$, will have a finite number (possibly zero) of finite cycles on $\mathbb{N}$, together with some 'incomplete' or 'partial' cycles. The key condition which we must preserve in the construction is the following.

(*) If $m \geq 5n$, then at step m we do not create any new finite cycles of $w_0, \ldots, w_n$.

The permutations built immediately before step k in the construction of x and y will be denoted x_k and y_k respectively. In the definitions below, we sometimes formally treat x_k and y_k as sets of ordered pairs. Pick a bijection $F : \mathbb{N} \longrightarrow \mathbb{N}^3$. Put $x_0 = y_0 = \varnothing$.

Step $5n$. Let $k := 5n$. If $n \in \operatorname{dom} x_k$, put $x_{k+1} := x_k$. Otherwise, choose $m \in \mathbb{N} \setminus (\operatorname{dom} x_k \cup \operatorname{ran} x_k \cup \operatorname{dom} y_k \cup \operatorname{ran} y_k)$ and put $x_{k+1} := x_k \cup \{(n, m)\}$.

Step $5n+1$. Let $k := 5n+1$. If $n \in \operatorname{ran} x_k$, put $x_{k+1} := x_k$. Otherwise, pick $m \in \mathbb{N} \setminus (\operatorname{dom} x_k \cup \operatorname{ran} x_k \cup \operatorname{dom} y_k \cup \operatorname{ran} y_k)$ and put $x_{k+1} := x_k \cup \{(m, n)\}$.

Step $5n+2$. Let $k := 5n+2$. If $n \in \operatorname{dom} y_k$, put $y_{k+1} := y_k$. Otherwise, pick $m \in \mathbb{N} \setminus (\operatorname{dom} x_k \cup \operatorname{ran} x_k \operatorname{dom} y_k \cup \operatorname{ran} y_k)$ and put $y_{k+1} := y_k \cup \{(n, m)\}$.

Step $5n+3$. Let $k := 5n+3$. If $n \in \operatorname{ran} y_k$, put $y_{k+1} := y_k$. Otherwise, pick $m \in \mathbb{N} \setminus (\operatorname{dom} x_k \cup \operatorname{ran} x_k \cup \operatorname{dom} y_k \cup \operatorname{ran} y_k)$ and put $y_{k+1} := y_k \cup \{(m,n)\}$.

We must check that each of these four kinds of steps preserves property $(*)$. In fact, *no* element w_i acquires a new finite cycle after one of these steps. For example, consider the case $k := 5n$, where x_{k+1} is a proper extension of x_k with $n^{x_{k+1}} = m$. Suppose that for some $l \in \mathbb{N}$ and $w \in \{w_i : i \in \mathbb{N}\}$, l^w is defined after Step k, but not before it. Then, by the choice of m, either $l = m$ and w begins with x^{-1}, or $l^w = m$ and w ends with x. Since w is of least length in its $\sim$-class, at most one of these can occur. After Step k, in the first case $l^{w^{-1}}$ is not defined, and in the second case l^{w^2} is not defined. Either way, l does not lie in a finite cycle of w, as required.

Step 5n+4. Put $k := 5n+4$. Also, let $F(n) = (p,q,r)$. If one of p,q lies in a finite cycle of w_r, or if p,q are in the same partial cycle of w_r, do nothing, that is, put $x_{k+1} = x_k$ and $y_{k+1} = y_k$. Otherwise, we join the partial cycle of w_r containing p to that containing q, as follows.

We first introduce some notation. If $w \in W$, let $l(w)$ be the sum of the absolute values of the exponents of x and y in w, that is, the *length* of w. Let $l_i := l(w_i)$ for each $i \in \mathbb{N}$. If $a \in \mathbb{N}$ and $w \in W$ with $l(w) = l$, then a *w-path at a* is a sequence $a_1, \ldots, a_l$ such that for each initial segment u of w, if $l(u) = j \geq 1$ then $a^u = a_j$. We shall from now on write w for w_r and l for l_r. When we say that a^u *is defined* for some $a \in \mathbb{N}$ and $u \in W$, we mean that it it defined *before* Step k.

Let $S \in \mathbb{N}$ be maximal such that p^{w^S} is defined, and put $p^{w^S} = a$. Let u_1 be a maximal initial segment of w such that a^{u_1} is defined, and put $a^{u_1} = a'$. Let $T \in \mathbb{N}$ be maximal such that $q^{w^{-T}}$ is defined, and put $b = q^{w^{-T}}$. Also, let v_2 be a maximal final segment of w such that $b^{v_2^{-1}}$ is defined, and put $b' := b^{v_2^{-1}}$. Let u_2, v_1 be the unique words in x, y, x^{-1}, y^{-1} such that $w = u_1 u_2 = v_1 v_2$ (so both u_2 and v_1 are non-empty, and could possibly equal w). Let $m_i = l(u_i)$ and $n_i = l(v_i)$ for $i = 1,2$. Finally, put

$$\Gamma := \mathbb{N} \setminus (\{p,q\} \cup \operatorname{dom} x_k \cup \operatorname{ran} x_k \cup \operatorname{dom} y_k \cup \operatorname{ran} y_k).$$

Put $N := (\operatorname{Max}\{l_0, \ldots, l_n\} + 2)^2$. The elements x_{k+1}, y_{k+1} will be the least extensions of x_k, y_k so that after Step k we have a sequence

$$c_1, \ldots, c_{m_2}, d_{1,1}, \ldots, d_{1,l}, d_{2,1}, \ldots, d_{2,l}, \ldots, d_{N,1}, \ldots, d_{N,l}, e_1, \ldots, e_{n_1}$$

from $\mathbb{N}$ with the following properties.

(a) The above sequence is a $u_2 w^N v_1$-path at a'.

(b) The $c_i, d_{i,j}, e_i$ are all chosen to be distinct elements of Γ apart from e_{n_1}, which equals b'.

We shall now verify that condition $(*)$ is not violated at this step. So let $w' \in \{w_0, \ldots, w_n\}$, with $l(w') = l'$. We must check that no new finite w'-cycle has been created, so suppose for a contradiction that there is a new finite w'-cycle. The key point in the argument below is the following observation.

(**) There is $f \in \mathbb{N}$ lying in the new finite cycle, and some subword v' of w' such that $f^{v'} \in \Gamma$; and furthermore, we may choose f and v' so that $f^{v'} = c_1$, and so that, if $L' \in \mathbb{N}$ is large enough, then the sequence

$$c_2, \ldots, c_{m_2}, d_{1,1}, \ldots, d_{1,l}, \ldots, d_{N,1}, \ldots, d_{N,l}, e_1, \ldots, e_{n_1}$$

is a path at c_1 for an initial subword of $w'^{L'}$ or of $(w'^{-1})^{L'}$.

This follows from the definition of Γ and the fact that w' is of minimal length in its $\sim$-class. The observation enables us, by the choice of the $c_i, d_{i,j}$, and e_i, to form equations involving w and w'.

Case (i). $w' = w$.

In this case, by (**), there is i with $1 \le i \le l$ such that the new cycle includes each $d_{j,i}$, and either $(d_{j,i})^w = d_{j+1,i}$ for each j $(1 \le j \le N-1)$ or $(d_{j,i})^w = d_{j-1,i}$ for each j $(2 \le j \le N)$. Suppose first that $i < l$. The new finite cycle of w includes $d_{2,i}$. Let u be the initial segment of w of length i (so $u \ne 1, w$), and suppose that $w = uv$. Then, as noted above, $(d_{2,i})^w \in \{d_{1,i}, d_{3,i}\}$. If $(d_{2,i})^w = d_{3,i}$ then the words uw and wu are equal, and if $(d_{2,i})^w = d_{1,i}$ then the words uw^{-1} and wu are equal. The latter is clearly impossible; for in this case, if u starts with $z \in \{x, x^{-1}, y, y^{-1}\}$ then so does w, so w^{-1} ends with z^{-1}, so u ends in z^{-1}. Repeating the argument with the second entry in u, and continuing in this way we find that u is not reduced. In the former case it follows that w is equal (as an element of F_2) to a proper power of u, contradicting the minimality of w in its $\sim$-class.

It follows that $i = l$. Now $(d_{j,i})^w = d_{j+1,i}$ for each $j = 1, \ldots, N-1$, and hence the new cycle includes p and q. However, before Step $5n+4$ p and q were in different w-cycles. Let M be least such that $q^{w^M} = p$ after Step k. Then if $L \in \mathbb{N}$ is largest such that $p^{w^{-L}}$ is defined before Step k, it follows as in (**) that $p^{w^{-(L+1)}} \in \Gamma$ (after Step k). Hence, using (**) and the last paragraph, there is $K < M$ such that, after Step k, $p^{w^{-K}} = d_{j,l}$ for some $j \in \{1, \ldots, N\}$. It follows that w does not define a partial permutation, since $d_{j,l}$ occurs twice in the cycle containing p.

Case (ii). $w' \ne w$.

By (**), some element of Γ occurs in the new w'-cycle. In fact, if $f \in \mathbb{N}$ is on this cycle and K is very large, then all elements of $\Gamma \setminus \{f\}$ occur in the union of a w'^K-path at f and a w'^{-K}-path at f. By the pigeonhole principle, as $N > (l'+2)^2$ there are distinct i, j with $1 \le i, j \le N$ and s with $1 \le s \le N$ and $M \in \mathbb{N}$ such that

$$(d_{s,i})^{w'^M} = d_{s,j}.$$

Let u be a proper initial segment of w of length s. Then $u(w')^M = w^{j-i}u$. It follows that $w \sim w'$, which is a contradiction.

This completes the construction of the x_k and y_k, for each step k. Now put $x := \bigcup(x_k : k \in \mathbb{N})$ and $y := \bigcup(y_k : k \in \mathbb{N})$. By the steps of form $5n, 5n+1, 5n+2, 5n+3,$

x and y are permutations of $\mathbb{N}$. We now verify the assertions of the theorem. If $w \in W$ and w is not the empty word, then w or w^{-1} is conjugate to a proper power of some w_r, so we may suppose that $w := w_r$, say. Then the finite cycles of w are just those defined before Step $5r$, so w has finitely many finite cycles. In particular, w induces a non-identity permutation, so the group generated by x and y is free. Furthermore, if Δ is the union of the finite cycles of w, and $p, q \notin \Delta$, then there is $n \in \mathbb{N}$ with $F(n) = (p, q, r)$. At Step $5n+4$, we ensure that p and q are in the same cycle of w_r. Hence w_r induces a single infinite cycle on $\mathbb{N} \setminus \Delta$, as required.

3 Background to the main theorem

Theorem 1.1 was motivated by the following collection of results, all due to Peter Neumann in response to the questions of Zimmer.

We adopt the following terminology. Let G be a permutation group on a set Ω. We say that (G, Ω) has *property Z* if Ω is infinite and every element of G of infinite order has only finitely many cycles (including cycles of length 1, that is, fixed points) in Ω. We say that it has *property ZTF* if Ω is infinite and every non-identity element has only finitely many cycles. Evidently, any group with property ZTF is torsion-free and has property Z; conversely, any torsion-free group with property Z has property ZTF. Furthermore, a subgroup of a group with either of these properties has that property. We remark too that if G has property ZTF then it is faithful on each of its infinite orbits. Observe that the permutation representation of F_2 in Theorem 1.1 has property ZTF.

The first lemma provides examples of groups with property Z.

Lemma 3.1 *Suppose that G is infinite and cyclic-by-finite. Then any faithful permutation representation of G with only finitely many orbits has property Z.*

Proof. Any element of G of infinite order generates a subgroup of finite index, and therefore has only finitely many orbits in each G-orbit.

In particular, the regular action of any infinite cyclic-by-finite group has popерty Z. The next lemma gives a partial converse to this.

Lemma 3.2 *If (G, Ω) has property ZTF and G acts regularly on Ω then G is infinite cyclic.*

Proof. Let g be any non-identity element of G. Since the cyclic group $\langle g \rangle$ has only finitely many orbits in Ω it has finite index in G. Therefore its core (the intersection of the G-conjugates of $\langle g \rangle$) has finite index, and so G is cyclic-by-finite. It now suffices to quote the following claim, which for completeness we prove.

Claim. Any torsion-free cyclic-by-finite group is cyclic.

Proof of Claim. Let H be a cyclic normal subgroup of finite index in the torsion-free group G. If $C := C_G(H)$ then the centre $Z(C)$ of C has finite index n, say, in C. The transfer homomorphism $T: C \longrightarrow Z(C)$ is given by the map $g \mapsto g^n$ (see [6] Exercise 3.5.8(c) and the discussion preceding it). Since G is torsion-free, T has trivial kernel, so C embeds in $Z(C)$. Hence C is a torsion-free abelian cyclic-by-finite group, so is itself cyclic. Replacing H by C we may suppose that $C_G(H) = H$. Now conjugation by elements of G induces a faithful action of G/H on H by automorphisms. Thus either $H = G$ or G/H is cyclic of order 2. In the latter case, if $g \in G \setminus H$ and $h := g^2$ then on the one hand $g^{-1}hg = h$ because h is a power of g, and on the other hand $g^{-1}hg = h^{-1}$ because $h \in H$. This would entail $h^2 = 1$, hence $g^4 = 1$, and then $g = 1$ since G was supposed to be torsion-free. But that contradicts the choice of g as an element not in H. Thus in fact we must have $H = G$, that is, G is cyclic.

The next lemma imposes an important restriction on the groups which can have property Z.

Lemma 3.3 *Suppose that Ω is infinite and (G,Ω) has property Z. If $g \in G$ and g has infinite order then its centraliser $C_G(g)$ is cyclic-by-finite.*

Proof. If g has k infinite cycles and if Ω_0 is the set of points of Ω permuted by g in finite cycles then $C_{\mathrm{Sym}(\Omega)}(g) \cong Z \,\mathrm{Wr}\, \mathrm{Sym}(k) \times C_0$, where C_0 is a subgroup of $\mathrm{Sym}(\Omega_0)$, and Z is the infinite cyclic group. Since Ω_0 is finite, so is C_0. Let C_1 be the intersection of $C_G(g)$ with the base group of the wreath product, so that the non-trivial orbits of C_1 are the infinite cycles of g. If Ω_1 is one of these then C_1 must act faithfully on Ω_1. Therefore C_1 is cyclic, and so $C_G(g)$ is cyclic-by-finite.

Recall that a permutation group G on a countably infinite set Ω is said to be *oligomorphic* if, for every positive integer k, G has finitely many orbits on Ω^k; equivalently, if, for every finite $\Gamma \subset \Omega$, $G_{(\Gamma)} := \{g \in G : g\restriction_\Gamma = \mathrm{id}\}$ has finitely many orbits on Ω.

The next result suggests that a non-cyclic group with property ZTF should be oligomorphic. However, this has not been proved, and the obstruction involves infinite Frobenius groups. Recall that a permutation group is a *Frobenius group* if it is transitive and every 2-point stabiliser is trivial. We shall call a Frobenius group *implausible* if

(a) the stabilisers are infinite cyclic; and

(b) the stabilisers have only finitely many orbits.

We call it *highly implausible* if in addition

(c) it has property ZTF.

We now give a little evidence in the direction that highly implausible groups do not exist. It is due independently to Károlyi, Kóvacs and Pálfy [2] and to Mazurov [4].

Proposition 3.4 ([2, 4]) *There is no 2-transitive permutation group whose one-point stabilisers are infinite cyclic.*

The next lemma is permutation group-theoretic folklore.

Lemma 3.5 *Let (G,Ω) be an infinite transitive permutation group, and suppose that G has finitely many orbits on Ω^2. Let $\alpha \in \Omega$, let Φ be the union of the finite orbits of G_α, and suppose that $|\Phi| > 1$. Then Φ is a block of imprimitivity in the action of G on Ω.*

Proof. This is left as an exercise. I know of no good reference (it is essentially Exercise 3(ii) of Section 2.2 of [1], and also was mentioned in a lecture course by Peter Neumann at Oxford in Autumn 1985).

Proposition 3.6 *Suppose that highly implausible Frobenius groups do not exist. If (G,Ω) has property ZTF then G is either cyclic or oligomorphic.*

Proof. Suppose this false, and let (G,Ω) be a counter-example, so that G has property ZTF but is neither cyclic nor oligomorphic. Since G has property ZTF it has only finitely many orbits, and since it is not oligomorphic there is some finite sequence whose stabiliser has infinitely many, so is the identity. Therefore we can choose a sequence $\alpha_1, \ldots, \alpha_k, \alpha_{k+1}$ which is such that $G_{\alpha_1,\ldots,\alpha_k}$ has only finitely many orbits on Ω, but $G_{\alpha_1,\ldots,\alpha_k,\alpha_{k+1}}$ has infinitely many. We may choose the sequence so that, in addition, for each $i = 1, \ldots, k$, the element α_{i+1} is in an infinite orbit of $G_{\alpha_1,\ldots,\alpha_i}$ (and α_1 is in an infinite G-orbit). Now $G_{\alpha_1,\ldots,\alpha_k}$ has property ZTF on its orbit containing α_{k+1}, which is infinite. It acts faithfully and regularly on this orbit and therefore by Lemma 3.2 it is cyclic.

For $0 \leq i \leq k+1$ define $G^{(i)} := G_{\alpha_1,\ldots,\alpha_i}$, so that $G = G^{(0)} > G^{(1)} > \ldots > G^{(k)} > G^{(k+1)} = \{1\}$. Since G is assumed not to be cyclic, $k \neq 0$. Furthermore, $G^{(k)}$ has infinite index in $G^{(k-1)}$, so $G^{(k-1)}$ is not cyclic. In addition, $G^{(k-1)}$ has property ZTF and is not oligomorphic. Therefore we may assume, without loss of generality, that $k = 1$. Let Γ be the G-orbit of α_1. Observe that $G^{(1)}$, being abelian, acts regularly (and faithfully) on each of its infinite orbits on Γ. Let Φ be the union of the finite orbits of $G^{(1)}$ which lie in Γ. Since $G^{(1)}$ has only finitely many orbits, Φ is finite. Furthermore, by Lemma 3.5, Φ is a block of imprimitivity of G on Ω. Let ρ be the corresponding congruence (G-invariant equivalence relation) on Γ. Any element of G which acts trivially on the quotient Γ/ρ fixes all the ρ-classes, and since these are finite, has infinitely many cycles, hence is 1. Thus G acts faithfully on Γ/ρ. Pick $\beta_1 \in \Gamma \setminus \Phi$ (so $G_{\alpha_1,\beta_1} = 1$). Let $\alpha := \rho(\alpha_1)$, let $\beta := \rho(\beta_1)$ thought of as elements of Ω/ρ, and let $H := G_{\alpha,\beta}$. In the action on Γ the sets $\rho(\alpha_1)$ and $\rho(\beta_1)$

are H-invariant, and so H_{α_1,β_1} has finite index in H. But $H_{\alpha_1,\beta_1} \leq G_{\alpha_1,\beta_1} = \{1\}$. Therefore H is finite, and so, since G is torsion-free, $H = \{1\}$. This means that G acts as a Frobenius group with cyclic stabiliser on Γ/ρ. Since, as is easy to see, $(G, \Omega/\rho)$ also has property ZTF, it is a highly implausible Frobenius group. This contradicts our assumption that these do not exist.

Remark. Let $\mathcal{X}$ be a subgroup closed class of groups, and suppose that no member of $\mathcal{X}$ has a faithful permutation representation as a highly implausible Frobenius group. If $G \in \mathcal{X}$ and (G,Ω) has property ZTF then G is either cyclic or oligomorphic. For, in the proof of Proposition 3.6 we only needed that subgroups of G had the same properties as G. In particular, if G is a non-abelian free group acting faithfully as a permutation group with property ZTF then G is oligomorphic (so the action in Theorem 1.1 is oligomorphic). Theorem 1.1 shows that the oligomorphic case in Proposition 3.6 really does occur.

Proposition 3.7 *Suppose that (G,Ω) is transitive and has property ZTF. Suppose furthermore that for every natural number k there exists a non-identity element of G that has at least k fixed points and just one infinite cycle. Then G is highly transitive on Ω.*

Proof. First, observe that if (H,Σ) is a transitive permutation group, and H contains an element h with at least one fixed point and with an infinite cycle which contains all but finitely many points of Σ, then H is primitive on Ω. This is a slight variation on a result from [3]. To prove it, let α be a fixed point of h, let Γ be the set of points on the infinite cycle of h, and let ρ be an H-congruence on Σ. If $\rho(\alpha) \cap \Gamma = \varnothing$ then, as $\Sigma \setminus \Gamma$ is finite, $\rho(\alpha)$ is finite, so all the ρ-classes are finite. In this case the restriction of ρ to Γ must be trivial, and, since only finitely many ρ-classes are disjoint from Γ, it follows that ρ is trivial. If, on the other hand, $\rho(\alpha) \cap \Gamma \neq \varnothing$, then, applying powers of h, we see that $\Gamma \subset \rho(\alpha)$. In this case $\rho(\alpha)$ has finite complement in Σ and is infinite, so is the whole of Σ, that is, ρ is the universal relation. Hence H is primitive.

Let (G,Ω) be as in the statement of the proposition, and let $\alpha \in \Omega$. By transitivity and our other assumption, there is a non-identity element g of G_α that has at least one fixed point, finitely many finite cycles, and one infinite cycle. By what we have just proved, G is primitive. Furthermore, G_α has one infinite orbit and finitely many finite orbits. By Lemma 3.5, as (G,Ω) is primitive, G_α has no finite orbits on $\Omega \setminus \{\alpha\}$, so G_α is transitive on $\Omega \setminus \{\alpha\}$. In particular, G is 2-transitive. Now G_α certainly satisfies the same hypotheses as G. We can therefore assume as inductive hypothesis that G_α is k-transitive, and get that G is $(k+1)$-transitive. Thus by induction G is k-transitive for every finite k, that is, it is highly transitive.

We record three questions on property ZTF.

Question 3.8 *Is there a (highly) implausible Frobenius group?*

Question 3.9 *Is every primitive group which has the property ZTF and is also oligomorphic necessarily highly transitive?*

Question 3.10 *What classes of groups other than free groups have oligomorphic actions with property ZTF?*

Finally, I remark that there is now quite a wide literature on possible cycle types of a particular permutation (rather than all permutations) in infinite permutation groups. For example, there are several naturally occurring permutation groups in which the realisable cycle types are classified, and in particular some of these have permutations acting as a single infinite cycle (see [5, 7, 8, 9] for more on this). Also, in [3] there are results restricting the possible cycle types in a primitive permutation group which is not highly transitive.

References

[1] P. J. Cameron, *Oligomorphic Permutation Groups*, London Math. Soc. Lecture Notes Series 152, Cambridge University Press, Cambridge, 1990.

[2] Gy. Károlyi, S. J. Kovács, P. P. Pálfy, *Doubly transitive permutation groups with abelian stabiliser*, Aequationes Math. **39**(1990), 161–166.

[3] H. D. Macpherson and C. E. Praeger, *Cycle types in infinite permutation groups*, J. Algebra, to appear.

[4] V. D. Mazurov, *Doubly transitive permutation groups*, Sibirsk. Mat. Zh. **31**(1990), 102–104 (in Russian).

[5] P. M. Neumann, *Automorphisms of the rational world*, J. London Math. Soc. **32**(1985), 439–448.

[6] W.R. Scott, *Group Theory*, Prentice-Hall, Englewood Cliffs, 1964.

[7] J. K. Truss, *The group of the countable universal graph*, Math. Proc. Cambridge Philos. Soc. **98**(1985), 213–245.

[8] J. K. Truss, *The group of almost automorphisms of the countable universal graph*, Math. Proc. Cambridge Philos. Soc. **105**(1989), 223–236.

[9] J. K. Truss, *Generic automorphisms of homogeneous structures*, Proc. London Math. Soc., to appear.

Automorphisms of Quotients of Symmetric Groups

J. L. Alperin

Department of Mathematics
University of Chicago
5734 University Avenue
Chicago Illinois 60637
U.S.A. *

Jacinta Covington

Department of Mathematics
Royal Melbourne Institute of Technology
GPO Box 2476V
Melbourne, Victoria 3001
Australia †

Dugald Macpherson

Department of Pure Mathematics
University of Leeds
Leeds LS2 9JT
England

Abstract

We describe the automorphism group of the quotient of the symmetric group on a countably infinite set by the group of finitary permutations. We show that the outer automorphism group is infinite cyclic. In the proof, we examine the structure of centralizers in this quotient.

1 Introduction

In this paper, S will denote $\mathrm{Sym}(\Omega)$, the full symmetric group on a countably infinite set Ω. For any $g \in S$, define the *support* of g to be $\mathrm{supp}(g) = \{\alpha \in \Omega : \alpha g \neq \alpha\}$ and the *fixed point set* to be $\mathrm{fix}\, g = \{\alpha \in \Omega : \alpha g = \alpha\}$. It is well-known (see Scott [6] or originally Baer [1]) that S has precisely two proper non-trivial normal subgroups. These are $\mathrm{FS}(\Omega) := \{g \in S : |\mathrm{supp}(g)| < \aleph_0\}$ and $\mathrm{Alt}(\Omega)$, the subgroup of index two in $\mathrm{FS}(\Omega)$ consisting of even permutations. Throughout this paper, we let $G := S/\mathrm{FS}(\Omega)$. This group is simple. The main goal of this paper is to describe $\mathrm{Aut}(G)$.

*This research was partially supported by a grant from the National Science Foundation.
†This research was partially supported by ARC grant A68830987.

W. C. Holland (ed.), Ordered Groups and Infinite Permutation Groups, 231–247.

Theorem 1.1 *Let* $G = \mathrm{Sym}(\Omega)/\mathrm{FS}(\Omega)$, *where* Ω *is a countably infinite set. Then the outer automorphism group* $\mathrm{Out}(G)$ *is infinite cyclic.*

To see why G admits an infinite cyclic group of outer automorphisms, observe that if Ω is identified with $\mathbf{N}$, then the shift $n \mapsto n+1$, acting on G 'by conjugation up to finitely many mistakes', induces an outer automorphism of infinite order.

We give in Section 2 a more formal description of the infinite cyclic group of outer automorphisms. In Section 3 we describe in Proposition 3.2 and Theorem 3.3 the structure of centralizers in G, and thereby show that certain conjugacy classes of G are invariant under $\mathrm{Aut}(G)$. We believe that these results are of independent interest, as the centralizers do not have as simple a structure as might first be imagined. In Section 4, we show that there are no outer automorphisms other than those described in Section 2, and thereby prove Theorem 1.1.

We frequently refer to elements of S by their cycle type. The set of *cycle lengths* is the set of cardinals $\{n: 1 \leq n \leq \omega\}$, and a *nontrivial cycle* has length $n > 1$. If T is a set of cycle lengths, we say that g has cycle type $\prod(n^{a_n}: n \in T)$ if g has exactly a_n cycles of length n for each $n \in T$, $(0 \leq a_n \leq \omega)$, and no others. Observe that if $g, h \in S$ have cycle types $1^\omega.\prod(n^{a_n}: 1 < n \leq \omega)$ and $1^\omega.\prod(n^{b_n}: 1 < n \leq \omega)$ respectively, then $\mathrm{FS}(\Omega)g, \mathrm{FS}(\Omega)h$ are conjugate in G if and only if their cycle types differ finitely: that is, if

(a) $a_n = b_n$ for all but finitely many finite cycle lengths n;

(b) $a_\omega = b_\omega$;

(c) $a_n = \omega$ if and only if $b_n = \omega$, for $1 \leq n < \omega$.

It is important that here g and h have infinitely many fixed points. As a small warning, we remark that G has three conjugacy classes of involutions, corresponding to elements in S of the cycle types $1^\omega \cdot 2^\omega$ and 2^ω and $1^1 \cdot 2^\omega$.

If $g \in S$ then $\bar{g}$ will be the corresponding element $\mathrm{FS}(\Omega)g$ of G, and similarly, given $\bar{g} \in G$, we use g to denote any preimage of it under the natural surjection from S to G. We shall use $\triangle$ to denote symmetric difference, and write $U \equiv V$ if $U \triangle V$ is finite. Also, we extend the notion of support to elements $\bar{g} \in G$. If $g \in S$ has support $\Gamma \subseteq \Omega$, then $\bar{g}$ has support

$$\mathrm{supp}(\bar{g}) = \{\Delta \subseteq \Omega: |\Gamma \triangle \Delta| < \omega\}.$$

Thus $\mathrm{supp}(\bar{g})$ lies in the Boolean Algebra $\mathcal{B} = \mathrm{Pow}(\Omega)/\mathcal{I}$, where $\mathcal{I}$ is the ideal of finite subsets of Ω. Observe the distinction between $\mathrm{supp}(g)$, which is a subset of Ω, and $\mathrm{supp}(\bar{g})$, which is an element of $\mathcal{B}$. All wreath products whose top group is a permutation group of infinite degree will be unrestricted, and similarly all direct products will be unrestricted (that is, will be the full Cartesian product). Finally, recall that a moiety Γ of a set Ω is a subset satisfying $|\Gamma| = |\Omega \setminus \Gamma|$; for a countably infinite set Ω this is an infinite coinfinite subset.

The paper by J.K. Truss in the present volume sketches an alternative proof of Theorem 1.1, due to Truss and Shelah, in which the Boolean algebra $\mathcal{B}$ is shown to be first-order interpretable in G. Their approach seems to be more powerful, since they obtain some generalisations of Theorem 1.1 for symmetric groups of uncountable cardinality. The strongest results concerning the interpretability of $\mathcal{B}$ in G were obtained by M. Rubin in Theorem 4.2 of [5]. There, he showed the following. Let $\mu < \kappa \leq \lambda^+$ be infinite cardinals and $G_{\lambda\kappa\mu}$ denote the group of permutations (of support of size less than κ) of λ, modulo the normal subgroup of permutations of support of size less than μ, and $B_{\lambda\kappa\mu}$ be the corresponding Boolean algebra. Then $B_{\lambda\kappa\mu}$, its boolean algebra structure, and the action of $G_{\lambda\kappa\mu}$ on it, is first-order interpretable in the abstract group $G_{\lambda\kappa\mu}$, and furthermore this interpretation is uniform, that is, the formulas in the interpretation are independent of the choice of λ, κ, μ. It seems, though, that the step from such results to a description of $\mathrm{Aut}(()G_{\lambda\kappa\mu}$ is never immediate. Our approach yields en route the group-theoretic information about centralizers.

Acknowledgement. We would like to thank both Peter Neumann and John Truss for a number of very helpful remarks.

2 The group of near symmetries

In this section we describe an infinite cyclic group of outer automorphisms of $G = \mathrm{Sym}(\Omega)/\mathrm{FS}(\Omega)$, where Ω is a countably infinite set. In the rest of the paper, we will show that this group is the whole outer automorphism group.

Define the set of *near bijections* of Ω to be

$$\mathrm{NB}(\Omega) = \{f \colon f \text{ is a bijection } \Omega_1 \to \Omega_2, \text{ where } \Omega_1, \Omega_2 \text{ are cofinite subsets of } \Omega\}.$$

Clearly, $S \subseteq \mathrm{NB}(\Omega)$. Extending our notation for sets, we define an equivalence relation $\equiv$ on $\mathrm{NB}(\Omega)$ by the rule

$$f_1 \equiv f_2 \text{ if there is a cofinite } \Omega' \subseteq \Omega \text{ such that } f_1 |_{\Omega'} = f_2 |_{\Omega'} .$$

It is easily seen that $\equiv$ is indeed an equivalence relation. We define $\mathrm{NS}(\Omega)$, the set of *near symmetries* of Ω, to be the set of equivalence classes of this relation. This set naturally forms a group induced from composition of maps.

The natural projection from S to $\mathrm{NS}(\Omega)$ is a group homomorphism with kernel $\mathrm{FS}(\Omega)$. Hence we have an exact sequence

$$\mathrm{FS}(\Omega) \longrightarrow S \longrightarrow \mathrm{NS}(\Omega).$$

Thus $G \leq \mathrm{NS}(\Omega)$. The elements of G are the equivalence classes of permutations.

For $f \in \mathrm{NB}(\Omega)$ with Ω_1, Ω_2 as above, we define the *index* of f to be

$$\mathrm{ind}\, f = |\Omega - \Omega_2| - |\Omega - \Omega_1|.$$

Lemma 2.1 *The map* $\text{ind}: \text{NB}(\Omega) \to \mathbf{Z}$ *induces a mapping* $\text{ind}: \text{NS}(\Omega) \to \mathbf{Z}$ *which is a homomorphism onto* $(\mathbf{Z}, +)$ *with kernel* G.

Proof. First note that if Ω' is a cofinite subset of Ω then for any $f \in \text{NB}(\Omega)$ we have $\text{ind} f = \text{ind}(f|_{\Omega'})$. Take $f, g \in \text{NB}(\Omega)$ with $f \equiv g$. We want to show that $\text{ind} f = \text{ind} g$. There is a cofinite subset Ω' of Ω such that $f|_{\Omega'} = g|_{\Omega'}$. But then clearly

$$\text{ind} f = \text{ind}(f|_{\Omega'}) = \text{ind}(g|_{\Omega'}) = \text{ind} g.$$

Thus ind is a well-defined function on $\text{NS}(\Omega)$.

We now show that $\text{ind}(fg) = \text{ind} f + \text{ind} g$. Restricting the domains of f and g if necessary, we may assume that $\text{dom}\, g$ is equal to $\text{ran}\, f$, the range of f, that is, we have $f: \Omega_1 \to \Omega_2$ and $g: \Omega_2 \to \Omega_3$. Then

$$\begin{aligned} \text{ind}(fg) &= |\Omega - \Omega_3| - |\Omega - \Omega_1| \\ &= |\Omega - \Omega_3| - |\Omega - \Omega_2| + |\Omega - \Omega_2| - |\Omega - \Omega_1| \\ &= \text{ind} g + \text{ind} f. \end{aligned}$$

Now the kernel of this homomorphism consists of all bijections $f: \Omega_1 \to \Omega_2$ such that $|\Omega - \Omega_1| = |\Omega - \Omega_2|$. Any such map is $\equiv$-equivalent to a permutation. Thus $\ker(\text{ind}) = G$. □

From this lemma it follows that G is a normal subgroup of $\text{NS}(\Omega)$ and we have an exact sequence

$$1 \longrightarrow \text{FS}(\Omega) \longrightarrow S \longrightarrow \text{NS}(\Omega) \longrightarrow \mathbf{Z} \longrightarrow 1.$$

Lemma 2.2 $C_{\text{NS}(\Omega)}(G)$ *is trivial.*

Proof. Let h be a near-bijection representing an element of $C_{\text{NS}(\Omega)}(G)$. Then for every permutation $g \in S$, we have $h^{-1}gh \equiv g$. Now, for any $g \in S$, the sets $\text{supp}(h^{-1}gh)$ and $(\text{supp}(g))h$ have finite symmetric difference, where $\text{supp}(h^{-1}gh)$ is the the set of $\alpha \in \Omega$ such that $\alpha h^{-1}gh$ is defined and not equal to α. It follows that $\text{supp}(g) \equiv \text{supp}(g)h$ for any $g \in S$. Since any set $\Phi \subset \Omega$ is the support of some permutation, we have $\Phi h \equiv \Phi$ for all $\Phi \subseteq \Omega$. But this is possible only if h has finite support, that is, $h \equiv \text{id}$. □

Proposition 2.3 $\text{NS}(\Omega)$ *induces an infinite cyclic group of outer automorphisms of* G.

Proof. Since $G \trianglelefteq \text{NS}(\Omega)$, we see that $\text{NS}(\Omega)$ induces a group of automorphisms of G by conjugation. Since $C_{\text{NS}(\Omega)}(G)$ is trivial, the outer automorphism group induced by $\text{NS}(\Omega)$ is $\text{NS}(\Omega)/G$, which is isomorphic to $\mathbf{Z}$. □

Remark. Let $\Omega = \{\alpha_i : i < \omega\}$ be any enumeration of Ω. A typical generator of $\text{NS}(\Omega)/G$ is the equivalence class of the near-bijection $\sigma: \Omega \to \Omega$ defined by $\alpha_i \mapsto \alpha_{i+1}$ (for $i < \omega$).

3 Structure of centralizers in G

In this section we give partial information on the structure of centralizers in G. From this, it will follow that $\mathrm{Aut}(G)$ induces an automorphism of $\mathcal{B}$, and hence that $\mathrm{Aut}(G)$ fixes certain conjugacy classes of G.

For the first four lemmas we denote by $\bar{g}$ a fixed element of G, and choose a fixed $g \in S$ such that $\bar{g} = g\mathrm{FS}(\Omega)$. We suppose that g has cycle type $\prod(i^{n_i} : i \in T)$, where $T \subseteq \mathbf{N} \cup \{\omega\}$. We define an equivalence relation $\mathcal{F}$ on Ω whose equivalence classes are listed below:

- The union of all the infinite cycles of g is an $\mathcal{F}$-class F_ω.
- If $1 \leq i < \omega$ and g has infinitely many i-cycles, then the union of the i-cycles of g is an $\mathcal{F}$-class F_i.
- The remaining points of Ω form a single $\mathcal{F}$-class F_0, called the *dregs*.

We define the *strong centralizer* of $\bar{g}$ in G, to be

$$\mathrm{SC}_G(\bar{g}) := \{\bar{h} \in \mathrm{C}_G(\bar{g}) : h \text{ fixes each } \mathcal{F}\text{-class for some representative } h \text{ of } \bar{h}\}.$$

Note that, even though $\mathcal{F}$ depends on g, this definition does not depend on the choice of g. For, let g' be another preimage of $\bar{g}$, and let $\mathcal{F}'$ be the corresponding equivalence relation. Take $h \in S$ such that $\bar{h} \in \mathrm{SC}_G(\bar{g})$. Now, $\mathcal{F}'$ and $\mathcal{F}$ differ only on a finite set Δ of points. There is a finitary permutation k fixing each $\mathcal{F}$ class such that $hk \in S_{(\Delta)}$. Then hk fixes each $\mathcal{F}'$-class.

Lemma 3.1 *Let $g, \bar{g}$ be as above. Then the group $\mathrm{C}_G(\bar{g})/\mathrm{SC}_G(\bar{g})$ is free abelian (where the trivial group is regarded as free abelian).*

Proof. Let $\{F_i : i \in I\}$ be the set of $\mathcal{F}$-classes, and let $K := \bigoplus(\mathbf{Z}f_i : i \in I)$ be free abelian with $\{f_i : i \in I\}$ as basis. We define a homomorphism $\chi : \mathrm{C}_G(\bar{g}) \to K$ as follows: if $\bar{k} \in \mathrm{C}_G(\bar{g})$ then for each $i \in I$, we define

$$\chi_i(\bar{k}) = |F_i k \setminus F_i| - |F_i \setminus F_i k|.$$

Then $\chi_i(\bar{k})$ is well-defined, that is, it does not depend on the choice of the preimages g and k. Also put

$$\chi(\bar{k}) := \sum_{i \in I} \chi_i(\bar{k}) f_i.$$

Now g and $k^{-1}gk$ agree on all but finitely many points, so k fixes all but finitely many $\mathcal{F}$-classes and $\chi_i(\bar{k}) = 0$ for all but finitely many $i \in I$, so $\chi(\bar{k})$ does indeed lie in the direct sum K. It can be checked that χ is a homomorphism, its image is a subgroup of a free abelian group so is free abelian, and its kernel is $\mathrm{SC}_G(\bar{g})$. □

Problem 3.1 *Find an expression for the rank of* $C_G(\bar{g})/SC_G(\bar{g})$, *in terms of the cycle type of* g.

We will need a considerable amount of information about all centralizers in G. To find $C_G(\bar{g})$, we must describe all elements $h \in S$ such that $[g, h] \in FS(\Omega)$. We list five types of such elements, and then show that the images in G of elements of such types generate $C_G(\bar{g})$.

1. **Obvious Centralizer Type** These are elements of $C_S(g)$. Conjugation by these elements fixes g. Note that $C_S(g) \cong \prod_{i \in T}(C_i \operatorname{Wr} S_{n_i})$, where C_i denotes the cyclic group of order i.

2. **Finite/Finite Type** These elements convert one finite set of finite cycles of g into another. Let A, B be the unions of supports of finitely many finite cycles of g, where

   - $|A| = |B| > 0$;
   - No cycle in B has the same length as one in A;
   - A or B contains the support of a g-cycle of length i only if $n_i = \omega$, that is, g has infinitely many cycles of length i.

   Let $\chi: A \to B$ be a bijection and define a permutation h as follows:

   - h agrees with χ on A;
   - If A contains an i-cycle, then h maps $F_i \setminus A$ to F_i, respecting the cycle structure;
   - If B contains an i-cycle, then h maps F_i to $F_i \setminus B$, respecting the cycle structure;
   - h fixes all elements of all other $\mathcal{F}$-classes.

   Then g^h agrees with g except on the finite set B, so $[g, h] \in FS(\Omega)$.

3. **Finite/Infinite Transfers** Conjugation by these elements moves a finite cycle of g into an infinite cycle. Let $(\ldots a_{-2}a_{-1}a_0a_1a_2\ldots)$ be an infinite cycle of g. Let $(b_1b_2\ldots b_k)$ be a finite cycle of g from the infinite $\mathcal{F}$-class F_k. Define a permutation h as follows:

   - h fixes a_i for $i \leq 0$;
   - $a_ih = b_i$ for $1 \leq i \leq k$;
   - $a_ih = a_{i-k}$ for $i > k$;
   - h maps F_k to $F_k \setminus \{b_1, b_2, \ldots, b_k\}$, respecting the cycle structure;
   - h fixes all elements of all other cycles.

Then g^h agrees with g except on $\{a_0, b_k\}$, and so $[g, h] \in \mathrm{FS}(\Omega)$.

We refer to the inverses of such elements as infinite/finite transfers.

4. **Infinite/Infinite Transfer** These are elements which move a single point from one infinite cycle of g into another. Let

$$(\ldots a_{-2}a_{-1}a_0a_1a_2\ldots) \text{ and } (\ldots b_{-2}b_{-1}b_0b_1b_2\ldots)$$

be two infinite cycles of g, and let h be the permutation of Ω whose only non-trivial cycle is

$$(\ldots b_{-2}b_{-1}b_0a_0a_1a_2\ldots).$$

Then the conjugate g^h differs from g only in that a_0 lies in the b-cycle of g^h. Hence g^h agrees with g except on the set $\{a_{-1}, a_0, b_0\}$, so $[g, h] \in \mathrm{FS}(\Omega)$.

5. **Infinite/Infinite Recombination** These are elements which cut two infinite cycles of g in half, then reattach the left side of one to the right side of the other. Let g have two infinite cycles as above, and let

$$h := (a_1b_1)(a_2b_2)\cdots$$

Then g^h agrees with g except on $\{a_0, b_0\}$, so $[g, h] \in \mathrm{FS}(\Omega)$.

Remark. Elements of types 1,4 and 5 lie in $\mathrm{SC}_G(\bar{g})$, but elements of types 2 and 3 do not.

Proposition 3.2 *(a) The group $\mathrm{SC}_G(\bar{g})$ is generated by elements of G represented by elements of S of types 1,4 and 5.*

(b) The group $\mathrm{C}_G(\bar{g})$ is generated by elements of G represented by the above types.

Proof. (a) Let $k \in S$ be such that $[k, g] \in \mathrm{FS}(\Omega)$, and assume that k fixes setwise each $\mathcal{F}$-class. We say that two infinite g-cycles

$$(\ldots a_{-2}a_{-1}a_0a_1a_2\ldots)$$

and

$$(\ldots b_{-2}b_{-1}b_0b_1b_2\ldots)$$

are *initially equal* if there are integers M and m such that $a_i = b_{i-m}$ for all $i < M$. *Finally equal* is defined similarly.

For all but finitely many of the infinite g-cycles, its conjugate under k is also a g-cycle. Let $(\ldots a_{-2}a_{-1}a_0a_1a_2\ldots)$ be any infinite g-cycle. Then

$$(\ldots a_{-2}k, a_{-1}k, a_0k, a_1k, a_2k, \ldots)$$

is initially equal to some g-cycle, and finally equal to some, perhaps different, g-cycle. Then we can use infinite/infinite recombinations to reduce to the case where

$(a_i k)$ is initially and finally equal to a particular infinite g-cycle. We then use a series of infinite/infinite transfers, together with finitary permutations to reduce to the case where k preserves the structure of all the infinite cycles. Using further finitary permutations, we may suppose that k preserves all finite g-cycles, and it follows that $\bar{k} \in \mathrm{C}_G(\bar{g})$.

(b) Let $k \in S$ with $[k,g] \in \mathrm{FS}(\Omega)$. Observe that $Fk \triangle F$ is finite for every $\mathcal{F}$-class F, and that all but finitely many of the $\mathcal{F}$-classes are fixed setwise by k. By adjusting k by a finitary permutation, we can reduce to the case where, for each $\mathcal{F}$-class F, either $F \subseteq Fk$ or $Fk \subseteq F$. Note that since the dregs class F_0 contains only finite cycles and has just finitely many cycles of any length, we must have $F_0 k = F_0$.

Now consider the classes F_i where $1 \leq i < \omega$. We wish to adjust k to ensure that $F_i k = F_i$ for each such i, for then also $F_\omega k = F_\omega$. So fix i with $1 \leq i < \omega$, and suppose that $F_i \subset F_i k$. (The case when $F_i \supset F_i k$ is similar.) If g has an infinite cycle then we can use a series of infinite/finite transfers to move cycles of $F_i k \setminus F_i$ into F_ω, then a combination of finitary permutations on F_ω and finite/infinite transfers to move appropriate elements of F_ω back into F_i. If g has no infinite cycle, we multiply k by a single element of finite/finite type to reduce to the case where k fixes all $\mathcal{F}$-classes setwise. We can then use a finitary permutation to ensure that k respects the cycle structure of finite cycles. After these reductions we have $k \in \mathrm{SC}_G(\bar{g})$, so can apply part (a).

□

We now give a partial structure theorem for centralizers in G. We first define some normal subgroups of $\mathrm{C}_G(\bar{g})$. Let $g \in S$ be as above. Let

$$\mathrm{OC}(\bar{g}) := \mathrm{C}_S(g) \cdot \mathrm{FS}(\Omega)/\mathrm{FS}(\Omega) = \{\bar{h} : h \in S, [g,h] = 1\}$$

the group of images of elements of obvious centralizer type. Also, define

$$N_1(\bar{g}) := \{\bar{h} \in \mathrm{OC}(\bar{g}) : h \text{ fixes setwise the support of each infinite } g\text{-cycle}\},$$

$$N_2(\bar{g}) := \{\bar{h} \in \mathrm{SC}_G(\bar{g}) : |Dh \triangle D| < \aleph_0, \text{ if } D \text{ is the support of an infinite } g\text{-cycle}\},$$

and

$$N_3(\bar{g}) := \langle N_2(\bar{g}), \mathrm{OC}(\bar{g}) \rangle.$$

These definitions are independent of the choice of the representative g.

Theorem 3.3 *(a) The groups*

$$N_1(\bar{g}), \mathrm{OC}(\bar{g}), N_2(\bar{g}), N_3(\bar{g}) \quad \textit{and} \ \mathrm{SC}_G(\bar{g})$$

are each normal subgroups of $\mathrm{C}_G(\bar{g})$, *with*

$$N_3(\bar{g}) \leq \mathrm{SC}_G(\bar{g}) \leq \mathrm{C}_G(\bar{g}).$$

(b) $N_2(\bar{g}) \cap \mathrm{OC}(\bar{g}) = N_1(\bar{g})$. *Hence we have the isomorphisms* $N_3(\bar{g})/\mathrm{OC}(\bar{g}) \cong N_2(\bar{g})/N_1(\bar{g})$ *and* $N_3(\bar{g})/N_2(\bar{g}) \cong \mathrm{OC}(\bar{g})/N_1(\bar{g})$.

(c) (i) *The groups* $\mathrm{C}_G(\bar{g})/\mathrm{SC}_G(\bar{g})$ *and* $N_2(\bar{g})/N_1(\bar{g})$ *are free abelian, and the latter has rank* n *if* g *has* $n+1$ *infinite cycles (*n *finite) and* $\aleph_0$ *if* g *has infinitely many infinite cycles.*

(ii) *The quotient* $\mathrm{OC}(\bar{g})/N_1(\bar{g})$ *is isomorphic to the symmetric group on the set of infinite cycles of* $\bar{g}$.

(iii) *The quotient* $\mathrm{SC}_G(\bar{g})/N_3(\bar{g})$ *is isomorphic to the finitary symmetric group on the set of infinite* g*-cycles.*

(d) *If* g *has more than one infinite cycle then the five normal subgroups listed in (a) are distinct, and if* g *has at most one infinite cycle then they are all equal.*

Proof. All parts of this are easy, so we omit details. To show that $N_2(\bar{g})/N_1(\bar{g})$ is free abelian, we argue much as in the proof of Lemma 3.1. This quotient is generated by infinite/infinite transfers, modulo $N_1(\bar{g})$. By Proposition 3.2(a), the finitary symmetric group $\mathrm{SC}_G(\bar{g})/N_3(\bar{g})$ is generated by infinite/infinite recombinations (modulo $N_3(\bar{g})$). □

We now define, for any $h \in S$, the group

$$G(\bar{h}) := \{\bar{k} : \mathrm{supp}(k) \subseteq \mathrm{fix}(h)\}.$$

Observe that this definition is independent of the choice of the element $h \in S$ such that $\bar{h} = h \cdot \mathrm{FS}(\Omega)$. Also, clearly $G(\bar{g}) \trianglelefteq \mathrm{C}_G(\bar{g})$.

Lemma 3.4 (a) *Suppose that* $g \in S$ *has infinitely many fixed points. Then* $G(\bar{g})$ *is isomorphic to* G, *and is the only normal subgroup of* $\mathrm{C}_G(\bar{g})$ *which is isomorphic to* G.

(b) *Suppose that* $g \in S$ *has just finitely many fixed points. Then there is no normal subgroup of* $\mathrm{C}_G(\bar{g})$ *isomorphic to* G.

(c) *Let* $g, h \in S$ *and* $\varphi \in \mathrm{Aut}(G)$, *and suppose* $\bar{g}\varphi = \bar{h}$. *Then* g *has infinitely fixed points if and only if* h *has infinitely many fixed points, and if they do, then* $G(\bar{g})\varphi = G(\bar{h})$.

Proof. (a) Let $\Gamma := \mathrm{fix}(g)$. Clearly

$$G(\bar{g}) = \mathrm{Sym}(\Gamma) \cdot \mathrm{FS}(\Omega)/\mathrm{FS}(\Omega) \cong \mathrm{Sym}(\Gamma)/\mathrm{FS}(\Gamma) \cong G.$$

Now let $\bar{N}$ be any normal subgroup of $\mathrm{C}_G(\bar{g})$ isomorphic to G. We shall show that $\bar{N} = G(\bar{g})$. First observe that $G(\bar{g})$ is simple, so for any normal subgroup M of

$C_G(\bar{g})$, either $G(\bar{g}) \leq M$ or $G(\bar{g}) \cap M = 1$. In particular, by Lemma 3.1, as $\bar{N}$ is non-abelian it must lie in $SC_G(\bar{g})$. Also, by Theorem 3.3, as G is not embeddable in a free abelian group or a finite or finitary symmetric group, we must have $\bar{N} \leq OC(\bar{g})$.

Define $N := \{h \in S : \bar{h} \in \bar{N}\}$. Observe that

$$C_S(g) = \prod (C_i \operatorname{Wr} S_{n_i} : i \in T),$$

and that $1 \in T$ and n_1 is infinite, and that $OC(\bar{g}) = C_S(g) \cdot FS(\Omega)/FS(\Omega)$. Put

$$K := C_1 \operatorname{Wr} S_{n_1} \times \prod (C_i^{n_i} : i \in T, i > 1),$$

and

$$L := \{k \in S : \operatorname{supp}(k) \subseteq \operatorname{fix}(g)\} \cong S_{n_1} = S.$$

Then $L \leq K$. Put $\bar{K} := K \cdot FS(\Omega)/FS(\Omega)$ and $\bar{L} := L \cdot FS(\Omega)/FS(\Omega)$. Then $\bar{L}$ and $\bar{K}$ are normal subgroups of $OC(\bar{g})$ and $\bar{K}/\bar{L}$ is abelian. The proof will be complete if we can show that $\bar{N} \leq \bar{L}$ (since $\bar{L}$ is simple), so we shall suppose that this is false. It follows that $\bar{N} \cap \bar{K} = 1$; for otherwise $\bar{N} \leq \bar{K}$ which is impossible as $\bar{K}/\bar{L}$ is abelian and $\bar{N}$ is non-abelian and simple.

Now pick $h \in N \setminus L \cdot FS(\Omega)$. Then $h \notin K \cdot FS(\Omega)$, so, up to a finitary permutation, h induces a permutation of infinite support on the set of non-trivial cycles of g. It is now easy to choose $k \in K$ so that $[h, k] \notin L \cdot FS(\Omega)$. Now $[h, k] \in N \cap (K \cdot FS(\Omega) \setminus L \cdot FS(\Omega))$, which contradicts the last paragraph.

Part (b) follows from the proof of uniqueness in (a), and (c) follows from (a) and (b).

□

We will adopt the notation $G(\bar{g})$, $G(\bar{h})$ later in the paper. Also, from now on φ will denote an element of $\operatorname{Aut}(G)$.

Lemma 3.5 *Let $\varphi \in \operatorname{Aut}(G)$, let $\bar{g}, \bar{h}, \bar{k} \in G$, and assume that g, h, k each have infinitely many fixed points. Then*

(a) $\operatorname{supp}(\bar{g}) = \operatorname{supp}(\bar{h})$ *if and only if* $\operatorname{supp}(\bar{g}\varphi) = \operatorname{supp}(\bar{h}\varphi)$,

(b) $\operatorname{supp}(\bar{g}) < \operatorname{supp}(\bar{h})$ *if and only if* $\operatorname{supp}(\bar{g}\varphi) < \operatorname{supp}(\bar{h}\varphi)$,

(c) $\operatorname{supp}(\bar{k}) = \operatorname{supp}(\bar{g}) \wedge \operatorname{supp}(\bar{h})$ *if and only if* $\operatorname{supp}(\bar{k}\varphi) = \operatorname{supp}(\bar{g}\varphi) \wedge \operatorname{supp}(\bar{h}\varphi)$.

Proof. To see (a), observe that $\operatorname{supp}(\bar{g}) \leq \operatorname{supp}(\bar{h})$ if and only if $G(\bar{h}) \leq G(\bar{g})$, and by Lemma 3.4(c), this holds if and only if $G(\bar{h}\varphi) \leq G(\bar{g}\varphi)$. But, again by Lemma 3.4(c), we have $G(\bar{g}\varphi) = G(\bar{g})\varphi$ and $G(\bar{h}\varphi) = G(\bar{h})\varphi$.

Part (b) follows directly from (a). Part (c) then follows since

$$\operatorname{supp}(\bar{k}) = \operatorname{supp}(\bar{g}) \wedge \operatorname{supp}(\bar{h})$$

if and only if $\text{supp}(\bar{k}) \leq \text{supp}(\bar{g}) \wedge \text{supp}(\bar{h})$ and $\text{supp}(\bar{\ell}) \leq \text{supp}(\bar{k})$ for all $\bar{\ell} \in G$ such that $\text{supp}(\bar{\ell}) \leq \text{supp}(\bar{g}) \wedge \text{supp}(\bar{h})$. □

It follows from this lemma that $\text{Aut}(G)$ embeds in $\text{Aut}(\mathcal{B})$ (via the action of each automorphism on the set of supports of elements with infinitely many fixed points). Observe that if $\text{supp}(\bar{g}) \wedge \text{supp}(\bar{h}) = 0$ then there are representatives g of $\bar{g}$ and h of $\bar{h}$ with disjoint supports.

Next, we define the conjugacy classes in which we will be especially interested. For any subset $T \subseteq \mathbf{N} \setminus \{0\}$, let

$$C(T)_{\text{inf}} := \{\bar{g} : g \in S, g \text{ has cycle type } 1^{\omega} \cdot \prod(n^{\omega} : n \in T)\}.$$

Also, define

$$C(T) := \{\bar{g} : g \in S, g \text{ has cycle type } 1^{\omega} \cdot \prod(n^{1} : n \in T)\}.$$

Both $C(T)_{\text{inf}}$ and $C(T)$ are conjugacy classes of G. Our goal is to show that they are preserved by any automorphism of G.

For $T = \{2\}$ we shall write $C(2)_{\text{inf}}$ for $C(\{2\})_{\text{inf}}$.

Lemma 3.6 *Let $\varphi \in \text{Aut}(G)$, and let $T \subseteq \mathbf{N} \setminus \{0\}$, and $\bar{g} \in C(T)_{\text{inf}}$. Then $\bar{g}\varphi \in C(T)_{\text{inf}}$.*

Proof. First, observe that for any i with $1 < i < \omega$ and $h \in S$ with infinitely many fixed points, h has infinitely many i-cycles if and only if

$$\text{supp}(\bar{h}^{i}) < \bigwedge(\text{supp}(\bar{h}^{j}) : 1 \leq j < i, j|i).$$

By Lemma 3.5, it follows that if $k \in S$ with $\bar{k} = \bar{g}\varphi$, then for each i with $1 < i < \omega$, the element k has infinitely many i-cycles if and only if $i \in T$.

Next, again assuming $h \in S$ has infinitely many fixed points, note that h has a single infinite cycle and just finitely many non-trivial finite cycles if and only if $\text{supp}(\bar{h}) \neq 0$ and we cannot write $\bar{h} = \bar{h}_1\bar{h}_2$ with the supports of $\bar{h}_1$ and $\bar{h}_2$ disjoint and not equal to 0 or $1 \in \mathcal{B}$. Hence, given such an element h, there is $k \in S$ with a single infinite cycle and no other non-trivial cycles such that $\bar{k} = \bar{h}\varphi$. Also, if $h \in S$ is not of this form, then h has at least one infinite cycle if and only if there are $h_1, h_2 \in S$ such that $\bar{h} = \bar{h}_1\bar{h}_2$, the $\bar{h}_i$ have disjoint supports not equal to 0 or 1 in $\mathcal{B}$, and h_1 has a single infinite cycle and no other non-trivial cycles. It follows from Lemma 3.5 that if $k \in S$ with $\bar{k} = \bar{g}\varphi$, then k has no infinite cycles. In fact, this argument shows that if $k_1, k_2 \in S$ and $\bar{k}_1\varphi = \bar{k}_2$ and k_1 has infinitely many fixed points, then k_1 and k_2 have the same number of infinite cycles.

Finally, suppose that $h \in S$ has infinitely many fixed points and no infinite cycles. Then the dregs class F_0 of h is finite if and only if, whenever there are

$h_1, h_2 \in S$ such that $\bar{h} = \bar{h}_1\bar{h}_2$ with $\bar{h}_1, \bar{h}_2$ having disjoint non-trivial supports and h_1 having finitely many i-cycles for all i $(1 < i < \omega)$, the elements $\bar{h}_2$ and $\bar{h}$ are conjugate in G. Again, this condition is φ-invariant, so the lemma follows from this and the previous two paragraphs. □

Lemma 3.7 *Let $\varphi \in \mathrm{Aut}(G)$, and let T be an infinite set of nontrivial finite cycle lengths. Then $C(T)$ is φ-invariant.*

Proof. We will show that $C(T)$ is the only conjugacy class of $G \setminus \{1\}$ whose elements $\bar{g}$ satisfy all the following properties.

(a) there is $\bar{k} \in C(T)_{\mathrm{inf}}$ with $\mathrm{supp}\,\bar{k} \wedge \mathrm{supp}\,\bar{g} = 0$ such that $\bar{k}\bar{g} \in C(T)_{\mathrm{inf}}$;

(b) for any moiety R of T, there is no $\bar{k} \in C(R)_{\mathrm{inf}}$ with $\mathrm{supp}\,\bar{k} \wedge \mathrm{supp}\,\bar{g} = 0$ and $\bar{k}\bar{g} \in C(R)_{\mathrm{inf}}$;

(c) if $\bar{g} = \bar{h}\bar{k}$ where $\bar{h}, \bar{k} \in G\setminus\{1\}$, $\mathrm{supp}\,\bar{h} \wedge \mathrm{supp}\,\bar{k} = 0$ and $\mathrm{supp}\,\bar{h} \vee \mathrm{supp}\,\bar{k} = \mathrm{supp}\,\bar{g}$, then one of (a) or (b) fails when $\bar{g}$ is replaced by $\bar{h}$.

By Lemma 3.5, the conditions on supports are preserved by φ, and by Lemma 3.6, $C(T)_{\mathrm{inf}}$ is φ-invariant, so the conditions (a)-(c) are preserved by φ. Hence, given the above, $C(T)$ is φ-invariant, as required.

Clearly if $\bar{g} \in C(T)$ then it satisfies (a)–(c). Conversely, assume that $\bar{g} \in G \setminus \{1\}$ satisfies (a)–(c), and choose a representative g. Property (a) says that g has infinitely many fixed points, that all but finitely many of the cycles of g have lengths lying in T, and that g has no infinite cycles. Property (b) says that all but finitely many elements of T arise as lengths of cycles of g. Thus (a) and (b) imply that the set of cycle lengths of g is almost exactly T. If the cycle type of g differs more than finitely from $1^\omega \cdot \prod(n^1 : n \in T)$, then (c) fails when we let h be the product of exactly one n-cycle of g for each $n \in T$, and let k be the product of the remaining g-cycles. □

4 Proof of the Theorem

In this section, we prove Theorem 1.1, that is, that every outer automorphism of G is one of the outer automorphisms induced by a near symmetry as described in Section 2.

Let φ be any automorphism of G. We must find $f_\varphi \in \mathrm{NB}(\Omega)$ such that the action of φ on G is that of conjugation by f_φ. In the introduction, we defined the *support* of an element of G, and in Lemma 3.5 we showed that φ respects supports of elements of G and the 'containment' relation induced on supports by the subset relation on Ω. We also showed that certain conjugacy classes of G in which we are interested are respected by φ, that is, are not fused by φ. We now pick a particular conjugacy class $C(T)$ of G, a subset $\mathcal{D}$ of unordered pairs from $C(T)$, and an equivalence relation $\mathcal{E}$ on $\mathcal{D}$, and show that $C(T)$, $\mathcal{D}$ and $\mathcal{E}$ are invariant under φ. Elements of $\mathcal{D}/\mathcal{E}$ will be naturally identified with ordered moieties of Ω, and this enables us to define $f_\varphi \in \mathrm{NB}(\Omega)$ corresponding to φ. We show that f_φ and φ agree in their action on one non-trivial conjugacy class of G. Then since G is a simple group, this conjugacy class generates G and so the actions of f_φ and φ agree on the whole of G, as required.

Fix a set $T = \{n_i : i < \omega\}$ of finite nontrivial cycle lengths such that for all $i < \omega$ we have $n_i > 2\sum_{j<i} n_j$. We note without proof the following easy lemma.

Lemma 4.1 *Let T be as above. Then*

(a) *if $a, b, c \in T$ then $c \neq 2a + b - 2$;*

(b) *if $a, b, c \in T$ then $2a \neq 2b + c - 1$;*

(c) *if $a, b \in T$ then $a \neq 2b - 1$;*

(d) *if $a, b, c, d \in T$ with $a + b = c + d$ then $\{a, b\} = \{c, d\}$;*

(e) *if $a, b, c, d \in T$ with $2a + b = 2c + d$ then $a = c$ and $b = d$.*

We say that $\{g, h\}$ is a *matched $C(T)$-pair* if:

(a) g, h have cycle types $1^\omega \cdot \prod(i^1 : i \in T)$;

(b) $\mathrm{supp}(g) \cup \mathrm{supp}(h)$ is a moiety of Ω;

(c) for each $n \in T$, the n-cycle of g meets $\mathrm{supp}(h)$ in a singleton lying in the n-cycle of h.

Next, let $\mathcal{D}$ be the set of all unordered pairs $\{\bar{g}, \bar{h}\}$ for which there are corresponding elements $g, h \in S$ which form a matched $C(T)$-pair. We say that $\{g, h\}$ is a *matched cover* of $\{\bar{g}, \bar{h}\}$.

If $\{\bar{g}, \bar{h}\} \in \mathcal{D}$, we will write $\mathrm{supp}(g, h)$ for $\mathrm{supp}(g) \cup \mathrm{supp}(h)$. If $\{g, h\}$ is a matched cover of $\{\bar{g}, \bar{h}\}$, let the *spine* $\sigma(g, h)$ of $\{g, h\}$ be the ω-sequence from Ω whose i^{th} entry is the point in which the n_i-cycles of g and h intersect. Then the *ribcage* of $\{g, h\}$ is the set $\mathrm{supp}(g, h) \setminus (\mathrm{supp}(g) \cap \mathrm{supp}(h))$, that is the elements of the supports of g and h which are not in the spine.

Lemma 4.2 *The set $\mathcal{D}$ defined above is φ-invariant for any $\varphi \in \mathrm{Aut}(G)$.*

Proof. We show that if $\bar{g}, \bar{h} \in G$ then $\{\bar{g}, \bar{h}\} \in \mathcal{D}$ if and only if the following conditions all hold:

(a) $\bar{g}, \bar{h} \in C(T)$;

(b) $\mathrm{supp}(\bar{g}) \vee \mathrm{supp}(\bar{h}) \neq 1$;

(c) for all $\bar{k} \in C(2)_{\text{inf}}$ such that $\text{supp}(\bar{k}) = \text{supp}(\bar{g}) \wedge \text{supp}(\bar{h})$ we have $\bar{g}\bar{k}, \bar{h}\bar{k} \in C(Q)$ for some subset Q of $\{n_i + n_j : n_i, n_j \in T\}$;

(d) $\bar{g}\bar{h} \in C(R)$, where $R = \{2n - 1 : n \in T\}$.

Here, part (b) says that $\text{supp}(g) \cup \text{supp}(h)$ is a moiety of Ω. Part (c) ensures that for some $g, h \in S$ representing $\bar{g}, \bar{h}$, every non-trivial cycle of g intersects $\text{supp}(h)$ in at most a singleton, and similarly with g, h interchanged. For, if the n_i-cycle of g intersects $\text{supp}(k)$ in a singleton, which k pairs with a singleton from the n_j-cycle of g, then gk contains an (n_i+n_j)-cycle. On the other hand, if, say, there are infinitely many cycles of g which each intersect $\text{supp}(h)$ in more than one point, then k can be chosen so that gk contains an infinite cycle. We leave the details to the reader.

If the m-cycle of g meets the k-cycle of h in a singleton, then they form an $(m+k-1)$-cycle in the product gh. By Lemma 4.1(d), if $m, k, n \in T$ then $m+k-1 = 2n-1$ if and only if $m = k = n$. Thus part (d) ensures that these cycles are matched up correctly, that is, the n-cycle of g meets the n-cycle of h.

By Lemmas 3.5, 3.6 and 3.7, the conditions in (a) to (d) are preserved by any automorphism. So $\mathcal{D}$ is φ-invariant. □

Next, we define an equivalence relation $\mathcal{E}$ on $\mathcal{D}$. If $\{\bar{g}_1, \bar{h}_1\}, \{\bar{g}_2, \bar{h}_2\} \in \mathcal{D}$, then we say that they are equivalent under $\mathcal{E}$ if they have matched covers $\{g_1, h_1\}$ and $\{g_2, h_2\}$ respectively, such that $\sigma(g_1, h_1) = \sigma(g_2, h_2)$.

Lemma 4.3 *Let $\varphi \in \text{Aut}(G)$. Then the equivalence relation $\mathcal{E}$ is φ-invariant.*

Proof. We say that $g_1, g_2, g_3, g_4 \in S$ form a *matched quadruple* if

(a) $\{\bar{g}_i, \bar{g}_j\} \in \mathcal{D}$ and has $\{g_i, g_j\}$ as a matched cover for all $1 \le i < j \le 4$;

(b) all sequences $\sigma(g_i, g_j)$, $(1 \le i < j \le 4)$ are equal;

(c) $\bigcup\{\text{supp}(g_i) : 1 \le i \le 4\}$ is a moiety of Ω;

Claim: The set of matched quadruples in G is φ-invariant.

Proof of Claim: Note that $\{\bar{g}_1, \bar{g}_2, \bar{g}_3, \bar{g}_4\}$ is a matched quadruple if and only if $\{\bar{g}_i, \bar{g}_j\} \in \mathcal{D}$ for all $1 \le i < j \le 4$ and $\text{supp}(\bar{g}_i) \wedge \text{supp}(\bar{g}_j) = \bigwedge\{\text{supp}(\bar{g}_k) : 1 \le k \le 4\}$ and $\bigvee\{\text{supp}(\bar{g}_k) : 1 \le k \le 4\} \ne 1$. These conditions are clearly preserved by φ, which yields the claim.

Now take $\{\bar{g}_1, \bar{h}_1\}, \{\bar{g}_2, \bar{h}_2\} \in \mathcal{D}$ such that $\{\bar{g}_1, \bar{h}_1\}\mathcal{E}\{\bar{g}_2, \bar{h}_2\}$. Let $\{g_1, h_1\}$ and $\{g_2, h_2\}$ be matched covers. We can find $\{\bar{g}_3, \bar{h}_3\} \in \mathcal{D}$ and $\{\bar{g}_4, \bar{h}_4\} \in \mathcal{D}$ with matched covers $\{g_3, h_3\}$ and $\{g_4, h_4\}$ such that

$$\text{supp}(g_1, h_1) \cup \text{supp}(g_3, h_3),\ \text{supp}(g_3, h_3) \cup \text{supp}(g_4, h_4) \text{ and } \text{supp}(g_2, h_2) \cup \text{supp}(g_4, h_4)$$

are moieties, the spines $\sigma(g_i, h_i)$ all agree and the ribcages of $\text{supp}(g_1, h_1)$ and $\text{supp}(g_3, h_3)$ are disjoint, as are the ribcages of $\text{supp}(g_3, h_3)$ and $\text{supp}(g_4, h_4)$, and the ribcages of $\mathrm{supp}(g_4, h_4)$ and $\mathrm{supp}(g_2, h_2)$. Then each of $\{\bar{g}_1, \bar{h}_1, \bar{g}_3, \bar{h}_3\}$, $\{\bar{g}_3, \bar{h}_3, \bar{g}_4, \bar{h}_4\}$ and $\{\bar{g}_4, \bar{h}_4, \bar{g}_2, \bar{h}_2\}$ is a matched quadruple of G. Hence, by the claim, their images under φ are all matched quadruples of G. Thus

$$\{\bar{g}_1\varphi, \bar{h}_1\varphi\}\mathcal{E}\{\bar{g}_3\varphi, \bar{h}_3\varphi\}\mathcal{E}\{\bar{g}_4\varphi, \bar{h}_4\varphi\}\mathcal{E}\{\bar{g}_2\varphi, \bar{h}_2\varphi\},$$

as required. □

It follows from the last lemma that φ has a natural induced action on the set of almost-equality classes of ω-sequences which enumerate moieties of Ω. We are now in a position to construct a near-bijection f_φ corresponding to φ. Let $\Omega = \Omega_0 \cup \Omega_1$ be a partition of Ω into moieties, and let $\{\bar{g}, \bar{h}\}, \{\bar{k}, \bar{l}\} \in \mathcal{D}$ have matched covers $\{g, h\}$, $\{k, l\}$ respectively such that $\sigma(g, h)$ is an enumeration of Ω_0 and $\sigma(k, l)$ is an enumeration of Ω_1. Since φ preserves the Boolean algebra $\mathcal{B}$, we see that if $\{g', h'\}$ and $\{k', l'\}$ are matched covers of $\{\bar{g}\varphi, \bar{h}\varphi\}$ and $\{\bar{k}\varphi, \bar{l}\varphi\}$ respectively, then $\sigma(g', h')$ and $\sigma(k', l')$, regarded as sets, have finite intersection and their union is cofinite in Ω. There is a near-bijection f_φ of Ω which induces a map taking $\sigma(g, h)$ to $\sigma(g', h')$ and taking $\sigma(k, l)$ to $\sigma(k', l')$ (except on a finite set).

The next lemma shows essentially that this action preserves subsequences. (Note that these subsequences are not necessarily increasing subsequences; that is, they may involve permutations of the elements of the sequence, so may not induce the same ordering.)

Lemma 4.4 *Let $\varphi \in \mathrm{Aut}(G)$. Suppose that $\{\bar{g}, \bar{h}\}, \{\bar{k}, \bar{l}\} \in \mathcal{D}$ have matched covers $\{g, h\}$, $\{k, l\}$ respectively, and that $\sigma(g, h)$ meets $\mathrm{supp}(k) \cap \mathrm{supp}(l)$ in a subsequence of $\sigma(k, l)$, that is, there is $A \subseteq \omega$ and an injection $\chi: A \to \omega$ such that if $i \in A$ then the ith element of $\sigma(g, h)$ is the $\chi(i)^{\text{th}}$ element of $\sigma(k, l)$, and if $i \notin A$ then the ith element of $\sigma(g, h)$ does not lie in $\sigma(k, l)$. Then there are matched covers $\{g', h'\}$ and $\{k', l'\}$ of $\{\bar{g}\varphi, \bar{h}\varphi\}$ and $\{\bar{k}\varphi, \bar{l}\varphi\}$ respectively such that if $i \in A$ then the ith element of $\sigma(g', h')$ is the $\chi(i)^{\text{th}}$ element of $\sigma(k', l')$, and if $i \notin A$ then the ith element of $\sigma(g', h')$ does not occur in $\sigma(k', l')$.*

Proof. By replacing $(\bar{k}, \bar{l})$ by a pair $\mathcal{E}$-equivalent to it, and using Lemma 4.3, we may suppose that

$$\mathrm{supp}(k) \cap \mathrm{supp}(l) \cap \mathrm{supp}(g) \cap \mathrm{supp}(h) = \mathrm{supp}(g, h) \cap \mathrm{supp}(k, l).$$

It follows that we may choose matched covers $\{g', h'\}$ of $\{\bar{g}\varphi, \bar{h}\varphi\}$ and $\{k', l'\}$ of $\{\bar{k}\varphi, \bar{l}\varphi\}$ so that

$$\mathrm{supp}(k') \cap \mathrm{supp}(l') \cap \mathrm{supp}(g') \cap \mathrm{supp}(h') = \mathrm{supp}(g', h') \cap \mathrm{supp}(k', l').$$

Hence there is a subset B of ω and an injection $\chi' : B \longrightarrow \omega$ such that if $i \in B$ then the ith element of $\sigma(g', h')$ is the $\chi'(i)^{\text{th}}$ element of $\sigma(k', l')$, and if $i \notin B$ then the ith element of $\sigma(g', h')$ is not in $\sigma(k', l')$, for almost all $i \in \omega$. We must show that $A = B$ and $\chi = \chi'$ almost everywhere (for then, by finite adjustments of g', h', k', l', we may arrange that equality holds everywhere).

It is clear that ghk has cycle type $C(S)$ where

$$S = \{2n_i - 1 : i \notin A\} \cup \{2n_i + n_{\chi(i)} - 2 : i \in A\} \cup \{n_i : i \notin \chi(A)\}.$$

Similarly, $g'h'k'$ almost has cycle type

$$S = \{2n_i - 1 : i \notin B\} \cup \{2n_i + n_{\chi'(i)} - 2 : i \in B\} \cup \{n_i : i \notin \chi'(B)\}.$$

It follows by Lemma 4.1(a),(b),(c) and (e) that $A \triangle B$ is finite and $\chi(i) = \chi'(i)$ for almost all $i \in \omega$, as required. □

Lemma 4.5 *Let f_φ be as defined above. Then the $\equiv$-class of f_φ is independent of the choice of Ω_0, Ω_1 and of $\bar{g}, \bar{h}, \bar{k}, \bar{l}$.*

Proof. Pick another partition $\Omega = \Omega_0^* \dot\cup \Omega_1^*$ into moieties, and let $\{\bar{g}^*, \bar{h}^*\}, \{\bar{k}^*, \bar{l}^*\} \in \mathcal{D}$ have matched covers $\{g^*, h^*\}, \{k^*, l^*\}$ so that $\sigma(g^*, h^*)$ is an enumeration of Ω_0^* and $\sigma(k^*, l^*)$ is an enumeration of Ω_1^*. Define a corresponding near-bijection f_φ^* of Ω. We must show $f_\varphi \equiv f_\varphi^*$. This, however, follows almost immediately from Lemma 4.4. □

To complete the proof of the theorem, we must prove the following lemma.

Lemma 4.6 *Let f_φ be as above, and let $\bar{f}_\varphi$ be the element of* NS(Ω) *corresponding to f_φ. Then for all $\bar{g} \in C(2)_{\text{inf}}$ we have $\bar{f}_\varphi^{-1} \bar{g} \bar{f}_\varphi = \bar{g}\varphi$.*

Proof. Let $g := \prod_{i \in \omega} (a_{2i} a_{2i+1})$. Choose permutations $x, y \in S$ such that $\{\bar{x}, \bar{y}\} \in \mathcal{D}$, the set $\{x, y\}$ is a matched $C(T)$-pair, and such that $\sigma(x, y)$ is the sequence $(a_i : i \in \omega)$. Then $\text{supp}(\bar{g}) = \text{supp}(\bar{x}) \triangle \text{supp}(\bar{y})$, so $\text{supp}(\bar{g}\varphi) = \text{supp}(\bar{x}\varphi) \triangle \text{supp}(\bar{y}\varphi)$. Also, $\{\bar{x}\varphi, \bar{y}\varphi\} \in \mathcal{D}$, so we may choose a corresponding matched $C(T)$-pair $\{x', y'\}$. Let $\sigma(x', y') = (b_i : i < \omega)$. Then by Lemma 4.5, $a_i f_\varphi = b_i$ for all but finitely many $i \in \omega$. In particular, $f_\varphi^{-1} g f_\varphi$ and $\prod_{i<\omega} (b_{2i} b_{2i+1})$ are $\equiv$-equivalent. Also, by Lemma 3.6 (or a direct argument), $\bar{g}\varphi \in C(2)_{\text{inf}}$, and if $g' \in S$ with $\bar{g}' = \bar{g}\varphi$, then the symmetric difference $\text{supp}(g') \triangle \{b_i : i \in \omega\}$ is finite. It remains to show that g' and $\prod_{i \in \omega} (b_{2i} b_{2i+1})$ differ by a finitary permutation.

Let $u := gx$. Then u has cycle type $1^\omega \prod((n_{2i} + n_{2i+1})^1 : i < \omega)$. This says that, for each $i < \omega$, g pairs up the n_{2i}-cycle of x with the n_{2i+1}-cycle of x. By Lemma 3.7, $\bar{u}\varphi$ and $\bar{u}$ are conjugate in G, so have preimages in S with cycle types differing finitely. If g' induces the transposition $(b_i b_j)$ then $x'g'$ has an $(n_{2i} + n_{2i+1})$-cycle. It now follows from Lemma 4.1(d) that g' differs finitely from $\prod_{i \in \omega} (b_{2i} b_{2i+1})$, as required. □

Proof of Theorem 1.1. The main theorem follows immediately from Lemma 4.6. For, since the conjugacy class $C(2)_{\text{inf}}$ generates G, conjugation by $\bar{f}_\varphi$ induces the action of φ on all elements of G.

References

[1] R. Baer, *Die Kompositionsreihe der Gruppe aller eineindeutigen Abbildungen einer unendlichen Menge auf sich*, Studia Math, **5**(1935), 15-17.

[2] John D. Dixon, Peter M. Neumann and Simon Thomas, *Subgroups of small index in infinite symmetric groups*, Bull. London Math. Soc. **18**(1986), 580-586.

[3] H. D. Macpherson and Peter M. Neumann, *Subgroups of infinite symmetric groups*, J. London Math. Soc. (2), **42**(1990) 64-84.

[4] Fred Richman, *Maximal subgroups of infinite symmetric groups*, Canad. Math. Bull. **10**(1967), 375-381.

[5] M. Rubin, *On the automorphism groups of homogeneous and saturated Boolean algebras*, Algebra Universalis **9**(1979), 54–86.

[6] W. R. Scott, *Group Theory*, Prentice Hall, New Jersey, 1964.

[7] Stephen W. Semmes, *Infinite symmetric groups, maximal subgroups, and filters*, Preliminary report, Abstracts Amer. Math. Soc., **69**(1982), 38.

[8] J. K. Truss, *On recovering structures from quotients of their automorphism groups*, this volume.

Other *Mathematics and Its Applications* titles of interest:

J.-F. Pommaret: *Partial Differential Equations and Group Theory. New Perspectives for Applications.* 1994, 473 pp. ISBN 0-7923-2966-X

Kichoon Yang: *Complete Minimal Surfaces of Finite Total Curvature.* 1994, 157 pp. ISBN 0-7923-3012-9

N.N. Tarkhanov: *Complexes of Differential Operators.* 1995, 414 pp. ISBN 0-7923-3706-9

L. Tamássy and J. Szenthe (eds.): *New Developments in Differential Geometry.* 1996, 444 pp. ISBN 0-7923-3822-7

W.C. Holland (ed.): *Ordered Groups and Infinite Permutation Groups.* 1996, 255 pp. ISBN 0-7923-3853-7